The Cortical Neuron

The Cortical

New York   Oxfor

Neuron

Edited by

MICHAEL J. GUTNICK, DVM, Ph.D.

ISTVAN MODY, Ph.D.

OXFORD UNIVERSITY PRESS 1995

Oxford University Press

Oxford New York Toronto
Delhi Bombay Calcutta Madras Karachi
Kuala Lumpur Singapore Hong Kong Tokyo
Nairobi Dar es Salaam Cape Town
Melbourne Auckland Madrid

and associated companies in
Berlin Ibadan

Published by Oxford University Press, Inc.
198 Madison Avenue, New York, NY 10016

Library of Congress Cataloging-in-Publication Data
The cortical neuron / edited by Michael J. Gutnick, Istvan Mody.
p. cm. Includes bibliographical references and index.
ISBN 0-19-508330-X
1. Cerebral cortex. 2. Neurons. 3. Neural circuitry.
I. Gutnick, M. J. (Michael Joseph). II. Mody, Istvan, 1957–
[DNLM: 1. Neurons—physiology. 2. Cerebral Cortex—physiology.
3. Synaptic Transmission—physiology.
WL 102.5 C829 1995]
QP383.C66 1995 612.8925—dc20 DNLM/DLC
for Library of Congress 94–24546

9 8 7 6 5 4 3 2 1

Printed in the United States of America
on acid-free paper

To the memory of Hans-Dieter Lux, a pioneer of the physiology of cortical neurons.

Preface

One of the major challenges of neurosciences is to understand the functioning of the cerebral cortex. Cortical neurons and circuits have been the subject of numerous volumes. Most of these, however, have focused largely on the anatomy of the cortex (*e.g.*, cortical areas, cell types, local connectivity) or on the function of particular cortical subsystems (*e.g.*, sensory systems and receptive fields, or the motor system and related output fields). While a wiring-diagram approach to the description of the cortex is undoubtedly important, understanding a functioning cortical circuit requires an appreciation of its dynamic physiological characteristics, at both the single-cell and network levels. Recent years have seen dramatic advances in our understanding of cerebral cortical function. Much of the progress has stemmed from a more profound insight into the basic cellular and molecular physiology of cortical neurons: their membrane characteristics, their synaptic responsiveness, their functional connectivity and the mechanisms of their responses to injury. This book consists of contributions by many of the neuroscientists who have been instrumental in making these advances—anatomists, physiologists, cell biologists, pharmacologists, developmental biologists and neurologists. Often, several experts in the same field have teamed up as co-authors of a single-chapter. In general, our goal has been to provide a balanced, broad and up-to-date overview of normal and pathological cortical neuronal function, and to identify the critical issues and controversies facing researchers in this exciting area.

Although this book is not a "proceedings" volume, the idea for it emerged from a scientific meeting. In May, 1993, students, trainees, collaborators and colleagues of David Prince held a symposium on "The Cortical Neuron" in

Asilomar, California, on the occasion of his 60th birthday. In the words of Dennis Choi, "This book is a special project. David Prince has had a major role in furthering our knowledge of cerebral cortical and hippocampal biology, especially in relation to epilepsy. He has accomplished this both through his own substantial scientific investigations, and through the recruitment and training of other neuroscientists and neurologists." Indeed, for this project, David has provided a unique critical analysis of normal and abnormal cortical cellular physiology, based on insights from three decades of productive investigation.

The symposium that led to this book was sponsored by the American Epilepsy Society, the Morris Research Fund, the Office of Naval Research, Axon Instruments and Ciba-Geigy. We would like to express our appreciation to all of those who helped putting together the symposium and the book, especially Myrna Canizares, Ted Brooks, and Becky Hilger. We also thank the authors for their time and for their valuable contributions to this volume.

Michael J. Gutnick
Istvan Mody

Contents

Contributors

Gregory W. Albers, M.D.
Stanford Stroke Center
Department of Neurology and Neurological Sciences
Stanford University School of Medicine
Stanford, California

Yael Amitai, M.D.
Department of Physiology
Faculty of Health Sciences
Ben-Gurion University of the Negev
Beer-Sheva, Israel

Larry S. Benardo, M.D., Ph.D.
Department of Pharmacology
Department of Neurology
State University of New York
Health Science Center at Brooklyn
Brooklyn, New York

Julie A. Bennett
Department of Pharmacology
Emory University School of Medicine
Atlanta, Georgia

Paul C. Bush
Howard Hughes Medical Institute
Salk Institute
LaJolla, California

Dennis W. Choi, M.D., Ph.D.
Department of Neurology and
Center for the Study of Nervous System Injury
Washington University School of Medicine
St. Louis, Missouri

Barry W. Connors, Ph.D.
Department of Neuroscience
Brown University
Providence, Rhode Island

Douglas Coulter, Ph.D.
Department of Neurology
Medical College of Virginia
Richmond, Virginia

Wayne E. Crill, M.D.
Department of Physiology and Biophysics
University of Washington School of Medicine
Seattle, Washington

Rudolf A. Deisz, M.D.
Clinical Neuropharmacology
Max-Planck-Institute for Psychiatry
München, Germany

Raymond J. Dingledine, Ph.D.
Department of Pharmacology
Emory University School of Medicine
Atlanta, Georgia

Michael J. Gutnick, DVM, Ph.D.
Department of Physiology
Faculty of Health Sciences
Ben-Gurion University of the Negev
Beer-Sheva, Israel

John R. Huguenard, Ph.D.
Department of Neurology and
Neurological Sciences
Stanford University School of
Medicine
Stanford, California

Edward G. Jones, M.D., Ph.D.
Department of Anatomy and
Neurobiology
University of California
Irvine, California

Peter Kellaway, Ph.D.
Department of Neurophysiology
Baylor College of Medicine
Houston, Texas

John E. Kraus
Department of Neurobiology
Duke University Medical Center
Durham, North Carolina

Arnold R. Kriegstein, M.D., Ph.D.
Department of Neurology
College of Physicians and Surgeons
of Columbia University
New York, New York

Jo LoTurco, Ph.D.
Department of Genetics
Harvard Medical School
Boston, Massachusetts

Heiko J. Luhmann, Ph.D.
Institute of Neurophysiology
Humboldt University
Berlin, Germany

Daniel V. Madison, Ph.D.
Department of Molecular and
Cellular Physiology
Stanford University School
of Medicine
Stanford, California

Robert C. Malenka, M.D., Ph.D.
Departments of Psychiatry and
Physiology
University of California
San Francisco, California

Leona M. Masukawa, Ph.D.
Department of Neurology
The Graduate Hospital
Philadelphia, Pennsylvania

David A. McCormick, Ph.D.
Section of Neurobiology
Yale University School of Medicine
New Haven, Connecticut

James O. McNamara, M.D.
Departments of Pharmacology
and Neurology
Duke University Medical Center and
Department of Veteran Affairs
Medical Center
Durham, North Carolina

Istvan Mody, Ph.D.
Department of Neurology
Department of Physiology
University of California
Los Angeles, California

Jeffrey L. Noebels, M.D., Ph.D.
Neuroscience and Molecular Genetics
Baylor College of Medicine
Houston, Texas

David A. Prince, M.D.
Department of Neurology and Neurological Sciences
Stanford University School of Medicine
Stanford, California

Howard Schulman, Ph.D.
Department of Molecular Pharmacology
Stanford University School of Medicine
Stanford, California

Philip A. Schwartzkroin, Ph.D.
Department of Neurological Surgery
University of Washington School of Medicine
Seattle, Washington

Terrence J. Sejnowski, Ph.D.
Howard Hughes Medical Institute
Salk Institute
La Jolla, California

Charles F. Stevens, M.D., Ph.D.
Howard Hughes Medical Institute
Salk Institute
La Jolla, California

Scott M. Thompson, Ph.D.
Brain Research Institute
University of Zürich
Zürich, Switzerland

Robert K.S. Wong, Ph.D.
Department of Pharmacology
State University of New York
Health Science Center at Brooklyn
Brooklyn, New York

I

General overview

1

Thirty years among cortical neurons

DAVID A. PRINCE

For Reva

"We know too well that unforeseen facts can modify or reverse our scientific conjectures from one day to the next. All we can hope is that some principles of our conception . . . will remain" (Santiago Ramon y Cajal, 1909)

I approached writing this chapter with some trepidation, in part because the title suggested by the editors raises expectations of a historical-philosophical review that I will not deliver. Also, in these days of neurobiological reductionism, it has become easier and easier to dwell on the outcomes of exquisitely focused experiments, and progressively more difficult to place their results into the context of a functioning cortical neuron, let alone the framework of a circuit or a brain. During the past three decades an increasingly complex picture of the operation of even a single neuron has emerged as avalanches of information have been released by application of new investigative approaches to old problems. The use of intracellular labeling, immunocytochemistry, molecular biological and patch-clamp techniques, and a large variety of other methodologies has had an astounding impact on our view of the cortex, and has moved studies of cortical neurons to levels of inquiry previously possible only in less complex systems. As a result, major advances have occurred in many areas including those presented in this book. Witnessing this evolution of brain science evokes both elation at what has been accomplished and speculation about issues that are important for a better understanding of cortical function. It would be impossible to mention even briefly the many areas in which major progress has been made; rather, I have selected a few general concepts to discuss.

The first part of this chapter deals with diversity as a key attribute of cortical neurons, and the importance of understanding interactions that occur among molecular, cellular, and circuit level functions in predicting the effects of naturally occurring or experimentally induced alterations. The functional implications of intrinsic burst generation in subsets of neurons are used to illustrate the potentially unexpected results that may ensue when translating molecular (membrane) properties into the language of the cortex. In subsequent sections I use the results of experiments on epileptogenesis to emphasize the important contributions that work on abnormal cortical function has made to understanding the normal operation of the cortex. Diversity and the plastic changes in the cortex produced by activity and injury are discussed as they relate to the pathophysiology of epilepsy. The study of epileptogenesis is a pseudonym for the study of brain function. The core issues are the same as those that form some of today's challenges in neuroscience.

LIVING WITH AND LEARNING FROM DIVERSITY

> "To be reductive is seductive. . . . There is then a great temptation to concentrate on smaller and smaller bits of the whole, hoping perhaps that no surprises would be in store if, eventually, one really tried to put it all back again in an integrative approach" (Noble and Boyd, 1993).

The structural complexity and heterogeneity of cortical neurons have been obvious since the earliest days of light microscopy. In fact, if there is one word that captures many of the key attributes of cortical structures, it is *diversity*. As new techniques have evolved, an amazing array of physiological, biochemical and molecular variables have been superimposed upon increasingly complex anatomical descriptions. To what extent do the patterns of connectivity and varieties of neuronal size, shape, transmitter phenotype, pre- and postsynaptic receptors, and voltage-gated channels have specific functional purposes that provide a basis for understanding the design of the brain?

When potential combinations of these variables are taken into account, numerous subgroups of cortical neurons emerge. One interesting hypothesis derived from such findings is that particular combinations of properties tend to be expressed within a given subgroup of neurons; that is, significant structural differences between cell types may always signal functional (physiological) differences. According to this prediction, each of the subtypes of anatomically distinct cortical pyramidal cells or interneurons will have different functions and connections within the cortical circuit and unique combinations of intrinsic properties, receptors, and the like, that will be best suited to these functional roles. One corollary is that differences in intrinsic cell behavior initially detected electrophysiologically will indicate anatomical and other differences in neuronal subclasses (Chagnac-Amitai et al., 1990; Kawaguchi, 1993; Kawaguchi and Hama, 1988; Kawaguchi and Kubota, 1993; Livsey et al., 1993; Mason and Larkman, 1990; McCormick et al., 1985; Prince and Huguenard, 1988; Tseng and Prince, 1993). Some of the earliest observations that support this notion came from studies of cortical pyramidal tract neurons where differences based on axonal diameters, conduction velocities, and biophysical properties (Calvin and Sypert, 1976; Koike et al., 1968; Takahashi, 1965) led to the definition of "slow" and "fast" subtypes that had tonic and phasic discharge properties, respectively (Humphrey and Corrie, 1978), and were correlated with particular soma-dendritic structures (Deschenes et al., 1979). Differences in the structure of intrinsically- bursting versus regular-spiking pyramidal neurons have also been described (Chagnac-Amitai et al., 1990; Kasper et al., 1994a; Mason and Larkman, 1990), and these subsets of cells also seem to have different cortical and subcortical targets. More recent data suggest further diversity in firing behaviors and structure among pyramidal tract neurons (Tseng and Prince, 1993), although a detailed neuroanatomical analysis (e.g., a quantitative analysis of dendritic arbors [Ohara and Havton, 1994]) has yet to be done. Anatomically different subsets of γ-amino-butyric-acid-containing (GABAergic) inhibitory interneurons in neocortex (Kawaguchi, 1993; Kawaguchi and Kubota, 1993) and hippocampus (Kawaguchi and Hama, 1988; Lacaille and Schwartzkroin, 1988a) also have different intrinsic signaling properties (e.g. fast-spiking versus the generation of low-threshold spikes and bursts) and different local projections. The physiological consequences of these differences in signaling and connectivity are discussed by the authors of Chapter 11. It is my bias that each will be found to have important implications for processing information within cortical circuits. For example, in the case of interneurons, evidence suggests that different presynaptic cells innervate $GABA_A$ and $GABA_B$ receptors (Bernardo, 1994b; Mueller and Misgeld, 1990; Segal, 1990) and that interneurons with intrinsic burst-generating capacities may evoke $GABA_B$ receptor-mediated inhibitory postsynaptic potentials (IPSPs) in postsynaptic cells (Kawaguchi, 1992; Lacaille and Schwartzkroin, 1988; Williams and Lacaille, 1992). Some speculations regarding the functions of bursting in cortical neurons are presented below. Since the operation of the cortex depends heavily on the potency and distribution of GABA receptor-mediated inhibition (Gilbert, 1993; Jones, 1993; Sillito, 1975), these and other new findings related to the nature of intracortical inhibitory activities (Jones, 1993; Otis and Mody, 1992a and b; Otis et al., 1991, 1993; see Chapters 8 to 10) will eventually have to be incorporated into models that emulate complex cortical functions (Douglas et al., 1991; Worgotter and Koch, 1991).

Recent experiments on thalamic neurons (see Chapter 11) provide another excellent example of functional diversity at the molecular (membrane) level that has a very specific role within a circuit (Huguenard and Prince, 1992a). Thalamic relay neurons are glutamatergic elements designed to transmit information from the periphery to the cortex, whereas nucleus reticularis (nRt) cells release GABA onto relay neurons, hyperpolarizing them and thereby leading to deinactivation of a low-threshold Ca^{2+} current and the generation of low-threshold Ca^{2+} spikes (LTSs; Deschenes et al., 1982; Llinás and Jahnsen, 1982). The deinactivation of the low-threshold current is voltage and time dependent, so that prolonged hyperpolarizations that bring the membrane potential close to −90 to −100mV are extremely effective in evoking rebound LTSs (Coulter et al., 1989a and b; Jahnsen and Llinas, 1984). Two features of relay neurons and nRt cells seem to be designed to maximize the efficacy of GABA in evoking LTSs. First, as a consequence of fundamental differences in the biophysical properties of the underlying low-threshold Ca^{2+} current in the two cell types (Huguenard and Prince, 1992a), bursts of spikes during LTSs are much longer in nRt cells than in relay cells, resulting in enhanced postsynaptic GABA release. Second, the equilibrium potential for $GABA_A$ receptor-mediated events (E_{Cl}) is very negative in the relay neurons (ca −90mV), probably due to active Cl^- extrusion (Huguenard and Prince, 1994b), so that activation of these receptors, as well as of $GABA_B$ receptors (Crunelli and Leresche, 1991), is very effective in evoking the LTS.

Functional Consequences of Neuronal Diversity; Speculation About the Potential Significance of Intrinsic Burst Generation

Variability in the electrical properties of single cells provides a rich repertory of possible network behaviors that are not predictable from the skeleton of a circuit diagram or even from a detailed molecular and biophysical characterization of the membrane receptors and ion channels of each participating neuron. The potential roles within the circuit of neurons generating intrinsic bursts of action potentials are considered here as one example of the functionally diverse effects that might occur as a consequence of a particular set of membrane properties. The capacity to generate intrinsic bursts is phenotypically determined in specific subclasses of neurons and is dependennt on the type, density, and distribution of Ca^{2+}, Na^{+}, and K^{+} channels in their membranes. Bursting in an excitatory presynaptic cell can induce a larger, more prolonged synaptic response and can increase the probability of firing in its postsynaptic target (Wong et al., 1986). In neocortex, burst-generating capacities are not distributed at random; deep-lying pyramidal cells are the predominant class of intrinsically bursting neuron (Chagnac-Amitai et al., 1990; Connors and Gutnick, 1990; Connors et al., 1982; Mason and Larkman, 1990). This innate capacity is apparently present in layer V pyramidal neurons very early in cortical development (Silva et al., 1991a) and is functionally expressed about 2 weeks after birth in the rat (Franceschetti et al., 1993; Hoffman and Prince, 1995; Kasper et al., 1994b). Burst generation may allow these cells to amplify signals and, by virtue of their intracortical connections (Burkhalter, 1989; Schwark and Jones, 1989), to synchronize neurons within the cortical circuit. These and other results have led to the hypothesis that bursting cells may be pacemakers for interictal epileptiform discharges (Connors, 1984; Gutnick et al., 1982; Schwartzkroin and Prince, 1978). However, in addition to increasing signal amplitude and the input-output function of synapses, there are a number of mechanisms by which repetitive presynaptic discharge may actually alter the *quality* of postsynaptic response.

In the case of glutamatergic synapses made by pyramidal neurons (Thomson et al., 1989), the enhanced transmitter release and depolarization evoked by a burst of presynaptic spikes may relieve the Mg^{2+} block of N-methyl-D-aspartate (NMDA) receptor channels and thus result in both a prolongation of the postsynaptic response and a change in the ionic basis of the postsynaptic current so that there is enhanced Ca^{2+} entry (Mayer et al., 1987; Regehr and Tank, 1990), with resultant long-term consequences for the regulation of cell excitability (Collingridge, 1987; Muller et al., 1988). Intrinsic burst discharges (Kawaguchi and Kubota, 1993; Lacaille and Schwartzkroin, 1988) or steep relationships between depolarization and spike frequency in GABAergic neurons (McCormick et al., 1985) might also have an important effect on the types of receptors that are activated, quite apart from the possible selective innervation of particular receptor subtypes by "$GABA_A$" or "$GABA_B$" presynaptic interneurons mentioned above. Spontaneous inhibitory currents in hippocampal granule cells (Otis and Mody, 1992a) and in neocortical pyramidal neurons (P.A. Salin and D.A. Prince, unpublished observations) do not have $GABA_B$ receptor-mediated components, although the latter may be evoked by electrical stimulation or during epileptiform discharges (Otis and Mody, 1992b). One explanation for these findings would be differences in the sensitivity of the postsynaptic $GABA_A$ and $GABA_B$ receptors to GABA (Newberry and Nicoll, 1984; see Otis and Mody [1992b] for discussion). Burst generation or high-frequency firing in GABAergic neurons might thus lead to enhanced GABA release and $GABA_B$ receptor-mediated IPSPs on postsynaptic neurons (Lacaille and Schwartzkroin, 1988; Williams and Lacaille, 1992), whereas $GABA_A$ IPSPs would be the predominant response when GABAergic neurons generate single action potentials. The generation of $GABA_B$ postsynaptic IPSPs can have important consequences, such as a reduction of the spontaneous spike firing rate and the responsiveness to threshold excitation (Connors et al., 1988). Enhanced GABA release produced by high-frequency presynaptic discharges might also determine the extent to which presynaptic auto- or heteroreceptors will be activated (Isaacson et al., 1993) and thus regulate the induction of long-term changes in excitability by "spill-over" effects at presynaptic $GABA_B$ autoreceptors (Davies et al., 1991; Olpe et al., 1992). In thalamic relay neurons, the long duration and negative reversal potential of these events (ca -100 mV $\cong E_K$) are tailor-made to deinactivate a potent low-threshold calcium conductance (Deschenes et al., 1982; Llinás and Jahnsen, 1982), leading to development of an oscillatory burst firing mode that affects excitability in the entire thalamo-cortical circuit (Steriade and Llinás, 1988).

Another more speculative consequence of burst generation in both excitatory and inhibitory neurons is raised by studies in the peripheral nervous system where co-release of a neurotransmitter and a neuropeptide occurs at higher frequencies of discharge, whereas the traditional neurotransmitter alone is released at lower frequencies (Kupfermann, 1991). Both cortical GABAergic interneurons and pyramidal cells contain a variety of peptides (Hendry et al., 1984a and b; Peters et al., 1983), so that intrinsic bursting or synaptically evoked high-frequency firing in these cells could produce a significant qualitative change in the nature of postsynaptic responses within the circuit. Frequency-dependent intrinsic peptide release has recently been shown in the hippocampus (Weisskopf et al., 1993). The long-term effects of peptidergic actions, such as those that alter postsynaptic potassium conductances via G-protein-coupled mechanisms (Raynor and Reisine, 1992) and presynaptically modulate transmitter release (Colmers et al., 1987; Weisskopf et al., 1993), would again exert important influences on cortical behavior.

Neuronal discharge frequency may also have an important influence on the *distribution* of activities within circuits by affecting propagation of action potentials through the fine axonal arbors within the cortex. In addition to extensive local branching, both pyramidal cells and interneurons have remarkable long-range complex axonal distributions (Kawaguchi and Kubota, 1993; Kisvarday et al., 1993; Kritzer et al., 1992; Ojima et al., 1991). The completeness with which these axonal arborizations are invaded by the action potential is a critical factor determining the amount of transmitter released and the dispersion of excitation or inhibition throughout the postsynaptic population. Interactions among ion channel densities, diameters of "parent" and "daughter" axons at branch points, the presence of synaptic boutons or other axonal enlargements, changes in extracellular potassium concentration ($[K^+]_o$) near bifurcations, and, most critically, the frequency of spike generation all determine the extent of invasion of a nerve impulse through an axonal arborization (Grossman et al., 1979a,b; Manor et al., 1991; Smith, 1980a,b; Westerfield et al., 1978). The widespread, complex arbors of cortical axons thus may be a site at which the strength and topographic distribution of synaptic activities can be regulated through influences of firing frequency. Such a complex integrative aspect of structure and function in the cortex might be missed with techniques focused exclusively on biophysical and molecular characterization of neural membranes and agonist-activated receptors.

Further diversity in the effects of burst generation on cortical function may occur because the intrinsic mechanisms underlying bursting vary among neurons. Bursts arise from low-threshold calcium spikes in some cells (Foehring and Wyler, 1990; Friedman and Gutnick, 1987; Kawaguchi and Kubota, 1993; Montoro et al., 1988; Tseng and Prince, 1993), whereas in others they are generated by summated depolarizing afterpotentials (Calvin and Sypert, 1976; Chagnac- Amitai et al., 1990; Mason and Larkman, 1990; Tseng and Prince, 1993) that may be due to underlying Ca^{2+} (Friedman et al., 1992; Wong and Prince, 1978, 1981), and Na^+ conductances (Friedman et al., 1992; Konnerth et al., 1986b). In still other cells, slow depolarizations mediated by a non-inactivating Na^+ conductance (Montoro et al., 1988; Stafstrom et al., 1985) contribute to burst generation. These differences in the underlying conductances mean that the conditions under which burst generation will occur (e.g., the level of depolarization), the firing patterns within bursts, and the capacity to generate repetitive bursts will differ in different neurons (Montoro et al., 1988; Tseng and Prince, 1993).

Not only are the mechanisms underlying bursts variable but so too are the portions of the cell membrane that are capable of generating burst discharges in different neurons, and the types of action potentials within the burst (Masukawa and Prince, 1984; Wong and Prince, 1979; Wong and Stewart, 1992; Wong et al., 1978). In neocortical pyramidal neurons, bursts with low thresholds (Chagnac-Amitai et al., 1990; McCormick et al., 1985) are presumably somatically generated. However, recent studies

(Amitai et al., 1993; Kim and Connors, 1993) show that bursts of Na^+ and Ca^{2+} action potentials also occur in dendrites. The presence of diverse burst-generating mechanisms at various sites on the neuronal membrane raises the possibility that each is activated under different circumstances, is regulated by a different set of transmitters/modulators, and is designed for a different functional outcome. The capacity of distal dendritic membranes to initiate action potentials in cortical pyramidal neurons may serve to increase the efficacy of remote EPSPs (Amitai et al., 1993; Deschenes, 1981; Kim and Connors, 1993), as originally suggested for fast prepotentials arising from dendrites in hippocampal pyramidal cells (Spencer and Kandel, 1961), whereas somatic bursts would be translated more directly into cell output, and spikes propagating from soma into dendrites (Stuart and Sakmann, 1994) might have quite a different effect on synaptic events. Burst discharges at dendritic (Wong and Prince, 1979) and somatic sites may be blocked by inhibitory postsynaptic potentials, raising the possibility that one or the other site might be selectively inactivated by proximal versus distal inhibitory inputs known to be derived from different subsets of inhibitory interneurons in neocortex (Kawaguchi and Kubota, 1993).

Without much more information, it is difficult to know which of the above speculative effects of neuronal discharge frequency and bursting actually occur in the cortex or how they will the affect performance of simplified model circuits in which input-output functions of neurons and connection strengths within the network are critically important (Lehky et al., 1992). Since anesthetics and slice procedures can have significant effects on inhibition and other variables that affect the dynamics of cortical function, one of the challenges over the next three decades will be to determine the validity of findings in cortical slices and anesthetized animals for normal cortical function *in vivo*.

An obvious consequence of the diversity and complexity described above is the difficulty in predicting the effects of experimentally or pathologically induced alterations, such as activation of a particular receptor subtype by a neurotransmitter; interference with an intracellular second-messenger signaling system; or "knock-out" of a gene controlling a particular channel, receptor, or messenger, on the behavior of a neuron or cortical circuit. In other words, neither alteration of a particular molecule or of a particular functional subclass of neuron will necessarily have a predictable effect on cortical function. In reality, cortical circuits are organized so that parallel activation of a number of interdependent systems controlling excitability is occurring with even simple operations. The behavior of single neurons must depend on concurrent activities in multiple classes of afferents, some of which leave traces as spontaneous EPSPs and

IPSPs, whereas the ongoing effects of others, such as remote dendritic synaptic activities or the tonic modulatory influences of extracortical ascending systems that release various neurotransmitters (McCormick et al., 1993), are harder to identify. The interaction of these agonist-activated conductances with a large number of voltage-dependent channels ultimately controls input-output relationships, action potential parameters, as well as the capacities of neurons to fire in various patterns. Therefore, to understand the behavior of individual neurons and to assess the potential effects of one experimental maneuver or another, it becomes critical to fit the new biophysical and molecular information into well-developed concepts and models of cellular function, just as predictions of the normal or abnormal operation of networks require detailed information regarding the properties of the constituent neurons. Without this step, we will be left with lists of items affecting neuronal properties, without a story line. "The truth is that a great intellectual challenge must be in understanding the overall logic of higher-level systems without which the lower-level detail may become mere cataloguing of information" (Noble and Boyd, 1993).

From the above, it is evident that prediction of the net effects of pharmacological agents on behavior within cortical circuits can also become a significant challenge. Experiments in which actions of drugs at single synapses are compared with effects in larger cortical networks can yield unexpected results. Small alterations in synaptic or membrane function that may be difficult to quantify electrophysiologically may be amplified by the network to produce large effects. For example, submicromolar concentrations of bicuculline that result in small decreases of *monosynaptic* IPSPs (about 10% to 20%; Chagnac-Amitai and Connors, 1989a), can have potent effects on the distribution of evoked excitatory events within neocortical slices and can lead to development of epileptogenesis during which *polysynaptic* IPSPs are actually exaggerated, probably because of the enhanced excitation of interneurons by disinhibited excitatory circuits (Chagnac-Amitai and Connors, 1989b; see Chapter 9). A general principle is suggested by these results: the larger and more complex the circuit, the smaller the alterations in synaptic or intrinsic properties required to produce substantial functional alterations. The results of recent experiments on the effects of ethosuximide on oscillations within intrathalamic circuits provide another example of this phenomenon. By causing small decreases in transient calcium currents in both nucleus reticularis and ventrobasal complex neurons, ethosuximide markedly decreases oscillatory activity in the circuit at doses that have only small effects on individual low-threshold calcium spikes at each site (Huguenard and Prince, 1994a).

EPILEPTOGENESIS AND STUDIES OF CORTICAL PHYSIOLOGY

"The Brain, within its Grove
Runs evenly and true—
But let a Splinter swerve
T'were easier for you—
To put a Current back—
When Floods have slit the Hills
And scooped a Turnpike for Themselves—
And trodden out the Mills"

Emily Dickinson (1862)

The long history of contributions of studies of human epilepsy to our knowledge of cortical function (Jackson, 1931; Penfield and Erickson, 1941; Penfield and Jasper, 1954) naturally led many of the pioneers of cortical intracellular recording techniques to perform experiments with models of epileptogenesis (Creutzfeldt, 1956; Goldensohn and Purpura, 1963; Kandel and Spencer, 1961; Li, 1955, 1959; Li and Jasper, 1953). Electrical stimulation or the application of convulsant drugs easily converted a "normally" functioning cortex to one that generated abnormally large and synchronous electrical signals similar to those found in human epileptic brain. Epileptogenesis could be induced in even phylogenetically primitive cortices (Servit and Strejckova, 1970a,b; Shen and Kriegstein, 1989), suggesting that such functional instability goes hand in hand with the evolution of complex vertebrate cortical circuits. The study of intracellular mechanisms underlying the generation of acute epileptiform activities began in earnest with work in the 1960s (for reviews see Ajmone-Marsan, 1969; Dichter and Ayala, 1987; Prince, 1978; Prince and Connors, 1986; Wong et al., 1986). The results of these and more recent experiments emphasize that delineating the cellular mechanisms of epileptogenesis has progressed hand in hand with our understanding of the behavior of normal cortical neurons.

The use of endogenous/spontaneous epileptiform signals as probes has made it possible to examine the properties of cortical synaptic networks, the influences of cortical output on distant structures, and the net effects of subcortical modulatory systems on cortical excitability. For example, *in vivo* experiments showed that the synchronous discharge of neurons in a cortical or hippocampal epileptiform focus could evoke a zone of "surround inhibition" (Dichter and Spencer, 1969a,b; Prince and Wilder, 1967), thus unveiling properties of cortical synaptic organization that are normally present (Krnjevic et al., 1966) and important in processing sensory signals (Sillito, 1975). Studies of projected epileptiform activities at more distant sites in contralateral cortex (Schwartzkroin et al., 1975b), thalamus (Gutnick and Prince, 1974; Noebels and Prince, 1978b; Schwartzkroin et al., 1975a) and brainstem sensory nuclei (Schwartzkroin et al., 1974) showed that the abnormal high frequency output of neurons within the epileptogenic focus could engage normal local circuitry in remote areas, resulting in widespread changes in excitability (Prince, 1976). The important roles of various subcortical structures in regulating cortical excitability were uncovered using models of epileptogenesis (McNamara et al., 1983; Mirski and Ferrendelli, 1986; Noebels, 1984a; Piredda and Gale, 1985; and Osorio, 1991). The ease with that epileptogenesis could be elicited by drugs that selectively decreased postsynaptic $GABA_A$ receptor-mediated IPSPs led to observations relevant to the normal role of GABAergic inhibition in regulating cortical neuronal excitability, such as control of intrinsic burst discharges (Wong and Prince, 1979), regulation of activity in polysynaptic excitatory circuits (Miles and Wong, 1987a,b), modulation of concurrent NMDA receptor-mediated excitatory synaptic events (Luhmann and Prince, 1990a; Staley and Mody, 1992), unusual and synchronous synaptic interactions among inhibitory interneurons (Aram et al., 1991; Michelson and Wong, 1991), and the potential for long-term changes in expression of inhibition due to activity (Merlin and Wong, 1993; Miles and Wong, 1987; Wong and Miles, 1993). Effective models of circuit and cellular activities in hippocampus (Traub and Jefferys, 1994; Traub and Wong, 1982) have also been derived from studies of acute epileptiform discharges.

Studies of epileptogenesis have uncovered several unusual types of cortical activity or response that might have been missed in experiments focused on normal cellular events. For example, indications that large shifts in $[K^+]_o$ and $[Ca^{2+}]_o$ can be induced by neuronal discharges were obtained from direct measurements in the course of experiments on cortical epileptogenesis (Benninger et al., 1980; Heinemann et al., 1977; Prince et al., 1973). It is now well accepted that increases in $[K^+]_o$ elicited by extracellular stimulation (Malenka et al., 1981) or by the discharge of single neurons (Hounsgaard and Nicholson, 1983; Yarom and Spira, 1982) can affect synaptic events and the excitability of adjacent cells. Nonsynaptic synchronization of neurons by large increases in $[K^+]_o$ (Yaari et al., 1986) and ephaptic interactions in hippocampus were described during epileptogenesis (Taylor and Dudek, 1984). Observations made during the course of studies of cortical epileptogenesis have also provided important (and at times unrecognized) leads for subsequent experiments. One example deals with the properties of cortical presynaptic terminals. Results of experiments on neuromuscular preparations treated with convulsant drugs and other agents had shown that nerve terminals can act as sites of impulse initiation independent of more proximal axonal action potentials (Dun and Feng, 1940; Grundfest, 1966). It was subsequently found that terminals of thalamocortical relay neurons in cortex could initiate bursts of action potentials during focal epileptiform discharges induced by convulsant drugs (Gutnick and Prince, 1972; Rosen and Vastola, 1971; Rosen et al., 1973), electrical stimulation (Noebels and Prince, 1978a), chronic epileptogenic lesions (Pinault and Pumain, 1985), or even under "normal" conditions (Pinault and Pumain, 1989). The mechanisms underlying these axon terminal discharges in cortex were not elucidated. Many subsequent studies have revealed that axonal terminals are important sites for modulation of the strength of synaptic events through activation of a large complement of agonist-activated hetero- and autoreceptors (for review see (Thompson et al., 1993a). It seems possible that activation of presynaptic receptors by locally released transmitters or modulators will turn out to be responsible for the "back-firing" phenomenon, perhaps with some contribution from changes in $[K^+]_o$ or $[Ca^{2+}]_o$ known to occur during cortical neuronal discharge (Benninger et al., 1980; Heinemann et al., 1977). Recent evidence suggests that electrically-induced epileptogenesis in hippocampal slices may also be associated with the onset of antidromic spikes in terminals of CA3 pyramidal neurons, which are dependent on activation of NMDA and GABA receptors (Stasheff et al., 1993a,b). Other data indicate that presynaptic terminals possess a variety of voltage-dependent channels (Bielfeldt et al., 1992; Robitaille and Charlton, 1992; Stanley, 1993; Uchitel et al., 1992), and that terminal membranes may be sites of genetically determined alterations in the efficacy of transmitter release (Wu et al., 1983). It is interesting to speculate whether the increases in membrane input resistance and time constant and the decreases in spike frequency accommodation that have been detected in somata of cortical neurons after injury (Prince and Tseng, 1993; Tseng and Prince, 1991) are also reflected in axons and terminals where they might have important effects on transmitter release. The above examples emphasize that there is much to learn about the function of normal cortical neurons and circuits from studies of pathophysiological events.

FUNCTIONAL DIVERSITY AND EPILEPTOGENESIS

Since epilepsy is a syndrome with a multitude of etiologies extending from genetic defects to chronic brain injury, such as might be produced by a stroke, tumor, infectious process, or cortical trauma, it is unlikely that some common mechanism underlies all types of seizure disorders. The diverse processes controlling neuronal excitability provide many avenues by which pathological events can lead to the development of epileptiform activities in cortex. Previous studies of animal models have catalogued these potential mechanisms (Dichter and Ayala, 1987; Prince and Connors, 1986). The application of molecular neurobiological techniques now promises to add to this list (Bergold et al., 1993; Casaccia-Bonnefil et al., 1993) and provide more complete information about already established themes, such as manipulations that decrease the efficacy of GABAergic neurotransmission, enhance synaptic excitation, or alter membrane channels to make neurons intrinsically more excitable. Cellular and molecular alterations detected in the course of pathological processes or produced by seizures per se will depend upon a complex set of variables within the cortical network for expression as "epileptogenic."

Initial animal experiments were often conducted *in vivo*, using acute epileptiform foci produced by applications of convulsant drugs to the normal cortical surface, and later in hippocampal and neocortical slices exposed to similar agents. Results (for reviews see Dichter and Ayala, 1987; Prince, 1978; Prince and Connors, 1986) provided information about some of the basic cellular phenomenology underlying acute epileptogenesis and emphasized the importance of both intrinsic membrane properties (Connors, 1984; Gutnick et al., 1982; Schwartzkroin and Prince, 1978, 1980a; Wong and Prince, 1979) and the cortical synaptic network (Gutnick et al., 1982; Hablitz and Johnston, 1981; Johnston and Brown, 1981; Traub and Wong, 1982) in initiating, synchronizing, and amplifying signals. Since most of these early studies were done following cortical exposure to agents that depressed $GABA_A$ receptor-mediated inhibition without effects on neuronal membrane properties (but see Hotson and Prince, 1981; Schwartzkroin and Prince, 1980b), it was reasonable to conclude that epileptogenesis was an emerging property of a cortical network that might be due to a change in balance between excitation and inhibition; it did not depend on an identifiable "intrinsic abnormality" of individual "epileptic" neurons. Although neurons with unusual subtypes of GABA or glutamate receptors, or with alterations in the density or distribution of such receptors, might certainly qualify as having an "intrinsic abnormality," this term has been taken to mean an abnormality of voltage-dependent or passive membrane properties. These models and conclusions were thus, in a sense, tautological. When other preparations have been used in which chronic cortical injury is a prominent feature, such as delayed epileptogenesis following status epilepticus or epilepsy as a consequence of cortical trauma, intrinsic abnormalities that increase excitability in neurons (Ashwood et al., 1986; Franck and Schwartzkroin, 1985; Prince and Tseng, 1993) and damage affecting inhibitory synaptic circuits (Bekenstein and Lothman, 1993; Franck and Schwartzkroin, 1985; Obenaus et al., 1993; Ribak and Reiffenstein, 1982; Sloviter, 1991a) have been discovered.

Studies of epileptogenesis that begin with the production of a specific alteration in excitability, followed by assessment of whether or why seizure activity occurs, essentially represent "bottom-up" (or inductive) approaches. Such experiments have revealed mechanisms by which epilepsy *might* occur as a result of some acquired or inherited human brain pathology. Results serve as a template against which to examine mechanisms underlying chronic models of epileptogenesis in animals and epilepsy in man. The alternative "top-down" (or deductive) approach is very much more difficult and involves starting with a model in which epileptogenesis occurs for unknown reasons (e.g. genetically determined, post-traumatic, kindling, etc.) and then attempting to determine which of many possible mechanisms are operative. Both strategies are necessary and interdependent, whatever the level of inquiry. "Top-down" experiments in the human epilepsies may ultimately yield solutions to questions of pathogenesis and specific approaches to prophylaxis or therapy.

EPILEPSY AND THE CHANGING BRAIN

> "The work on the cortex has seen recently the almost complete end of one and the beginning of a new phase. What might be termed the anatomical era has come to a close, and the physiological era has begun" (von Bonin, 1960).

It seems that work on the cortex, in whatever subdiscipline of neuroscience, will be never ending. Progress in understanding the changing brain illustrates this point. Results of experiments over the past three decades have revealed that a surprising degree of both anatomical and functional plasticity is present in both the developing and mature brain (Antonini and Stryker, 1993; Brown et al., 1990; Cusick et al., 1990; Jones, 1993; Matthews et al., 1976; Pons et al., 1988; Schlaggar et al., 1993; Shatz, 1990; Ulas et al., 1990). Multiple mechanisms probably exist by which properties of cortical neurons and circuits may be changed over time during normal development, as a response to repetitive activation, or as a consequence of cortical injury, as discussed in Sections IV and V. Epileptogenesis can be regarded as a striking result of maladaptive cortical plasticity, as well as a process that, once established, might induce long-term changes in cortical function. In this context, three issues relevant to epilepsy and the changing brain have generic importance in studies of the cortex and represent major investigative challenges.

Activity-Related Changes in Gene Expression

> "Many problems are raised by these results, and a temptation that has to be resisted is to load them with more interpretation than they should be made to bear" (Phillips, 1959).

In situ hybridization techniques have been used to demonstrate changes in mRNAs following seizures. Changes in the levels of message affecting a large variety of postsynaptic receptors (Gall et al., 1990; Wong et al., 1993), trophic factors (Ernfors et al., 1991; Gall and Lauterborn, 1992; Merlio et al., 1993), immediate early genes (Herdegen et al., 1993; Lanaud et al., 1993; Morgan et al., 1987; Shin et al., 1990) and transmitters (Marksteiner et al., 1990; Rosen et al., 1992; Sonnenberg et al., 1991) have been reported. The extent of such alterations seems only limited by the number of mRNAs studied; changes in more than 25 mRNAs had been reported by the time this chapter was prepared. The answers to a number of important questions are required to interpret these findings. Do any of these changes bring about a fundamental reorganization in the structure or function of cortical circuits? If functional alterations do result from activity-induced up- or downregulation of mRNAs, are they adaptive (i.e. homeostatic) or maladaptive when viewed in the context of the spectrum of parallel shifts in many other factors that can alter excitability of neurons and circuits? Since changes in levels of many of these mRNAs occur after a variety of nonepileptogenic stimuli (Dragunow et al., 1990; Ruzdijic et al., 1993), what determines the specificity of the activity-induced alterations that would be relevant to epileptogenesis versus other processes? Are seemingly transient changes part of a broader pattern that initiates hyperexcitability, or are they part of the "noise" in any complex system that is irrelevant to any long-term processes?

Criteria are obviously needed for establishing a cause-and-effect relationship between an activity-induced alteration in gene function and some appropriate change in the brain. Similar issues in the field of infectious diseases led to the development of Koch's postulates to prove that a pathogen (the tubercle bacillus) caused a disease (tuberculosis). A tentative schema along these lines is shown in Figure 1-1. The hypothesis to be tested in this illustration is that the increase in mRNA for protein "X," which is induced by an event that elicits a seizure, has some expected effect on the structure or function of cortical neurons (the "index response"). For example, suppose one finds an increase in brain-derived neurotrophic factor (BDNF) mRNA in dentate granule cells as the result of seizures (Isackson et al., 1991) and hypothesizes that this increase contributes to mossy fiber sprouting within the dentate gyrus. It would first be necessary

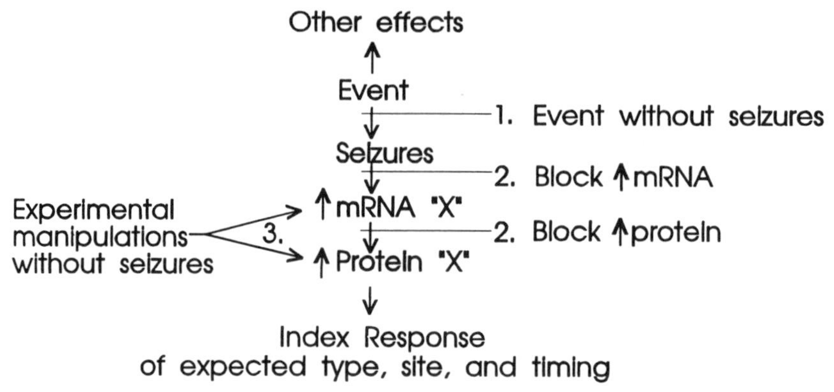

FIGURE 1-1. Seizures and gene expression. Koch's postulates adapted for epilepsy-induced changes in gene expression. An event, such as drug administration or an electrical stimulus, evokes a seizure and has various other effects. It is postulated that the seizure results in some appropriate structural or functional change in the brain ("index response") via the sequence of steps shown. If this formulation is correct, then other manipulations should elicit or block the response: (1) uncoupling the "event" from seizures, (2) blockade of the seizure-induced increase in mRNA or the translation of mRNA to protein, and (3) manipulations that increase the mRNA or protein without seizures.

to show that the increased mRNA results in an increase in BDNF (protein) temporally appropriate to the onset of sprouting. To show that the activation of the mRNA was the critical signal, rather than another event associated with either the stimulus that evoked the seizure or the seizure itself, some other experimental manipulation would have to be performed that would selectively overexpress the gene and increase BDNF in granule cells in the absence of seizures and result in mossy fiber sprouting (pathway 3 of Figure 1-1). Also, the "index response" should be blocked if manipulations are used to prevent the seizure-induced increase in mRNA, the translation of mRNA to protein, or the activation of BDNF receptors (sites 1 and 2 of Figure 1-1). If the investigator then wanted to focus on the functional implications of the sprouting response, other questions would emerge: Is the sprouting associated with the formation of functional synaptic connections and, if so, to which elements? Does the new connectivity contribute to net changes in excitability within the circuit, or is it homeostatic?

This is obviously a simplified schema and experiments might not yield clear-cut results. For example, the occurrence of the sprouting response in the face of a blockade of increases in mRNA and protein would not rule out a contribution of BDNF under normal conditions, but clearly would indicate that it was not the sole necessary factor. In addition, the occurrence of sprouting associated with overexpression of the gene and its product would not prove a direct effect on the sprouting mechanism; indirect effects through stimulation or suppression of another gene might be present. The diagram in Figure 1-1 would also become more complex if one were dealing with an immediate early gene, such as c-fos, or another gene whose product had multiple potential effects in the cortex. As yet, proof that seizure-induced changes in mRNA have some meaningful functional outcome (along the lines of the scheme of Figure 1-1) is not availabe for any single alteration.

There is certainly a variety of evidence showing that seizure activity can lead to excitability changes in cortical circuits. Persistent epileptiform activity can be induced through an NMDA receptor-dependent mechanism acutely *in vitro* by electrical stimulation of slices (Stasheff et al., 1989) or by exposing them to ionic manipulations or convulsant drugs (Hoffman and Haberly, 1989). Several alterations occur during the kindling paradigm that may effect changes in long-term excitability (Köhr et al., 1993; Mody et al., 1992; Yeh et al., 1989; and others; see McNamara et al., 1993 for review). Cell death and circuit reorganization within the hippocampus also occur during the course of limbic system kindling (Cavazos and Sutula, 1990; Cavazos et al., 1991) and in the human epileptic temporal lobe (Isokawa et al., 1993; Masukawa et al., 1992; Strowbridge et al., 1992). This raises the possibility that epilepsy-induced long-term changes in excitability may be in part consequences of injury, rather than purely activity-related.

Effects of Epileptogenesis on Brain Development

A second link between epileptogenesis and long-term changes in cortical activities relates to potential interactions between seizure activity and brain development. Such interactions are a critical issue in the human epilepsies because of the frequent onset of seizures in the immature brain (Aicardi, 1986). Many of the features of developing cortical neurons described in Section IV affect the capacity of immature cortex to generate seizure activity. Another important aspect of this problem has to do with the effects of abnormal signals (i.e. epileptiform discharges) on normal brain development. The establishment of normal connections within the cerebral cortex depends in part on the level and pattern of activity within the circuit (Antonini and Stryker, 1993; Schlaggar et al., 1993; Shatz, 1990); neurotransmitters (Mattson and Kater, 1989) and intracellular levels of Ca^{2+} (Rehder and Kater, 1992) may play an important role as well. The specificity of connections between structures can be influenced by glutamatergic excitation at NMDA receptors (Debski et al., 1990; Simon et al., 1992) and the strength of connections, once formed, by concurrent excitation of pre- and postsynaptic neurons (Brown et al., 1990; Malenka and Nicoll, 1993). Excessive connectivity exists between the hemispheres and between cortex and subcortical structures early in development, which is subject to subsequent structural and functional reorganization (LaMantia and Rakic, 1990; O'Leary, 1992). Certainly these processes could be influenced by the increases in interacellular calcium concentration ($[Ca^{2+}]_i$) that presumably occur in the neurons of epileptogenic cortex (Heinemann et al., 1977, 1986), as well as excessive transmitter release during seizures.

Does the generation of abnormal epileptiform patterns of signals within cortex during development interfere with the establishment of normal functional or structural connectivity, just as deafferentation does in sensory systems (Callaway and Katz, 1991; Jones, 1993)? Available results strongly suggest that this is the case. Experiments by Chow and colleagues (Baumbach and Chow, 1981; Chow et al., 1978; Crabtree et al., 1981, 1983) showed that acute focal interictal epileptiform discharges in immature rabbit striate cortex during the critical period produced abnormal development of neuronal receptive field properties in cortex and lateral geniculate resembling those seen in visual deprivation experiments. These effects were long lasting, were selective in that they did not involve neurons of the superior colliculus, and were not associated with gross lesions in the visual cortex. The mechanisms underlying these alterations have not been elucidated. Results of a recent anatomical study of the effects of cortical epileptiform activity on callosal connections between visual cortices of postnatal rabbits suggest that abnormal discharges tend to stabilize immature interhemispheric projections that are normally eliminated during development (Grigonis and Murphy, 1994). Obviously much more information is needed to determine whether epileptiform activity during cortical development produces long-lasting changes of intrinsic cortical circuitry or alterations in other, more distant projections. The suggestion from these results, however, is that abnormal activity may stabilize connections that ordinarily would be eliminated during development and thus may affect the intracortical connectivity that is necessary for processing sensory information. Such changes could interfere with normal cortical functions and also serve to enhance seizure susceptibility.

The Injured Cortical Neuron and Epileptogenesis

One of the goals of understanding normal cortical function is to obtain information relevant to distortions that occur in the malfunctioning brain.

Abnormalities might be a direct result of injury produced by trauma or a disease process or be due to a sequence of changes within the cortex, set in motion by the precipitating events. In this sense, the frequent and often delayed development of epilepsy after a cortical lesion (Hauser et al., 1991; Salazar et al., 1985) is an overt and dramatic sign of malfunctional cortical plasticity that can be used as an index response to investigate the underlying mechanisms (see Chapters 20 and 21). There are numerous changes in neurons and circuits after injury that might lead to cortical hyperexcitability and epileptogenesis, including alterations in neurotransmitters, trophic factors, and receptors (Anderson et al., 1991; Isackson et al., 1991; Pumain et al., 1986); initiation of axonal sprouting responses with presumed development of new functional circuitry (Masukawa et al., 1992; Salin et al., 1993; Sutula et al., 1988), loss of subclasses of neurons that may be selectively vulnerable (Obenaus et al., 1993; Ribak et al., 1989; Sloviter, 1991b) and altered membrane properties (Prince and Tseng, 1993; Titmus and Faber, 1990; Tseng and Prince, 1991).

Comparatively little is known about the nature of the changes in intrinsic membrane properties of chronically injured cortical cells. Recent studies have assessed the alterations in cortical neurons after distal axotomy (Tseng and Prince, 1990) or following direct cortical trauma that results in epileptogenesis (Prince and Tseng, 1993; Salin et al., 1993). Chronically injured neurons in both experiments had significant increases in membrane input resistance, prolongation of the slow membrane time constant, and a decreased capacity to generate slow afterhyperpolarizations following trains of spikes. The last finding, also reported in hippocampal CA1 pyramidal neurons surviving kainic acid-induced seizures (Ashwood et al., 1986; Franck and Schwartzkroin, 1985), may account for the development of a steeper relationship between firing frequency and intracellular current application (f-I slopes) in these cells. These alterations in cell properties, which would all tend to make chronically injured neurons more excitable, are similar to those found in axotomized mammalian CNS neurons that project to peripheral targets (Gustafsson, 1979; Heyer and Llinás, 1977; Laiwand et al., 1988). In spite of these changes, there was no evidence of evoked epileptiform activities in *in vitro* slices obtained from motor cortex up to 1 year after axotomy of corticospinal neurons in the cervical cord, perhaps because the axotomized cells are the minority of neurons in layer V, or because distal axotomy is an insufficient stimulus to induce intracortical sprouting and reorganization of synaptic connections. However, in the case of direct cortical trauma produced by lesions that partially isolated islands of cortex (Prince and Tseng, 1993), epileptogenesis developed in 1 to 3 weeks (Hoffman et al., 1994). Intracellular labeling showed evidence for sprouting within axonal arborizations of layer V pyramidal neurons, suggesting that a reorganization of intracortical connections had occurred (Salin et al., 1993).

A variety of techniques have been used to induce chronic cortical epileptogenesis (see (Purpura et al., 1972) for review), including recent approaches that produce injury through drug- or electrically-induced seizures (Franck and Schwartzkroin, 1985; Lancaster and Wheal, 1984; Mello et al., 1993; Tauck and Nadler, 1985), prolonged kindling protocols (Sutula et al., 1986), or lesions that partially isolate islands of cortex ((Prince and Tseng, 1993; see also Halpern, 1972). The combination of multiple models of chronic epileptogenesis and a list of plastic changes in cortex that result from injury raises a number of critical questions and challenges for the future. Do all chronic models in which there is cortical injury have qualitatively similar underlying mechanisms (e.g. sprouting that enhances excitatory more than inhibitory circuits), or will seizures that follow direct trauma, stroke, drug-induced status epilepticus, etc. each have a different pathogenesis? Only a minority of cortical injuries result in seizures. There must be either qualitative or quantitative differences in the abnormalities within injured but nonepileptogenic cortex versus epileptogenic cortex. Strategies to determine those differences include attempts to assess the time course of development of various anatomical, electrophysiological, or other abnormalities in relation to the development of hyperexcitability in the injured tissue. Also, as has been previously emphasized (Jasper, 1970), examination of mechanisms underlying chronic epileptogenesis must use injured but nonepileptogenic cortex as controls. Such approaches might provide an opportunity to identify differences between adaptive and maladaptive cortical plasticity and lead to the development of interventions that would prevent epileptogenesis following injury without interfering with normal mechanisms of restitution of function.

CONCLUSIONS

Not even those with the most vivid imaginations 30 years ago would have been able to picture the face of neuroscience today. As likely as not, the same will be said of this era 30 years hence. In Cajal's words, "All we can hope is that some principles of our conception . . . will remain." The rich diversity of all aspects of cortical neurons is not unexpected, when one considers the tasks they accomplish and the multifaceted disturbances in their behavior that are the focus of the clinical neurosciences. The real challenge will be in understanding the implications of neuronal diversity that will lead to a more complete description of cortical function. Examples of circuit operations that can be explained on the basis of the complex intrinsic properties of single neurons and their interactions with synaptic events are starting to emerge. Studies of the cellular and molecular basis of cortical malfunction have made remarkable progress and will continue to contribute substantially to our understanding of normal cerebral function and the brain as a changing organ. One cannot help but marvel at what has been achieved in our quest to fathom the workings of cortical neurons and wonder what the future will hold.

> "The Brain—is wider than the Sky
> For—put them side by side—
> The one the other will contain
> With ease—and You—beside"
>
> (Emily Dickinson, 1862).

Acknowledgments. The work reported in this chapter was supported in part by NIH Grants NS06477 and NS12151 from the NINDS, and the Morris Research Fund. I thank Myrna Canizares and Heather Harris who helped prepare the manuscript, John Huguenard for his valuable comments and discussion, and Reva Prince for her editorial assistance. Studies in my laboratory over the past 30 years could not have taken place without the achievements of a remarkable group of pre- and postdoctoral fellows and colleagues who have made this volume possible. I will always be indebted to them for their friendship and contributions to my life as an investigator.

II

A cellular and synaptic perspective of the cortical neuron

2

Cortical synaptic transmission: an overview

CHARLES F. STEVENS

A microliter of brain contains approximately a billion synapses, about two-thirds of which are excitatory. This neurobiological "factoid" underlines the importance of synapses in the operation of brains and explains the difficulty in studying how synapses work. Clearly synapses are vital because the brain uses so many of them, but their close packing makes the investigation of individual synapses, a necessity for a complete elucidation of their function, very difficult.

Our concept of how synapses work is based largely on work from the Katz school on a particular synaptic model, the frog neuromuscular junction (Katz, 1969). Because this is a specialized synapse for communication between a motoneuron and a muscle cell, one would expect that some aspects of its structure and function are not typical of synapses in general, whereas other properties might be characteristic of all synapses. The trick is discovering where the neuromuscular junction is a valid model for central synaptic transmission and where it breaks down. Various technical advances are helping the analysis of central synapses reach the Katz school's depth of understanding of the neuromuscular junction.

This chapter briefly summarizes the picture of transmission at the neuromuscular junction, defines some of the unresolved issues relating to the function of this model synapse, compares transmission at central synapses, and explores similarities and differences between the neuromuscular junction and central synapses.

TRANSMISSION AT THE NEUROMUSCULAR JUNCTION

Two ideas dominate our concept of neuromuscular transmission: its quantal nature and the fact that it operates probabilistically. The neurotransmitter at the neuromuscular junction, acetylcholine, is released in integral multiples of a smallest transmitter packet, usually identified as a single synaptic vesicle-full. Because these packets of neurotransmitter are nearly the same size, a packet is termed a *quantum*. When a nerve impulse arrives, the release of quanta is probabilistic. In the original Katz view, a synapse possesses N sites at which quanta can be released, and each site releases its quantum with a probability p. This view is formally the same as a coin-flipping problem. If N coins are flipped, each with a probability p of giving heads, the number of heads is distributed according to the binomial distribution. Similarly, Katz and his co-workers showed that the number of quanta released follows a binomial distribution. Each release site corresponds to a coin that is flipped, and the probability of a release (heads) is p, so the distribution of releases is the same as the distribution of heads. From coin flipping theory we know that the average number of quanta released is Np and that the variance in the number released is $Np(1-p)$, so that the coefficient of variation c in the number released is

$$c = \frac{\sqrt{Np(1-p)}}{Np} = \frac{\sqrt{1-p}}{\sqrt{Np}}$$

Because the coefficient of variation specifies the fractional variation in a random process, the relative precision of synaptic transmission becomes greater as p approaches 1, or as N increases. Since N is a very large number for the neuromuscular junction, transmission there is very reliable with little variability (relative to the mean) from one impulse arrival to the next. This is as it should be, because we want a muscle twitch to occur each time the motoneuron sends out a command.

Katz and his colleagues originally identified quanta because nerve terminals spontaneously release individual ones at infrequent intervals. Katz realized that this low spontaneous release rate could be reconciled with the processes of evoked release by assuming that quantal release occurs randomly at a rate that increases sharply but continuously, just after a nerve impulse arrival. His laboratory mapped out the release rate as a function of time after a nerve impulse arrival by counting the number of quanta released in each time bin (Katz and Miledi, 1965). Barrett and Stevens (1972a,b) then demonstrated that the release follows a Poisson process and used a different method to determine the rise and fall of quantal release after the activation of the neuromuscular junction.

Although the Katz theory of synaptic function is generally accepted for transmission at the neuromuscular junction, several difficulties with it are widely recognized. The theory assumes that each release site is identical and independent so that a single release probability is adequate to specify the quantal release process. Direct estimates of regional release probabilities, however, reveal that this quantity is not uniform and that the binomial distribution provides only an approximation, of unknown accuracy, to the actual case (Robitaille and Tremblay, 1991). Extending the Katz theory to include nonidentical release sites presents no theoretical difficulties, but the number of free parameters needed for such a theory ensures that it will always fit the data, whether it is true or not.

However, the most serious difficulty with the Katz theory is the unknown physical (molecular) identity of the release sites, the quanta, and the release probability. Quanta have always been identified with synaptic vesicles, but this identification has never been experimentally verified, and some workers have questioned the correspondence (Kriebel and Gross, 1974). Nor is an independent means available for recognizing release sites, so we really do not understand to what molecular entity N corresponds. Further, the release probability p in reality reflects a complex transient that describes the release rate following a nerve impulse arrival, which means that a complete theory of release must specify this release rate function and provide a mechanistic explanation for it; we are very far from such a theory. However, recent work has identified many of the molecules involved in exocytosis (Bennett and Scheller, 1993; Südhof et al., 1993), so we can anticipate that advances in our understanding the molecular details of release should occur in the next years.

CENTRAL SYNAPSES

In contrast to the general agreement about the adequacy of the Katz theory for neuromuscular transmission, our understanding of central synaptic function is much less advanced. The reason for this is that central synapses are more difficult to study, and appropriate techniques for studying them have become available only relatively recently. The main problem presented by central neurons is that, in most cases, synapses are distributed over a broad spreading dendritic tree so that the investigator has no way to know the precise site at which a particular synaptic current originated. Because synaptic currents are attenuated and distorted by flowing through a length of cable, the underlying time course and amplitude of any synaptic current at its site of origin are uncertain. Testing the Katz theory for a central neuron depends on recognizing the size of a central quantum and then using the appropriate coin flipping theory to predict the fluctuations in the distribution of synaptic current amplitude. Yet, disentangling inherent fluctuations in synaptic current size from variations that arise from differences in site of origin is very difficult. In fact, when an amplitude distribution of synaptic currents is made, one generally does not see the discrete quantal peaks that are typical for the neuromuscular junction, so that the fitting of theories with so few constraints (such as independent knowledge of quantal size) becomes hard to believe. As a consequence, the extent to which the Katz theory applies to central synapses is still being debated (Edwards, 1991; Korn and Faber, 1991).

Three technical advances should ultimately permit an understanding of central synaptic transmission. The first is the advent of whole-cell recording (Hamill et al., 1981). One of the main technical problems has been that quanta in central neurons are small so that the signal-to-noise ratio provided by traditional microelectrode recording is inadequate. With modern recording techniques (Blanton et al., 1989; Edwards et al., 1989), however, spontaneous synaptic currents are easily resolved.

The second technical advance has been the development of brain slices from various brain regions (Dingledine, 1984). The use of brain slices permits good control of the extracellular medium composition, a necessity for the experimental analysis of synaptic transmission, and provides good access to neurons and afferent pathways. By selecting particular cell types—cerebellar granule cells (Silver et al., 1992) or hippocampal mossy fiber synapse (Jonas et al., 1993)—problems associated with the distribution of synapses through the dendritic tree can be minimized.

Finally, the use of primary neuronal cultures (Banker and Goslin, 1991) provides even better control over synaptic contacts. Individual synapses can be studied with microscopic as well as electrophysiological methods, and contacts between defined cell pairs can be investigated. All conclusions derived from cell cultures must, of course, be confirmed in slices (and ultimately *in vivo*) because the extent to which cells in primary culture (or in slices) are typical is still uncertain.

Three main conclusions about central synaptic transmission are widely accepted, in addition to some classical results relating to excitatory and inhibitory mechanisms. First, central synapses, like the neuromuscular junction, release transmitter probabilistically (Redman, 1991), but the applicability of the Katz theory has been disputed (Edwards et al., 1990; Korn and Faber, 1991). As a result, no generally accepted theoretical structure is available for the description of central synaptic transmission. What is becoming clear, however, is that individual synapses are very unreliable, with a nerve impulse arrival failing to produce release more than half of the time for many synapses (Hessler et al., 1993; Raastad et al., 1992;

Rosenmund et al., 1993). Precisely what fraction of synapses are unreliable is currently unknown, nor have the implications of this phenomenon for neuronal information processing been fully explored.

The second main conclusion is that quanta at central synapses are much more variable in size than are those at the neuromuscular junction (Bekkers and Stevens, 1989; Bekkers et al., 1990; Edwards et al., 1990; Jonas et al., 1993; Kullmann and Nicol, 1992; Malinow, 1991; Malgaroli and Tsien, 1992; Manabe et al., 1992; Raastad et al., 1992; Silver et al., 1992), so much so that the quantal nature of central synaptic transmission is—if it exists—obscured. The source of this quantal variability is still not understood. Some workers believe that a single vesicle-full of neurotransmitter is sufficient to saturate the postsynaptic receptors (Edwards et al., 1990; Jonas et al., 1993; Larkman et al., 1991), so that quantal variability must arise from different receptor numbers at different synapses. Each quantal size, on this theory, corresponds to a synaptic size. If these workers are correct in believing that a single vesicle can saturate the available postsynaptic receptors, then the Katz theory—in which synaptic current is a measure of the number of quanta released—cannot be applied except in the special circumstance that each synapse can release at most one quantum. Such a property of synapses has been proposed (Korn and Faber, 1991), but the evidence for mammalian central synapses is not yet compelling.

The alternative to the notion that quantal size variations arise from a population of synapses is that different amounts of neurotransmitter are released at a single synapse and that the postsynaptic receptors are not saturated; in this case synaptic currents reflect the quantity of transmitter applied. Bekkers, Richardson, and Stevens (1990) have reported that a single identified bouton exhibits as much variability in quantal size as do a population of synapses. This observation was made on primary cultured neurons and thus might not be typical of central synapses in general, and the other laboratories yet have reported repeating a single bouton experiment. If the Bekkers, Richardson, and Stevens result were accepted and were it generally applicable, it would provide decisive support for the notion that much of the quantal variability arises at individual synapses and that the postsynaptic receptors are not saturated by a single quantum.

The third conclusion is that individual synapses are multifunctional. Since the original description of the unusual properties of NDMA receptors (Mayer et al., 1984; Nowak et al., 1984), it has been apparent that two distance modes of excitatory neurotransmission are present in the brain. These two modes are now known to be present at individual synapses (Bekkers and Stevens, 1989), at least is some cases, so that glutamate released presynaptically has two different types of effects on the postsynaptic cell: through the non-NMDA receptors, it produces a conductance change analogous to the operation of the neuromuscular junction, but through the NMDA receptors, it produces a much longer-lasting effect, the magnitude of which depends, via the postsynaptic membrane potential, on what other inputs the cell is receiving. The computational significance of these dual modes of synaptic operation is not yet well understood.

Beyond the acetylcholine receptors, other postsynaptic features of neuromuscular transmission have received scant attention. Since most excitatory synapses in the cortex reside on spines, the neuromuscular junction offers little guidance for understanding the probably complex processes just beyond the postsynaptic glutamate receptors. However, some hints about what is going on in the spines are available. They are richly endowed with enzymes, such as CaM kinase II (Carlin et al., 1980; Kennedy et al., 1983; Ocorr and Schulman, 1991), and exhibit complicated calcium handling properties (Guthrie et al., 1991; Muller and Connor, 1991). Although these features of spines have been linked to synaptic plasticity, their precise role remains to be elucidated.

Considerable controversy still surrounds central synaptic transmission, but the tools for deciding many of the unresolved issues are at hand. Application of the best physiological methods, combined with exploitation of the recent advances in understanding the molecular basis for exocytosis, hold out the promise that we shall soon know how central neurotransmission is accomplished.

3

The cortical neuron as an electrophysiological unit

MICHAEL J. GUTNICK and WAYNE E. CRILL

For most neuroscientists, the image evoked by the term "neocortical circuit" is primarily an anatomical one: a diagrammatic abstraction of the magnificent histological drawings from the Golgi studies of Cajal (1899) and Lorente de Nó (1949). Icons and lines depict the pyramidal and nonpyramidal neurons of the different layers, interacting in an elaborate network of axodendritic and axosomatic connections. The only physiological diversity clearly expressed in such anatomy-based representations is a polar distinction between excitatory and inhibitory synapses. Functional neuronal characteristics are generally not depicted, leaving a tacit assumption that all cortical cells have similar, "integrate-and-fire" properties. However, cellular physiological studies from all areas of the brain have revealed that a rich assortment of ionic channels combine variously in the soma-dendritic membranes of different neurons to produce a diversity of firing types (Llinás, 1988); the cortex is no exception. To add a dynamic dimension to our concept of the neocortical circuit, we must consider the distinctive input-output characteristics of different circuit elements and the diverse ionic channel types that determine them. In this chapter, we briefly review the different firing types of neocortical neurons and summarize the properties of the ionic currents that have been identified in cortex. We also review mounting evidence that the elaborate dendrites of neocortical neurons are not merely passive accumulators of synaptic input but rather that they also possess functionally significant active properties.

INTRINSIC FIRING PATTERNS OF NEOCORTICAL NEURONS

Although it has long been clear from *in vivo* intracellular studies that neocortical neurons do not all have the same intrinsic physiological characteristics (Calvin and Sypert, 1975; Takahashi, 1965), it was only with the greater stability and experimental manipulability afforded by the brain slice technique that it became possible to systematically study cortical membrane characteristics. This study has led to the classification of neocortical neurons into three primary physiological types, based on spike characteristics and on response patterns during sustained, intracellularly applied current pulses: regular-spiking (RS), fast-spiking (FS) and intrinsically bursting (IB); (Connors and Gutnick, 1990; McCormick et al., 1985; Fig. 3–1). As is often the case when typology is based primarily on phenomenology, this classification is not definitive. However, although strong arguments can and have been made for additional subcategories in different specific cortical regions and/or in different species, there does seem to be general agreement in the literature about these three broad classes. Results of intracellular staining studies indicate that each morphological cell type probably has a consistent physiological correlate.

The most commonly encountered neuron is the RS cell, which is found in all subpial laminae (Fig. 3–1A). During sustained depolarization, RS neurons undergo profound frequency accommodation. Differences in the time course of accommodation led Agmon and Connors (1992) to distinguished between two types of RS neurons in mouse barrel cortex: In RS_1 cells, which were found in all cortical layers, the firing frequency was initially high, but fell precipitously after a few impulses to a much lower maintained rate, whereas in RS_2 cells, which were found only in the deeper layers, firing frequency decreased steadily throughout the intracellular stimulus. In RS cells, individual action potentials are relatively prolonged

→

FIGURE 3–1. Intrinsic firing patterns of neocortical neurons. **A**. In response to an intracellular depolarization, RS neurons show initial high frequency spiking and rapid adaptation to a sustained lower frequency. **B**. FS neurons fire at a high, sustained frequency throughout the course of the stimulus. **C**. IB neurons respond to a threshold stimulus with an all-or-none burst. **D–F**. Three different neurons in the infragranular layers of the guinea pig cortex, illustrating a spectrum of RS firing patterns. Each row shows activity of a different neuron, and each successive column shows the response to a larger depolarizing pulse. The F-I graphs in the 4th column plot instantaneous frequency calculated from the first interval, as a function of stimulus current intensity (normalized to threshold intensity). Note that the during larger stimuli, the pattern of the initial response varies from cell to cell depending on the spike afterpotentials. (**A-C** from Connors and Gutnick, 1990; **D-F** from Friedman, 1991).

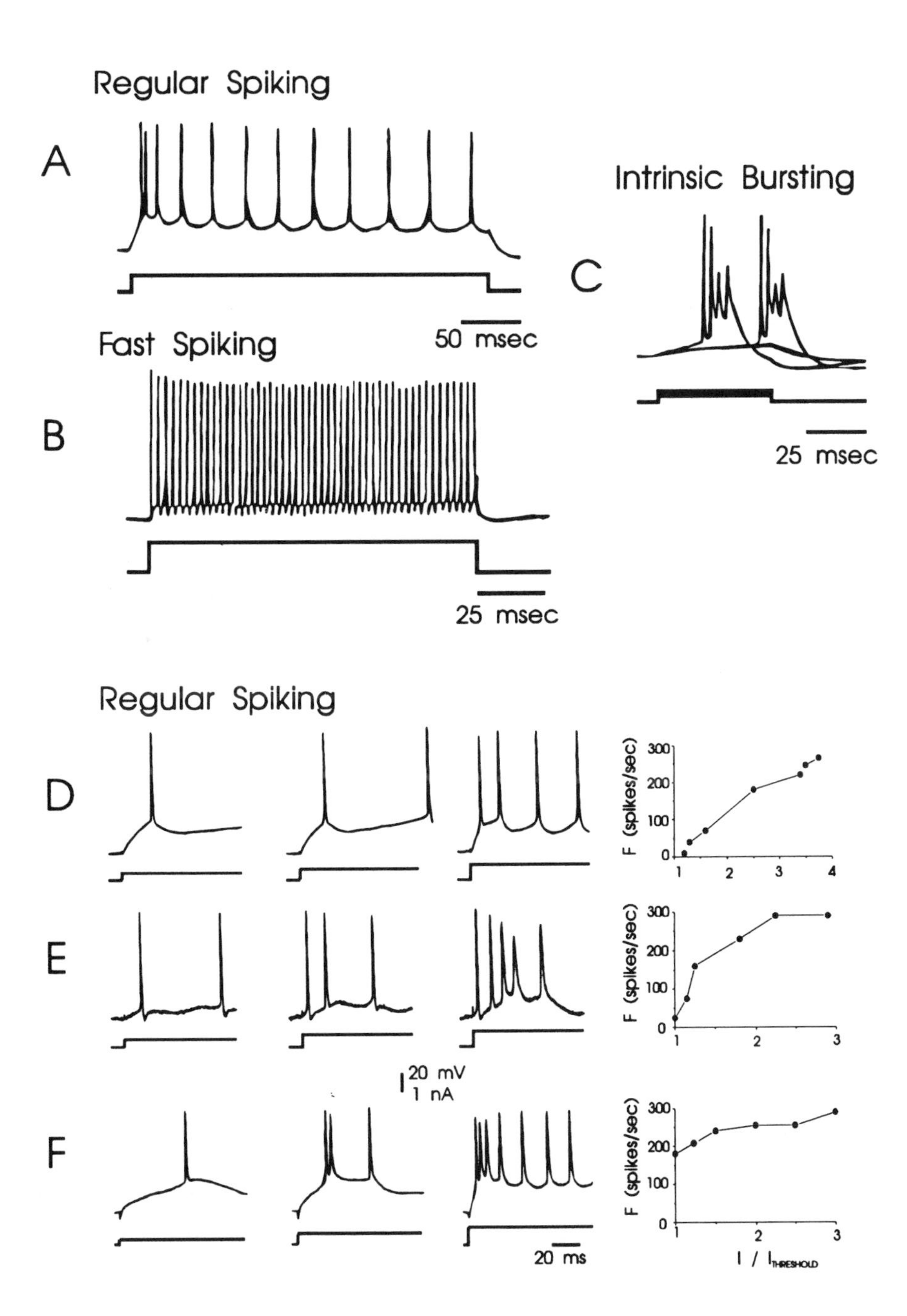
Regular Spiking
A
50 msec
Fast Spiking
B
25 msec
Intrinsic Bursting
C
25 msec
Regular Spiking
D
E
F
20 mV
1 nA
20 ms
F (spikes/sec)
I / I THRESHOLD

and are followed by a variable complex of depolarizing and hyperpolarizing afterpotentials. Thus, the initial discharge during an intracellular pulse can be quite variable from cell to cell (Fig. 3–1D–F). Moreover, a given cell's firing pattern can vary as a function of membrane potential, depending on the currents underlying its afterpotentials. The RS category thus seems to encompass a spectrum of behaviors, with some neurons firing an initial cluster of spikes as depolarizing afterpotentials sum, whereas in others, spike afterhyperpolarizations prevent initial bursting (Friedman, 1991). Since RS cells are the most frequently encountered neurons in "blind" electrophysiological experiments, it is not surprising that, in intracellular staining experiments, they are usually found to be pyramidal neurons. However, some spiny stellate cells have also been found to have RS characteristics (Gutnick and Lobel-Yaakov, unpublished observations).

FS neurons are distinguished by an ability to fire action potentials repetitively at high frequencies with little or no adaptation throughout the course of a sustained intracellular depolarization (Fig. 3–1B). The action potentials themselves are extremely brief (<0.8 msec), due to accelerated repolarization. Whenever they have been stained, FS neurons have always been found to be smooth or sparsely spiny nonpyramidal neurons (McCormick et al., 1985). Because this morphology is typical of neurons shown histochemically to contain either GABA or its associated enzymes, the evidence is strong that all FS neurons are GABAergic inhibitory cells. This conclusion was directly confirmed in a patch-clamp study in which the typical nonpyramidal morphology was first ascertained during the experiment using infrared-differential interference contrast video microscopy; then, subsequent analysis of the mRNA of single cells using primers specific for glutamic acid decarboxylase (GAD65) revealed mRNA for the GABA-related enzyme in three of four FS cells and in none of four RS pyramidal cells (Jonas et al., 1994).

Although it is evident that all FS neurons are GABAergic, the converse is apparently not the case—not all inhibitory interneurons are FS cells. Kawaguchi (1993) showed that many sparsely spiny stellate neurons in layer V of rat frontal cortex have a different firing pattern, which entails low-threshold spiking (LTS) at hyperpolarized potentials. GABAergic interneurons are among the most richly diverse cells in the cortex with regard to co-localization of classic neurotransmitters with various neuropeptides, especially calcium-binding proteins (DeFelipe, 1993). In a study that combined intracellular recording and staining with subsequent immunohistochemistry, Kawaguchi and Kubota (1993) found evidence that this chemical diversity may have physiological correlates; they found that parvalbumin-containing cells had FS characteristics, whereas calbindin$_{D28K}$-containing cells were of the LTS type.

IB neurons generate stereotyped clusters of three or more spikes; typically, the burst is the minimal response to a threshold depolarization (Fig. 3–1C). Responses to sustained depolarizations vary from cell to cell. Some fire an initial burst followed by repetitive single spikes, whereas others fire repetitive bursts, and it is not uncommon for an IB neuron to flip back and forth between regular spiking and repetitive bursting modes. IB neurons have a clear laminar distribution in the cortex. They were first noted in layer IV and upper layer V of somatosensory and cingulate cortex of the guinea pig (Connors et al., 1982). In subsequent studies in primary somatosensory cortex of rats (Chagnac-Amitai et al., 1990; Larkman and Mason, 1990) and mice (Agmon and Connors, 1989, 1992), they were found primarily in layer Vb. In studies involving intracellular staining, they were shown to be large pyramidal neurons with stereotyped dendritic and axonal morphologies (Chagnac-Amitai et al., 1990; Larkman and Mason, 1990). In recordings from retrogradely labeled neurons in somatosensory and visual cortex, many corticopontine and corticotectal neurons had IB characteristics (Wang and McCormick, 1993). Similarly, many identified cortico-spinal neurons of the rat motor cortex had firing patterns characteristic of IB cells (Tseng and Prince, 1993). It thus seems that most IB cells send axonal projections out of the cortex while at the same time providing important collateral feedback into the local circuit. Burst generation by such strategically placed circuit elements must have profound functional implications, both for output to subcortical areas and for normal and abnormal synchronization of activities within the local neuronal network (Gutnick and Friedman, 1986; Gutnick et al., 1982). These implications are discussed at length by Connors and Amitai in Chapter 9.

As noted above, in many RS neurons action potentials are followed by prominent DAPs, which can evoke additional spikes (ie. bursts) under appropriate conditions. Thus, the boundary between categories is somewhat indistinct, as neurons with very similar firing patterns are classified as IB by one author and RS by another (compare the IB neurons of Chagnac-Amitai et al. [1990] and Wang and McCormick [1993] with the "intermediate" cells of Silva et al. [1991b] and the RS_{DAP} cells of Tseng and Prince [1993]). This raises the question whether cortical neurons might be pluripotential, moving between RS and IB modes of firing depending on physiological conditions. On the one hand, Friedman and Gutnick (1989) demonstrated that virtually all neocortical RS neurons are transformed into bursters when intracellularly injected with a Ca^{2+} chelator, an effect they attributed to blockade both of Ca^{2+}-dependent K^+ conductance and of Ca^{2+}-dependent Ca^{2+} current inactivation. Moreover, Wang and McCormick (1993) reported that noradrenergic, cholinergic, and glutaminergic modulatory input can cause IB neurons to switch from bursting mode to repetitive firing mode by blocking voltage-dependent and Ca^{2+}-dependent K^+ conductances, thereby causing membrane depolarization. On the other hand, the tendency to burst seems to be a robust phenotype of certain neurons. Silva et al. (1991b) showed that in the mutant reeler mouse, in which there is a general inversion of the neocortical laminae and significant distortion of neuronal morphology, the depth distribution of intrinsic firing patterns is also inverted: IB and "intermediate" neurons, which were only found in layer V of the normal mouse, were restricted to the superficial layers of the reeler cortex.

Taken together, the evidence seems to indicate that in the infragranular layers of all neocortical regions of all species, there is a population of large pyramidal neurons, the axons of which project to various subcortical regions; these neurons have a tendency to intrinsic burstiness. Among these neurons, a spectrum of firing patterns is displayed in *in vitro* experiments, ranging from regular spiking with prominent DAPs to repetitive bursting. At any given moment, the actual firing mode of these neurons may vary as a function of such factors as membrane potential level and intracellular Ca^{2+} accumulation—functionally relevant parameters that are normally under synaptic and neuromodulatory control. Another population of pyramidal neurons with distinct RS characteristics is present in all cortical layers, such that they include virtually all the supragranular pyramidal neurons and a sizeable fraction of the infragranular cells; although these neurons may differ in their adaptation characteristics, they normally do not flip into a bursty firing mode.

VOLTAGE-DEPENDENT CURRENTS IN NEOCORTICAL NEURONS

The different membrane properties that determine how neocortical neurons transform synaptic inputs into spike trains must reflect the varied type, quantity, and modulation state of ionic channels. Unfortunately, most of our present knowledge about the ionic basis of repetitive firing derives only from a single neuron type, the RS cell. Clues about putative ionic mechanisms first came from examining the effects of pharmacological agents upon intracellularly recorded responses. Subsequently, with the development of single electrode and patch voltage-clamp techniques, whole-cell ionic current components were measured. The quantitative biophysical details of each of these ionic currents have been difficult to define because in slice preparations electrotonically remote portions of the neuron (dendrites) often are not isopotential with the soma during voltage clamp. Nonetheless, this approach has revealed many different ion channel types in neocortical neurons. Molecular biological and immunological investigations have provided us with even more information about the diversity and detail of channel molecules and the physiologically and pharmacologically identified groups of whole-cell currents must contain a mixture of molecular channel types.

Table 3–1 lists the channel types that have been identified so far in RS neurons from cat neocortex. We have grouped the ionic currents on the basis of their primary role in the transduction of synaptic signals into electrogenic responses. The transient sodium current, I_{Na}, provides the regenerative depolarizing current for the all-or-none action potential. Depolarizing currents activated negative to the spike threshold—I_H, $I_{Ca(T)}$ and I_{NaP}—dictate the firing pattern of neurons and can summate with incoming synaptic activity. The high-threshold (HVA) calcium currents primarily couple electrophysiological responses with intracellular chemical signaling mechanisms. The bevy of potassium currents rapidly repolarize the individual action potentials and control the pattern of adaptation to a sustained input by calcium-activated currents.

Depolarizing Currents

Sodium Currents. We have accepted the presence of voltage-dependent sodium channels of the Hodgkin-Huxley type in central neurons since the intracellular recording experiments of Sir John Eccles in motoneurons (Eccles, 1957). With the development of *in vitro* brain slice and tissue culture techniques, the specific sodium channel blocker tetrodotoxin (TTX) could be applied and, as expected, rapidly abolished the action potential of neocortical neurons (Connors et al., 1982). Huguenard et al. (Huguenard et al., 1988) measured the transient sodium currents in acutely isolated neurons from neonatal and developing neocortex. The threshold for transient sodium channel activation is about −50 to −60 mV. Fifty percent of the sodium channels inactivate at −60 mV. These measurements agree with a whole-cell spike threshold of −50 to −55 mV. Sodium channels are present in the soma and dendrites of neocortical neurons (Huguenard et al., 1988; Kim and Connors, 1993; Stuart and Sakmann, 1994).

Soon after the introduction of the slice technique, Connors et al. (1982) reported a TTX-sensitive, persistent, depolarizing electrogenic response in neocortical neurons, similar to the sodium-dependent subthreshold response previously recorded in cerebellar Purkinje cells (Llinás and Sugimori, 1980a). Stafstrom et al. (1982) identified it with a sodium-dependent, TTX-sensitive noninactivating current (I_{NaP}), which they

TABLE 3–1 Physiologically Identified Channel Types in Regular Firing Neocortical Pyramidal Neurons

Current	Blockers	Primary Function
		depolarizing currents
I_{Na}	TTX	Action potential depolarization
I_{NaP}	TTX	Constant threshold; adds to synaptic currents
I_H	Cesium	Keeps E_m near threshold; burst firing in some cell types
$I_{Ca(T)}$		? Burst firing
$I_{Ca(L)}$	Dihydropyridines	Intracellular signaling; activates K^+ currents
$I_{Ca(N)}$	ω-conotoxin	Intracellular signaling; activates K^+ currents
$I_{Ca(other\ ?P)}$	ω-Aga IVA toxin	Intracellular signaling; activates K^+ currents
		hyperpolarizing currents
I_{KTf}	TEA	Action potential repolarization
I_{Ks}	4-AP	Action potential repolarization
$I_{K(Ca)}$	Apamin	mAHP
$I_{K(Ca)}$	Muscarinic and β-adrenergic agonists	Slow AHP; adaptation
$I_{K(Na)}$	Muscarinic and β-adrenergic agonists	Slow AHP; adaptation

found in cat neocortical pyramidal cells (Fig 3–2B). I_{NaP} is also present in cells of the entorhinal cortex, hippocampus, and thalamus (Llinás, 1988).

The functional role of the persistent sodium current has been difficult to measure directly. Pharmacological blockade with preservation of the transient sodium current has not been possible. We can nonetheless infer some of the specific contributions of I_{NaP} to neocortical neuron behavior based upon its kinetics and voltage dependence. I_{NaP} activates about 10 mV negative to the transient current. It therefore provides an inward current at depolarized but subthreshold values of membrane potential. In neocortical neurons I_{NaP} is large enough to cause a net inward current that sets the spike threshold by depolarizing the cell to potential values where the transient current is activated (Stafstrom et al., 1982, 1985). Because the spike threshold is set by I_{NaP}, which has no detectable inactivation, the cell should show no accommodation of spike threshold to slow depolarizations, a property of neocortical cells that differs from axons, which have only the transient inactivating sodium current. I_{NaP} also supplies a constant depolarizing current that sums with synaptic currents. It has fast-activation kinetics that allow it to add to a brief synchronous excitatory synaptic current. I_{NaP} is required for the voltage-dependent

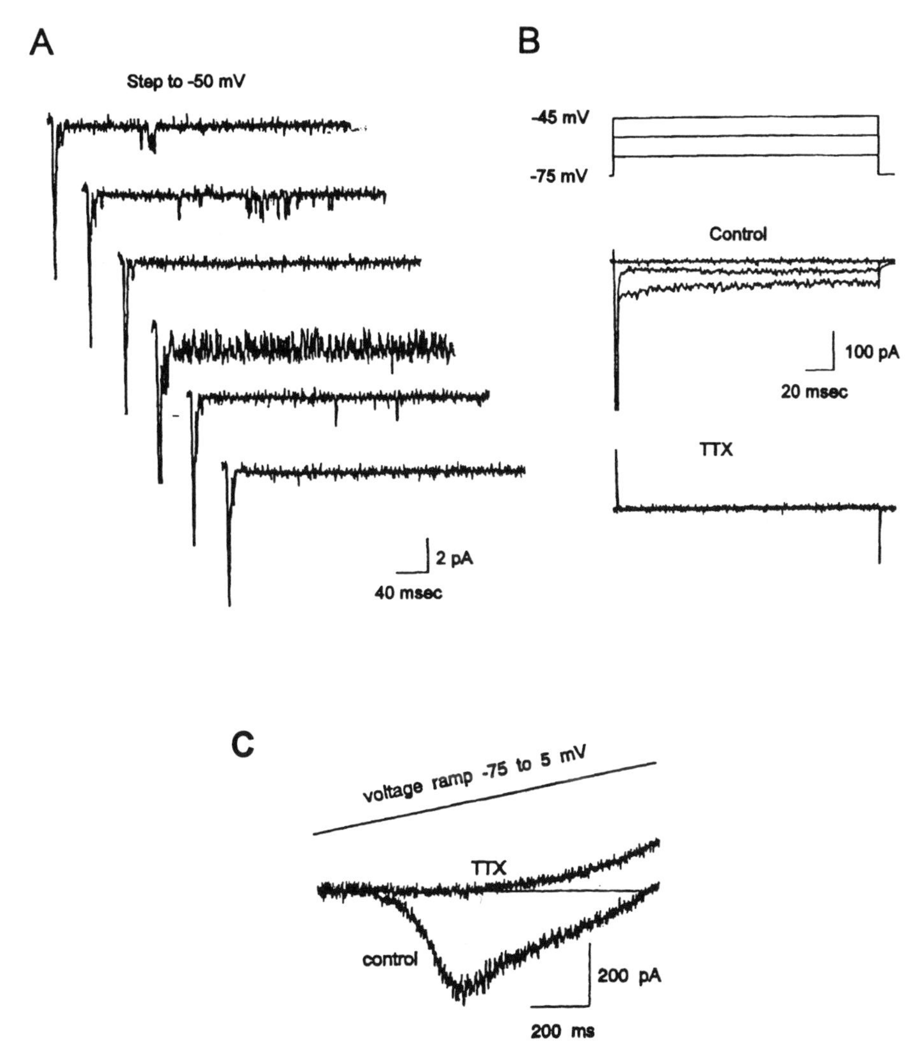

FIGURE 3–2. Multichannel patch recordings from acutely isolated rat neocortical neuron. **A**. Sodium channel activity recorded in the cell attached mode and evoked by a step to −50 mV from a holding potential of −100 mV. Cell bathed in high potassium to zero the membrane potential. The transient openings of sodium channels overlap at the beginning of the step and are partially truncated. Brief late openings appear on traces 1 and 2 from the top and persistent openings are present in trace 4. **B**. Whole cell currents measured in the absence and presence of TTX. Because the gain is high, the transient current is truncated. **C**. Effect of a slow voltage ramp from −75 to 5 mV, before and after exposure to TTX.

rhythmic oscillations present in entorhinal neurons (Alonso and Llinás, 1989).

Two different mechanisms for the persistent current have been proposed. In cardiac myocytes and hippocampal pyramidal neurons the conductance for the persistent current increases with depolarization in a Boltzman fashion (French et al., 1990; Saint et al., 1992). Channels with classical Hodgkin and Huxley properties will give a persistent window current that peaks near threshold because the steady-state activation and inactivation curves for the transient sodium current overlap in the spike threshold range of membrane potential. One of the subtypes of sodium channels might have different inactivation properties. That is, the current does not peak as expected from a window current. Because immunocytochemical techniques have identified one subtype of sodium channel in cell bodies (RI) and another subtype in axons (RII), Westenbroeck and co-workers (1989) postulated that these two antigenic types might underlie the persistent and transient sodium currents, respecitvely.

Alzheimer et al. (1993b) used on-cell patch recordings from visually identified pyramidal neurons in thin slices and acutely isolated neocortical pyramidal neurons to search for populations of sodium channels with different gating properties. Both preparations had a measurable persistent current with whole-cell recording (Fig. 3–2B & C). These studies revealed transient sodium channel openings and two modes of late openings following the transient openings. Brief late openings were clustered as minibursts lasting 10-40 msec (Fig. 3–2A trace 1 and 2). This type of late opening provided an open probability less than 0.01. In the second mode of late channel opening an individual channel produced a burst of openings that persisted throughout the depolarizing voltage clamp command lasting up to 400 msec (Fig. 3–2A trace 4). Calculations revealed that the minibursts and sustained bursts could produce a small but significant persistent current. Both the unitary amplitude and slope conductance of transient and late openings were nearly identical.

Calcium Currents. Subthreshold depolarization opens transient or inactivating calcium channels (T channels), and larger depolarizations open the high-voltage-activated (HVA) calcium channels (L, N, and P channels; Tsien et al., 1991). L-type calcium channels are blocked by dihydropyridines and may be more concentrated on the cell soma (Westenbroek et al., 1990). N channels are blocked by ω-conotoxin GVIA (ω-CgTx) and show a moderate and variable amount of inactivation. Site-directed antipeptide antibodies to the α1 subunit of N channels stain primarily dendritic shafts and nerve terminals in neocortex (Westenbroek et al., 1992a). P channels were first identified by their selective sensitivity to funnel-web spider venom in cerebellar Purkinje cells, where they are located in the dendrites and somata (Hillman et al., 1991). P current is blocked by ω-Aga-IVA spider toxin (Llinás et al., 1989) which decreases the HVA calcium current in rat neocortical neurons by about 30% (Brown, et al., 1994). Low-threshold calcium electrogenesis can be prominent in neocortical neurons (Friedman and Gutnick, 1987; Sutor and Zieglgänsberger, 1987). $I_{Ca(T)}$ is relatively small in cortex as compared to the HVA channels (Sayer et al., 1990). In view of the importance of $I_{Ca(T)}$ in generation of voltage-dependent bursting in thalamic neurons (see Chapter 11), the low-threshold calcium conductance might be larger in the bursting type of neocortical neurons, but calcium current types have not been measured directly in IB cells.

The HVA calcium currents contribute little to the electrogenic response of regular spiking neurons since blocking calcium currents has no detectable effect upon action potential duration. The depolarization during the action potential still allows an appreciable calcium entry (Markram et al., 1994). The increase in $[Ca^{2+}]_i$ activates calcium-dependent conductances, as well as various second-messenger systems, and probably increases $[Ca^{2+}]_i$ further by calcium-induced calcium release from intracellular stores.

Hyperpolarizing Inward Rectification (I_H). Many types of excitable cells, including neurons, cardiac myocytes, and various receptors, have an inward cation current activated by hyperpolarization (I_H). In neocortical neurons (Fig. 3–3) the threshold for activation is about −55 mV and it is half- activated at −82 mV. A small fraction of the maximum I_H is activated at resting potential (−70 mV). The reversal potential is about −50 mV, and I_H ionic current is carried by sodium and potassium ions. The time course for activation is described by two exponentials with time constants of about 40 and 300 msec. I_H causes the sag in the hyperpolarizing response evoked by a constant current. Because the inward I_H resists any attempt to hyper-polarize the neuron, it keeps the resting membrane potential positive to E_K and within striking distance of the action potential threshold. I_H is not blocked by extracellular barium, but extracellular cesium causes a voltage-dependent block.

Spain et al. (1987) showed that the prominent I_H in the larger layer V neurons of the cat provides a memory for changes in membrane potential. A conditioning hyperpolarization (which activates I_H) will increase the excitability of the large pyramidal neurons to a depolarizing stimulus while I_H deactivates. A relatively modest 10-mV conditioning hyperpolarization can increase the subsequent peak firing threefold.

In neurons, both the biogenic amines noradrenaline (NA) and serotonin (5-HT) modulate I_H (McCormick and Pape, 1990b). Such a finding was expected since a similar current in heart, I_f, is modulated by both acetylcholine and NA (for review see DiFrancesco, 1993). In thalamic neurons activation of β-adrenergic receptors or serotonergic receptors (blocked by the 5-HT_1 and 5-HT_2 antagonist methysergide) increases the magnitude of I_H by shifting the voltage-dependent activation curve toward zero potential by about 6-8 mV. This might be the mechanism whereby brainstem orig-

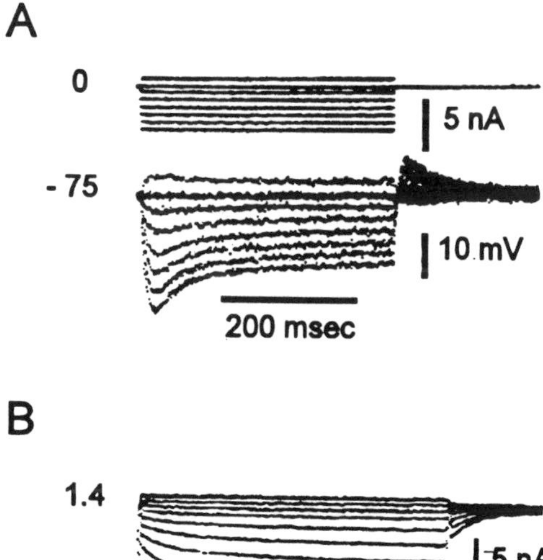

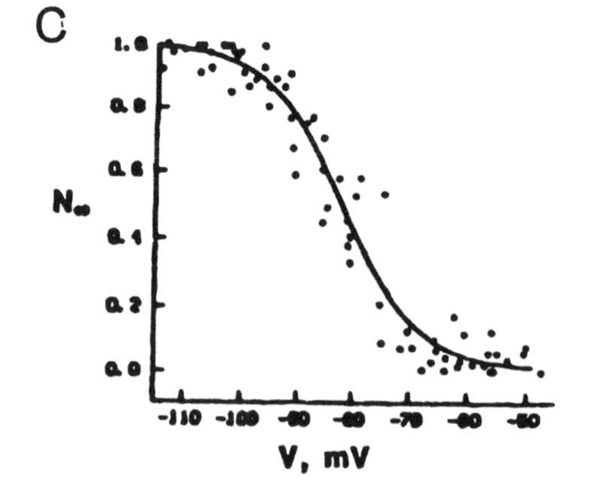

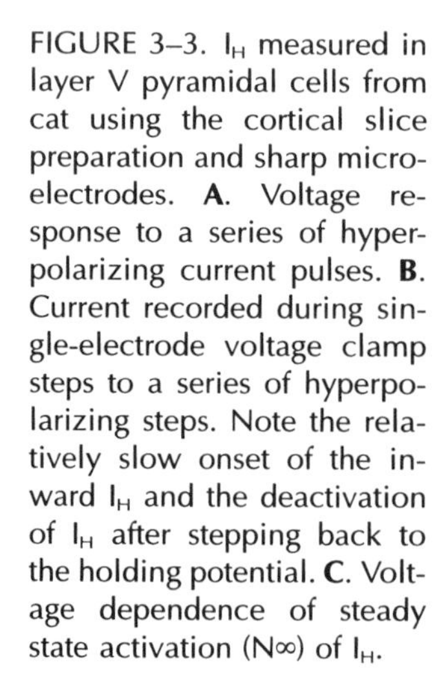

FIGURE 3–3. I_H measured in layer V pyramidal cells from cat using the cortical slice preparation and sharp microelectrodes. **A.** Voltage response to a series of hyperpolarizing current pulses. **B.** Current recorded during single-electrode voltage clamp steps to a series of hyperpolarizing steps. Note the relatively slow onset of the inward I_H and the deactivation of I_H after stepping back to the holding potential. **C.** Voltage dependence of steady state activation (N_∞) of I_H.

inating adrenergic pathways evoke the alerting response. Thalamic neurons are depolarized and show rhythmic firing, rather than the burst firing present at more hyperpolarized values of membrane potential. Serotonin also shifts the steady-state voltage-dependent activation of I_H in the pyramidal neurons of cat cortex, and as expected, it increases posthyperpolarization excitation in the cells with a large resting I_H (Spain, unpublished observations). In heart the increase in cAMP directly modulates I_H channels without an intermediary kinase. In the thalamus both the effects of β-adrenergic and serotonergic receptor activation are mediated by cAMP. Phosphodiesterase inhibitor, cyclic AMP analogs, and the adenyl cyclase stimulant, forskolin, all increase I_H.

Repolarizing Currents

The superfamily of voltage-dependent and second-messenger-gated potassium channels now includes over 50 members, many of which have been identified by molecular neurobiologists. The general characteristics of many of the potassium currents in neurons have been reviewed (Llinás, 1988; Rudy, 1988). The following discussion primarily focuses on the work of Crill, Schwindt, and colleagues, who, over the past years, have examined in detail the outward currents of cat neocortical neurons (Foehring et al., 1989; Schwindt et al., 1988, 1989, 1992a,b; Spain et al., 1991a,b).

Voltage-Dependent Potassium Currents. In cat neocortical neurons there are two physiologically and pharmacologically defined voltage-dependent potassium channels. Both types show rapid activation kinetics and are distinguished by the time course of inactivation and their sensitivity to blocking agents. The fast inactivating potassium current, I_{KTf}, (Fig. 3–4A) decays within msec and is 60% blocked by 1 mM extracellular tetraethylammonium. 4-aminopyridine (4-AP) has little effect upon I_{KTf}. A slowly inactivating transient potassium current, I_{KTs}, (Fig. 3–4B) is also present in neocortical neurons. I_{KTs} is not appreciably affected by tetraethylammonium, but is reduced by 4-AP. I_{KTf} begins to activate at about −60 mV whereas I_{KTs} activates at more depolarized potentials, about −50 mV. Both transient currents, I_{KTf} and I_{Ks}, are activated negative to the spike threshold in Betz cells, and both have an appreciable window current where the inactivation and activation curves overlap, providing a steady conductance over a window of membrane potentials. As originally noted by Connor and Stevens (1971), removal of inactivation of a transient potassium current (I_A) by hyperpolarization delays the generation of an action potential to an ensuing depolarizing stimulus. Similar behavior occurs in the smaller

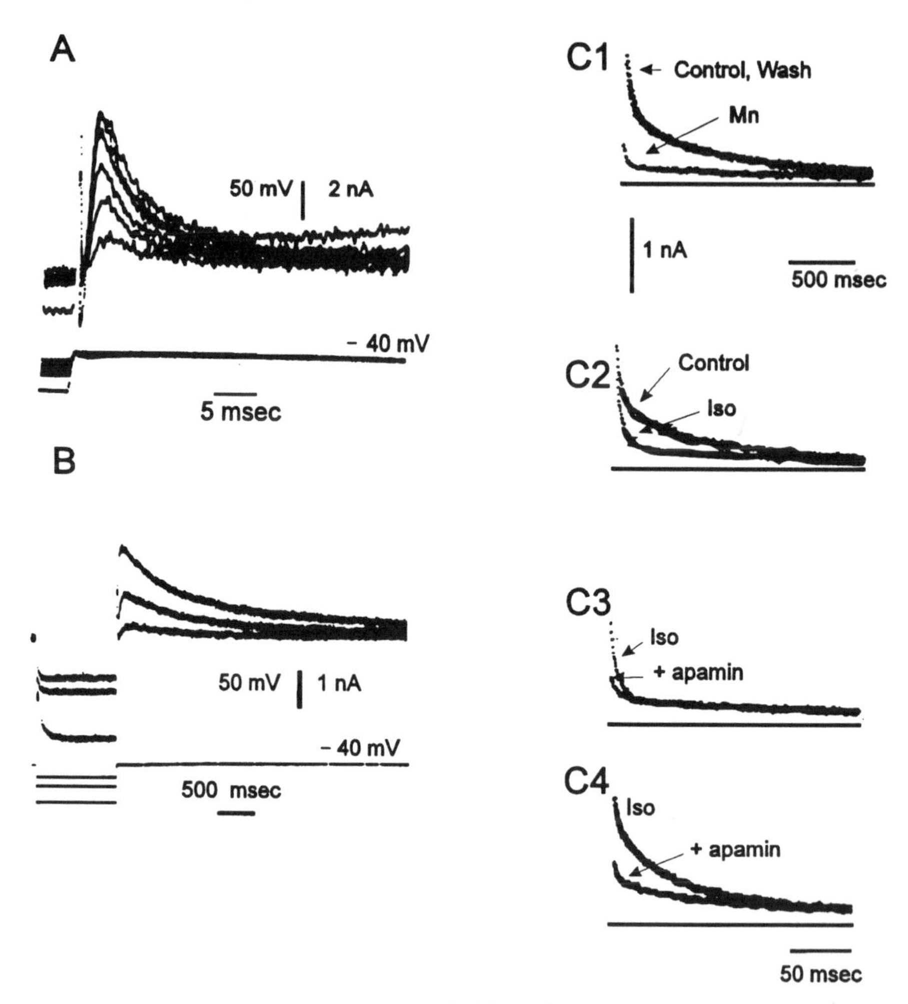

FIGURE 3–4. Potassium currents recorded from layer V pyramidal neurons in cat neocortical slices using the single electrode voltage clamp. **A**. Superimposed current traces of I_{KTf} evoked by stepping to −40 mV from different conditioning potentials. **B**. Superimposed current traces of I_{KTs} evoked by stepping to −40 mV from different conditioning voltages. Note the different time bases for A and B. C: Outward currents evoked by a train of 90 action potentials at 100 Hz. **C1**. Effect of substituting Mn^{2+} for Ca^{2+}. **C2**. Isoproteronol (iso) blocks the slow component of the tail current. **C3**. apamin blocks the remaining fast component. **C4**. same as C3 at faster time base.

layer V pyramidal cells of the neocortex (Spain et al., 1991b). Inhibition, caused by I_{Ks}, follows a hyperpolarizing conditioning current in these neurons. The two rapidly activated, transient, voltage-dependent potassium currents, I_{KTf} and I_{Ks}, produce a brief afterdepolarization (fAHP) lasting only a couple milliseconds after each action potential.

Ca- and Na-Activated Potassium Currents. Neurons in the mammalian central nervous system have several potassium currents activated by increases in intracellular calcium concentration. Some calcium-mediated potassium currents are also voltage dependent. I_C, a calcium-mediated potassium current first identified in frog sympathetic ganglia (Pennefather et al., 1985), is sensitive to tetraethylammonium and is blocked by charybdotoxin. I_C contributes to spike repolarization in rat hippocampal CA1 neurons but it is not present in large cat neocortical pyramidal neurons.

Neocortical pyramidal neurons have two calcium-mediated potassium currents. One has pharmacological properties similar to the small conductance calcium-mediated potassium channels (SK; Blatz and Magleby, 1986). It is selectively blocked by the bee toxin apamin and is not blocked by tetraethylammonium ions (1 mM) or muscarinic and β-adrenergic agonists (Fig. 3–4C traces 3 and 4). Repetitive firing in response to a constant current stimulus in the presence of apamin shows both the normal phases of rapid and slow adaptation, but the instantaneous rate is increased about 30% throughout the stimulus.

The second type of calcium-mediated potassium current in neocortical neurons decays slowly and causes the slow afterhyperpolarization (sAHP), which may last a second or two after repetitive firing. This Betz cell current is blocked by both β-adrenergic (Fig. 3–4C trace 2) and muscarinic neurotransmitter agonists and is responsible for the slow adaptation of repetitive firing that occurs in response to a prolonged constant stimulus. The transmitter-sensitive current is similar to I_{AHP} identified in other neurons of the central nervous system. Figure 3–4C shows tail currents measured with a single electrode voltage clamp following 90 spikes evoked at 100 Hz. Figure 3–4C2 shows that isoproterenol blocks a slow component of the tail current, and Figure 3–4C3 shows that apamin blocks a more rapidly decaying portion of the tail current.

The slow afterhyperpolarization includes a potassium current component that is present when calcium entry is blocked and when intracellular calcium is buffered with EGTA. This potassium current is blocked by tetrodotoxin and can be evoked by activating only the persistent sodium current. It behaves like a sodium-activated potassium current and it is also blocked by β-adrenergic and muscarinic agonists.

Summary

In sum, neocortical neurons, like other central neurons, have a large group of selectively permeable ion channels in their membranes. Evidently, neurons have used a diversity of channels with different biophysical properties to sculpture their transduction of synaptic inputs into spike trains and to control the entry of calcium ions that link action potential activity to intracellular messenger systems. Because many of these channels that dictate transduction are modulated by neurotransmitters, the old idea of a stationary response pattern is no longer tenable. It seems that the electrical responses of neurons can be explained by about a dozen channel types. Immunocytochemistry has taught us, however, that many of these channel types have a restricted spatial location on parts of the dendrites, soma, and axon (Westenbroek et al., 1992a). As yet, we know little about the electrophysiological significance of this observation. The spatial location of channels on the anatomically complex neocortical pyramidal neuron is the subject of the next section.

DISTRIBUTION OF IONIC CONDUCTANCES IN THE DENTRITIC MEMBRANE

Neocortical pyramidal neurons possess dendritic trees that are among the longest and most elaborate in the central nervous system. Classically, these dendrites have been considered as passive structures that linearly integrate synaptic input. However, from recent evidence it is clear that dendrites of cortical neurons, like those of hippocampal pyramidal cells (Benardo et al., 1982; Wong et al., 1978) and cerebellar Purkinje cells (Llinás and Sugimori, 1980b) possess voltage-dependent channels that invest them with complex, nonlinear properties.

In normal cortex, TTX-resistant slow spikes are readily generated in direct recordings from dendrites, either with sharp electrode penetrations (Amitai et al., 1993) or patch-clamp recordings (Kim and Connors, 1993; Fig. 3–5A). However, it is usually only possible to evoke high-threshold Ca^{2+} spikes with a somatic electrode if rectifying K^+ currents have been pharmacologically blocked (Connors et al., 1982). Under conditions of K^+ current blockade with a high concentration of TEA, and of Na^+ current blockade with TTX, brief somatic depolarizations evoke prolonged Ca^{2+} spikes in neocortical neurons, during which the membrane potential remains at a plateau level for seconds before rapid repolarization. Often, steps appear in the repolarization phase, as the membrane potential lingers at one or more intermediate plateau levels before returning to rest (Fig.

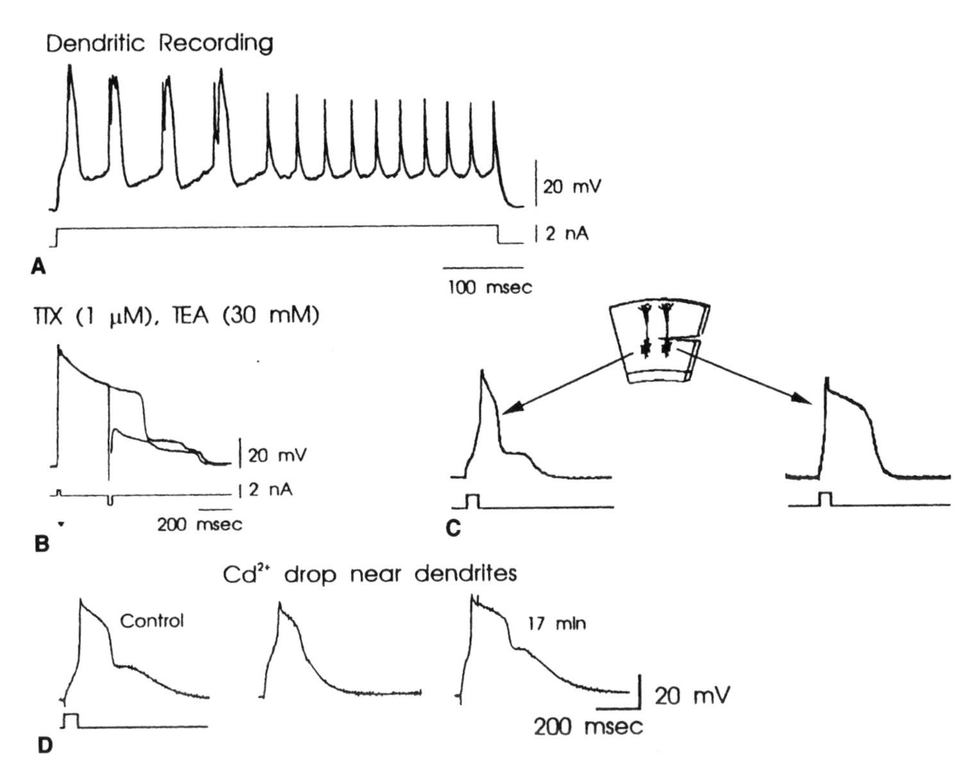

FIGURE 3–5. Active electrogenesis in neocortical dendrites. **A**. Fast and slow spikes in the apical dendrite of a layer III pyramidal cell. In this preparation, the fast spikes, which are blocked by TTX, reflect partial propagation of Na^+ spikes into the dendrite (Amitai et al., 1993). The slow spikes have a higher threshold, are Ca^{2+}-dependent, and are followed by a prominent AHP. **B-D**. In somatic recordings from layer V pyramidal cells, under conditions of blockade of Na^+ and K^+ currents, stepwise repolarization following Ca^{2+} plateaus indicates an electrotonically distal site of HVA Ca^{2+} electrogenesis. **B**. A brief (20 msec), intense (>1 nA) hyperpolarizing current pulse, applied during the Ca^{2+} spike, caused repolarization from the largest plateau without terminating the lower-amplitude plateau levels, since the latter were generated at a distance. **C**. A slit was cut in the slice, parallel to the pial surface, separating apical dendrites of deep neurons from their somas. Stepwise repolarization was recorded in neurons in the intact region of the slice, but was not seen in cells located beneath the cut, suggesting that the lower-amplitude plateau was generated in the apical dendrite. **D**. In a layer V pyramidal cell, focal drop application of 1 mM Cd^{2+} just beneath the pia (>1.5 mm from the somatic recording site) causes reversible blockade of the low amplitude plateau without impeding the large Ca^{2+} spike. (**B** modified from Reuveni et al., 1993; **C & D** modified from Yuste et al., 1994)

3–5B–D). Reuveni et al. (1993) studied stepwise repolarizations in detail, using computer simulations to help in the design of experiments and analysis of results. They concluded that, in addition to a patch of Ca^{2+} electrogenesis at or near the soma, Ca^{2+} spikes are also generated at dendritic "hot spots" that are separated from the proximal patch of excitability and from each other by dendritic zones through which spike propagation is passive. Their findings further indicated that, although HVA channels are not evenly distributed in the soma-dendritic membrane of neocortical pyramidal cells, all have the same voltage dependence. Further experiments on layer V pyramidal neurons (Yuste et al., 1994) showed that the distant plateau (smaller step) is associated with Ca^{2+} electrogenesis in the apical dendrites, since it was not present in cells whose apical trunk had been cut (Fig. 3–5C) and it could be reversibly removed by local application of a Ca^{2+} blocker near the subpial tuft of the apical dendrite (Fig. 3–5D). Using Ca^{2+} imaging techniques, Yuste et al. (1994) directly demonstrated that the distant spike is associated with a band of enhanced Ca^{2+} entry out on the apical dendritic trunk, in the region where it crosses layers II and III. They also showed, however, that HVA channels are present throughout the soma-dendritic membrane; this situation is readily simulated in a compartmental model if it is assumed that the Ca^{2+} current density in the apical band is five to six times that of the rest of the somadendritic membrane (Reuveni, Saar, and Gutnick, unpublished observations).

An important technical advance in the study of dendritic properties has been the introduction of infrared-differential interference contrast video (IRDICV) microscopy, which permits patch-clamp recording from dendrites in slices under direct visual control (Stuart et al., 1993). Coupling electrophysiological recordings with imaging, Markram and Sakmann (1994) have directly shown that excitatory synaptic potentials along the dendritic tree are associated with Ca^{2+} entry through voltage-dependent channels.

The use of IRDICV microscopy allowed Stuart and Sakmann (1994) to record simultaneously from one patch pipette on the dendrite and another on the soma of the same neuron. They found that voltage-dependent Na^+ channels are present throughout the apical dendritic membrane. Na^+ channels in dendrites of isolated neocortical neurons had previously been studied in detail (Huguenard et al., 1989), and their presence had been inferred in dendritic recordings in the slice preparation (Amitai et al., 1993; Kim and Connors, 1994). Moreover, local active boosting of distal synaptic potentials by dendritic Na^+ currents was evident in studies of layer I inputs to the apical tufts of layer V pyramidal cells (Cauller and Connors, 1992). Regehr et al. (1993) found that, when they patch clamped the somata of pyramidal cells, Na^+ spikes originated in the unclamped dendrites; they concluded that synaptic input generates dendritic Na^+ spikes that actively propagate to the soma. However, in simultaneous recordings from dendrites and soma that did not entail voltage clamping the soma region, it was possible to demonstrate directly that the Na^+ channel density is so much higher at the axon hillock that spikes invariably originate there and only then partially propagate back into the apical dendrite (Stuart and Sakmann, 1994). Such active signaling of neuronal output to input sites in the dendrites may be a critical feature of local circuit processing and synaptic plasticity in the cortex.

SUMMARY

In summary, it is now clear that neocortical dendrites possess a rich complement of active channels that render dendritic processing extremely nonlinear and complex. Voltage-dependent inward currents in dendrites undoubtedly serve multiple functions, the precise nature of which are still speculative. In addition to an obvious influence on dynamic control of the local concentration of the ubiquitous Ca^{2+} ion, they also undoubtedly serve to boost distant synaptic signals (Cauller and Connors, 1992) and to effect an intriguing retrograde communication between the sites of axonal output and synaptic input (Stuart and Sakmann, 1994). The evidence indicates that the distant dendrites can constitute an electrotonically distinct compartment, attached to the soma by the apical dendrite. This "stalk" has been shown to contain active inward currents, which enhance the proximity of the dendritic compartment. However, it is also the site of strategically placed inhibitory synaptic inputs. Moreover, it undoubtedly also possesses numerous outward conductances, the control of which by voltage, Ca^{2+}, and modulatory transmitters may make of the dendritic trunk an electrotonically extendable structure that dynamically determines the functional distance of the soma from the sites of dendritic input.

4

Signal transduction and protein phosphorylation

HOWARD SCHULMAN

Higher functions of the central nervous system are based on communication between functional units consisting of many neurons. Communication within and between functional units of neurons is largely based on the chemical transmission of signals with time courses ranging from milliseconds to seconds and minutes. Neurophysiologists first focused attention on fast transmission via ligand-gated ion channels, such as the nicotinic receptors, which can be activated and deactivated within 10 msec. Such a fast response time necessitates that all the components of the release machinery and of the receptor/transducer machinery be in very close proximity (Hille, 1992). In fact, the prototypical ligand-gated ion channel, the nicotinic receptor, contains all of its receptive and signal-transducing components in a single complex of five polypeptides that can rapidly transmit information about binding of acetylcholine into a conformational change that opens the cation-selective ion channel residing in the same complex. The biochemical components of such a signal transduction system are part of the "hard wiring" of the nervous system that makes rapid multisynaptic computations possible, but have a limited inherent capacity for plasticity or modulation.

We now realize that most chemical transmission requires a cascade of enzymatic steps that are relatively slow, but provide for essential modulation of fast transmission and of effects that are independent of ion channels. This typically involves receptors that are coupled to membrane-bound, GTP-binding proteins (G proteins); (Fig. 4–1). Most neurotransmitters in cortex have at least one receptor subtype that functions via a G protein (Nicoll et al., 1990).

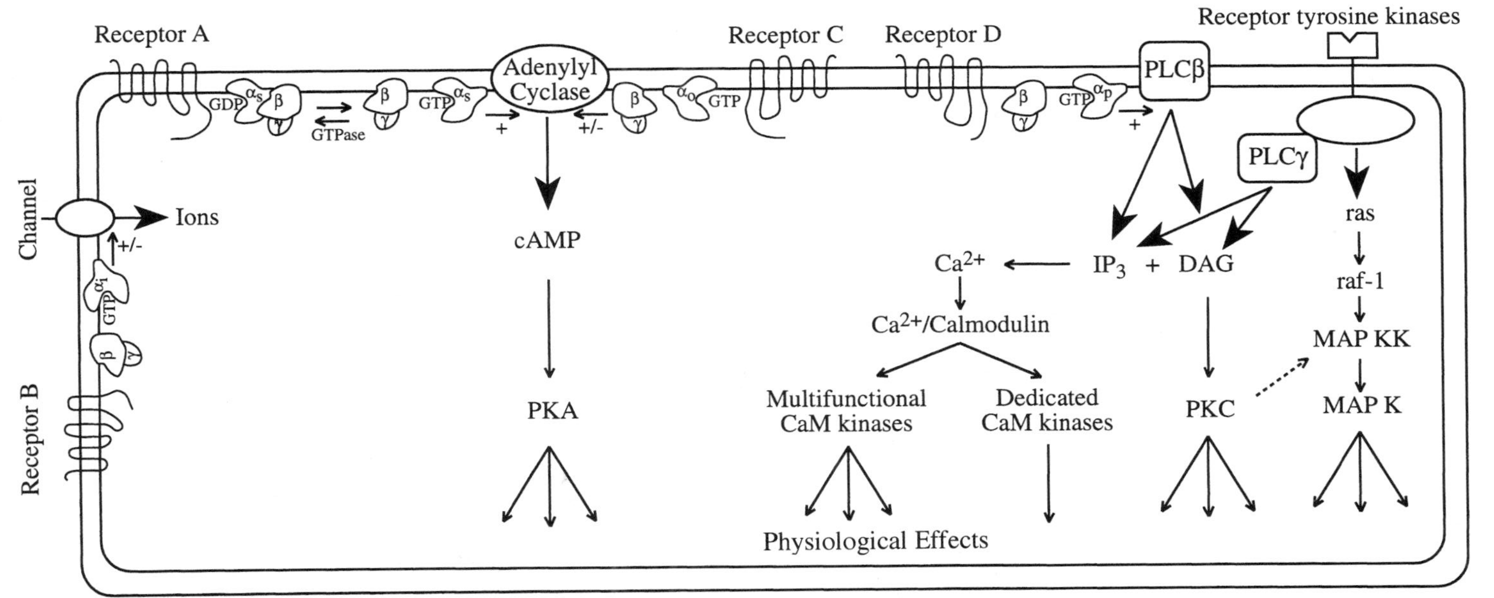

FIGURE 4–1. Generation of second messengers via G-protein- and tyrosine-kinase-linked receptors and their activation of protein kinases. Receptors belonging to the seven transmembrane domain family stimulate dissociation and activation of G proteins by replacement of GTP for GDP, a reaction illustrated for receptor A and that occurs with all G-protein-linked receptors. The GTP-bound α subunits activate effector enzymes (α_s stimulates adenylyl cyclase to make cAMP and α_p stimulates PLCβ to make DAG and IP_3 and to the release of stored Ca^{2+}). G proteins can also modulate ion channels directly by a membrane-delimited pathway, such as via receptor B and α_i. The β/γ dimer can also have direct effects on effector channels and effector enzymes, such as modulation of adenylyl cyclases via receptor C and the β/γ derived from G_o. The second messengers—cAMP, Ca^{2+}, and DAG—modulate PKA, multifunctional CaM kinase, and PKC to coordinate cellular changes in the cytosol and nucleus in response to neurotransmitter stimulation. Receptor-tyrosine kinases can utilize the PI signaling pathway by activation of PLCγ whereas the PI pathway may signal via tyrosine phosphorylation by activation of MAP kinase kinase (MAP KK) by PKC, which in turn activates MAP kinase (MAP K).

G-PROTEIN-COUPLED SIGNAL TRANSDUCTION

The G protein transduces information relayed by a neurotransmitter-bound receptor across the membrane to several types of effector molecules, including ion channels and enzymes (reviewed in Hepler and Gilman, 1992; Hille, 1992). For example, acetylcholine acting on an M_4-type muscarinic receptor, such as receptor B in Fig. 4–1, couples to a G protein that inhibits an N-type Ca^{2+} channel in a membrane-delimited process, i.e. a process not requiring a soluble second messenger. The involvement of several steps in the pathway, such as a distinct receptor, transducer, and effector, slows the response to 100 to 300 msec. This additional time enables the amplification of the signal which is an inherent attribute of G protein signaling (see below), and its spread to multiple ion channels at some distance from the initial site of interaction between transmitter and receptor. Fast synaptic release dependent on Ca^{2+} entry via N channels is thereby modulated over a relatively long period of time by the action of the slower signaling of the G-protein-coupled muscarinic receptor. Neuropeptides diffusing beyond release sites may affect a wide region of cortex by G-protein-coupled effects on such ion channels, thereby altering the receptive or reactive state of whole functional units.

G proteins can couple to enzyme effector molecules, as well as to channel effector molecules, thereby extending the action of neurotransmitters to a multitude of physiological processes (Fig. 4–1). These effectors are typically enzymes that synthesize second messengers. Two of the most common effector enzymes are adenylyl cyclase, which converts ATP to cAMP, and phospholipase C (PLC), which converts phosphatidylinositol (PI) to diacylglycerol (DAG) and inositol trisphosphate (IP_3). IP_3 in turn stimulates the release of intracellular stores of Ca^{2+}. A prominent effect of the three second messengers—cAMP, DAG, and Ca^{2+}—is stimulation of three multifunctional protein kinases, each of which is capable of phosphorylating and thereby changing the activity of many substrate proteins that underlie the "physiological agenda" of the neurotransmitter in that particular neuron (Hille, 1992). Before elaborating on one such protein kinase—multifunctional Ca^{2+}/calmodulin-dependent protein kinase (CaM kinase or CaM kinase II)—signal transduction via G proteins is presented in detail.

G proteins exist in two functional states based on the type of nucleotide bound to the α subunit of this trimeric protein consisting of α, β, and γ subunits (Fig. 4–1). In the inactive state the G protein has a tightly bound GDP whereas in the active state it has GTP bound to it. The function of the neurotransmitter (or agonist) is to convert the receptor to a form that can catalyze the dissociation of GDP from G-GDP and its replacement with GTP to form G-GTP. This leads to the dissociation of the G protein into α-GTP monomers and β/γ dimers (see receptor A, Fig. 4–1). Although the accepted dogma has been that the α subunit is the mediator of G protein action on ion channels and other effectors, there is now compelling evidence for an independent role for β/γ (see cross-talk below). At least 20 distinct G proteins have thus far been identified by purification and molecular cloning, and several distinct effector molecules for these have been identified (Hepler and Gilman, 1992). For example, activation of G_s by receptors for dopamine ($D\text{-}_1$) and norepinephrine (β) stimulates adenylyl cyclase, activation of G_p by a receptor for acetylcholine (M_1) stimulates a PI-specific phospholipase C, and activation of G_1 by receptors for epinephrine (α_2) and acetylcholine (M_2) stimulates a K^+ channel.

The G-protein-based signaling system provides several advantages over fast transmission. These advantages include amplification of the signal, modulation of cell function over a broad temporal range, diffusion of the signal to a large spatial range, cross-talk among signaling systems, and coordinated modulation of diverse cell functions.

- *Amplification*—each receptor can activate numerous G proteins before the neurotransmitter dissociates. A single G protein can, in turn, activate many effector molecules, each of which produces many second messengers that activate kinases that can phosphorylate many substrates before deactivation. Such a cascade can amplify the action of a single neurotransmitter molecule several thousand-fold.
- *Temporal range*—although receptors coupled to G proteins function more slowly than ligand-gated ion channels, transmission can occur within 100 to 300 msec. For example, maximal levels of cAMP and IP_3 in the rat olfactory system are attained within 50 msec and can decline due to feedback termination of the signal by protein kinases within 100 to 200 msec (Boekhoff and Breer, 1992). Membrane-delimited coupling of G proteins directly to ion channels is presumably even faster. Production of second messengers and their consequent action on protein kinases can also be more protracted, lasting many seconds, and provide a temporal range of action beyond that which fast transmission can accomplish.
- *Spatial range*—a slower time frame means that cellular processes that are quite distant from the receptor can be modulated. Diffusion of second messengers, such as IP_3, Ca^{2+}, and DAG, can extend neurotransmission through the cell body and to the nucleus to alter gene expression (Allbritton et al., 1992; Hille, 1992).
- *Cross-talk*—both the signal transduction machinery and the ultimate mediators of their responses, such as the protein kinases, are capable of cross-talk, as illustrated by the following three examples. First, neurotransmitters coupling to G_i can enhance the ability of other neurotransmitters that couple to G_s to increase cAMP (Bourne and Nicoll, 1993; Tang and Gilman, 1991). Such agents as GABA (or baclofen) that do not stimulate cAMP production on their own can release a large amount of β/γ from G_i (or G_o) and

conditionally stimulate type II (and type IV) adenylyl cyclase. These cyclases are only activated by β/γ (from G_i or G_o) during coincident stimulation by α_S (from G_S). This cross-talk is the basis for potentiation of β-adrenergic receptor-mediated inhibition of afterhyperpolarization in the hippocampus by $GABA_B$ and $5HT_{1A}$ receptor activation (Andrade, 1993). Stimulatory input to type II adenylyl cyclase by norepinephrine is enhanced by coincident stimulation of G_i by GABA or serotonin, leading to the significant release of β/γ from this abundant G protein. Thus, GABA and serotonin enhance norepinephrine-mediated neurotransmission by coincident stimulation of adenylyl cyclase via activators derived from two types of G proteins. Second, signals that elevate Ca^{2+} via the PI-linked signaling pathway can cross-talk and stimulate adenylyl cyclase and produce cAMP (Choi et al., 1993). Type I and type III adenylyl cyclases are primarily regulated by Gs, but are also activated by Ca^{2+}/calmodulin. Such cross-talk may be the basis for decoding of temporal asymmetry between cAMP and Ca^{2+} that underlies conditioning in *Aplysia* (Yovell and Abrams, 1992). Third, protein kinases activated by any of the three second messengers can regulate signaling proteins, including receptors, G proteins, and calmodulin-binding proteins, thereby mediating cross-talk among diverse inputs into a single cell.

- *Coordinated modulation*—the multifunctional protein kinases that are activated by G-protein-coupled signals phosphorylate proteins that modulate release, stimulate resynthesis of neurotransmitter (by stimulating synthetic enzymes or by increased gene expression), modulate cytoskeleton, and provide greater energy to the cell by regulating carbohydrate metabolism. These actions provide for a coordinated response to stimulation by neurotransmitters.

PROTEIN PHOSPHORYLATION

Protein phosphorylation is a ubiquitous mechanism for controlling cellular function. Most of the effects of the second messengers—cAMP, Ca^{2+}, and DAG—are due to activation of protein kinases. Phosphorylation can either increase or decrease the activity of enzymes or can modify protein-protein interactions, leading to changes in their intracellular localization or in their activity. It is this change in the functional state of the target that enables kinases and their cognate activators to control a wide range of physiological processes. Protein kinases can be classified as serine/threonine or tyrosine kinases based on the amino acid moiety phosphorylated on the target protein. G-protein-coupled signal transduction processes are mediated primarily by serine/threonine kinases, whereas signaling by growth factors and neurotrophic factors involve a cascade of protein kinases that prominently include tyrosine kinases.

The major protein kinases involved in signal transduction are illustrated in Figure 4–1. The kinases are often named based on the second messenger that activates them. Thus, PKA responds to changes in cAMP, CaM kinase to Ca^{2+}/calmodulin, and PKC to DAG (and Ca^{2+}). Ca^{2+} signaling is more heterogeneous than the other second messengers because it has many other targets, in addition to protein kinases. Furthermore, several protein kinases are activated by Ca^{2+} or by Ca^{2+}/calmodulin (Schulman, 1993). These include multifunctional CaM kinase (also referred to as CaM kinase II), as well as CaM kinase I and myosin light chain kinase, which are dedicated to the phosphorylation of only one or a few key substrates. PKA, PKC, and CaM kinase are all multifunctional protein kinases.

Figure 4–1 also shows a possible pathway for signaling by growth factors and neurotrophic factors. These factors often activate receptor-linked tyrosine kinases that contain an intrinsic tyrosine kinase activity on their cytoplasmic surface. Stimulation of these receptors results in autophosphorylation of their catalytic domain on tyrosine residues and initiates a cascade of tyrosine kinase phosphorylation that includes ras, raf, and MAP kinases. MAP kinases are capable of regulating multiple substrates in both the cytoplasm and nucleus. Access may be gained to the cascade by G-protein-based pathways that activate PKC since PKC may activate MAP kinase. Such neurotransmitters as glutamate, which activate PKC via metabotropic receptors, thereby activate MAP kinase to produce changes in gene expression (Bading et al., 1993). The autophosphorylated receptor tyrosine kinases also activate other signaling molecules, such as PLC γ, which binds to specific phosphotyrosine moieties on the kinase. This activation enables growth factor receptors to access the PI signaling pathway and coordinate an even wider range of intracellular events.

PKA, CaM kinase, and PKC are structurally homologous and regulated by a similar regulatory motif. The function of the second messenger in each case is to deinhibit the kinase by binding near, and displacing, an autoinhibitory domain from the kinase catalytic site (Fig. 4–2). Peptides corresponding to the inhibitory domains are often used as selective kinase inhibitors for functional studies. Interestingly, nature has taken advantage of this design by evolving mechanisms by which disruption of the inhibitory domain produces a kinase catalytic domain whose active state outlasts the presence of the second messenger (Schwartz and Greenberg, 1987). PKA is kept active during sensitization in *Aplysia* by proteolysis of its regulatory subunit, the regulatory domain of PKC can be cleaved by Ca^{2+}-stimulated proteases to form an active catalytic fragment, and CaM kinase undergoes an autophosphorylation that disables its inhibitory domain (see below).

Termination of phosphorylation-based signaling is ultimately reversed by phosphoprotein phosphatases. The primary phosphatases in the brain are phosphoprotein phosphatase 1, 2A, 2B (calcineurin; Ca^{2+}/calmodulin-dependent) and 2C. Phosphatase 1 and 2A, in particular, are responsible for much of the dephosphorylation of neurotransmitter signals in brain. Inhibition of these by such agents as okadaic acid leads to a general increase in phosphoprotein levels.

MULTIFUNCTIONAL CaM KINASE

Ca^{2+} is an essential informational molecule in the nervous system. Since activation of neurons typically results in increases in intracellular Ca^{2+}, this ion has evolved as a key index of the state of the neuron, and it is not surprising that protein kinases transmit this information to dozens of substrate proteins in cells. CaM kinase is the major protein kinase mediating Ca^{2+} action in cells (Hanson and Schulman, 1992; Schulman, 1993; Schulman et al., 1992). This kinase modulates synaptic release (via phosphorylation of synapsin I), catecholamine synthesis, cytoskeletal function, gene expression, ion channels (for Ca^{2+}, Cl^-, and K^+), and carbohydrate metabolism and has been implicated in LTP and neuronal memory.

Regulation of CaM kinase by autophosphorylation is very interesting since it provides a mechanism for potentiating its activity and may enable it to decode the frequency at which the neuron is being stimulated. How can a kinase potentiate the action of Ca^{2+} or function as a frequency detector? In the absence of its autoregulatory features CaM kinase responds faithfully to fluctuations in intracellular Ca^{2+}. It is quickly activated when Ca^{2+} is elevated above basal levels and quickly deactivates when Ca^{2+} levels decline again. In order to prolong kinase activity and respond

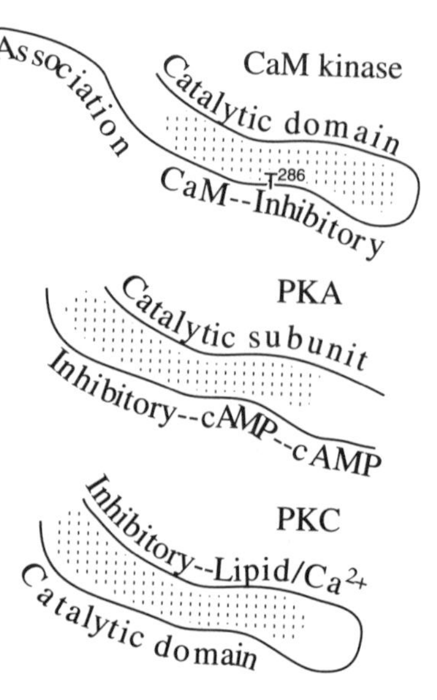

FIGURE 4–2. Second-messenger-regulated protein kinases are designed with a similar regulatory motif. Basal activity of PKA, PKC, and CaM kinase is kept low by an inhibitory domain that resides near the second-messenger binding site. The "T" in the inhibitory domain of CaM kinase is Thr^{286}, an important site of autophosphorylation that disables the inhibitory domain. The second messengers "displace" the inhibitory segment, thereby activating the kinases by deinhibiting them.

differentially to repetitive stimulation, each stimulus must elicit a modification of the kinase that persists until a second stimulus arrives. Furthermore, the sensitivity of a frequency detector would be enhanced if the process is cooperative. This section examines the regulation of CaM kinase and suggests how it may function to potentiate the effects of repetitive stimulation.

Structure of CaM Kinase

The kinase is a large multimeric enzyme encoded by four genes that give rise to at least nine closely related isoforms (Hanson and Schulman, 1992). Alternative splicing provides unique sequences that underlie major differences among the isoforms. The α and β isoforms are restricted to nervous tissue, whereas the γ and δ isoforms show a wide tissue distribution that includes the brain. The neuronal specific isoforms are highly abundant, constituting up to 2% of hippocampal protein and 10 to 20% of postsynaptic density protein. As shown schematically in Fig. 4–3, the N-terminal half of each isoform contains a catalytic domain with inherent kinase activity that is kept inactive in the basal state by the regulatory domain that follows. It consists of an autoinhibitory domain with an overlapping calmodulin-binding domain. The C-terminus contains the association domain by which the kinase associates into approximately a decameric structure and a variable region introduced by alternative splicing that may facilitate targeting the kinase to distinct sites in the neuron. Electron microscopic analysis suggests that the association domains of all the subunits assemble into a globular structure and that the catalytic/regulatory domains radiate out via a short tether (Fig. 4–3; Kanaseki et al., 1991). In essence, this structure produces a molecular machine with appropriate arrangement of subunits for rapid autophosphorylation among subunits (see below).

Regulation by Ca^{2+}/Calmodulin and Autophosphorylation

In the presence of physiological levels of Ca^{2+}, calmodulin undergoes a change in conformation that allows it to wrap around the calmodulin-binding domain of the kinase (Ikura et al., 1992; Meador et al., 1992). The inhibitory domain, which blocks access of both ATP and protein substrate to the active site of the kinase, cannot bind to both the catalytic site and to calmodulin so it is presumed that Ca^{2+}/calmodulin activates the enzyme by displacing the autoinhibitory domain and deinhibiting the kinase. Activation of the kinase results in autophosphorylation of a critical residue in the autoinhibitory domain Thr^{286} (numbering refers to the α

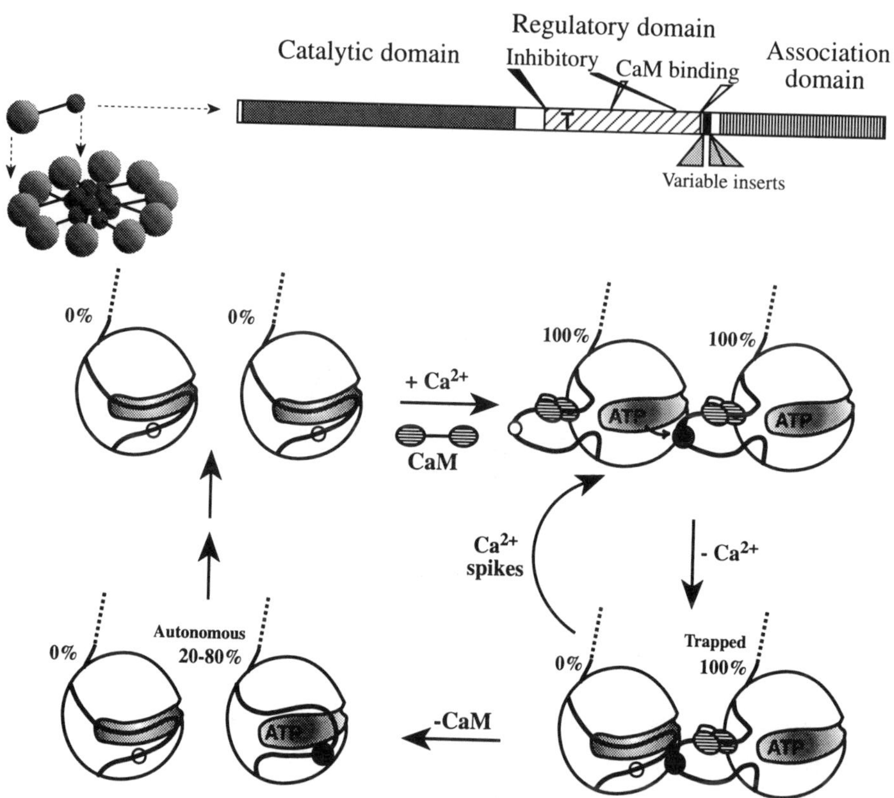

FIGURE 4–3. Molecular potentiation by autophosphorylation of CaM kinase. The domain structure of a typical CaM kinase subunit is shown. Subunits assemble into a decameric structure in which individual catalytic/regulatory domains radiate out from the central association domains. The schematic diagram shows a single autophosphorylation involving two subunits of the decamer. Thr^{286} is shown as an open circle in the dephosphorylated state and as a larger filled circle when phosphorylated. In the example above, calmodulin presents an inhibitory domain of one subunit for autophosphorylation by a second nearby subunit that is activated by calmodulin. As Ca^{2+} declines, calmodulin quickly dissociates from the nonphosphorylated subunit, leading to its deactivation (0% activity), whereas calmodulin is trapped on the autophosphorylated subunit, keeping that subunit fully active (100%). Repetitive stimulation (Ca^{2+} spikes) sets up a feedforward reaction in which increasing amounts of calmodulin are recruited to kinase subunits. At a low frequency of stimulation calmodulin dissociates during the interspike interval when Ca^{2+} is at basal levels, leaving a kinase that retains partial activity (20% to 80% of maximal activity) until it is fully deactivated by dephosphorylation.

subunit), which disrupts its inhibitory function and locks the kinase in the activated state. Autophosphorylation of this "autonomy" site increases the affinity of the kinase for calmodulin several hundredfold and reduces the rate of dissociation of calmodulin at either high or low Ca^{2+}. Even after calmodulin dissociates, the phosphokinase remains partially Ca^{2+} independent or autonomous. The steps of this process and their consequences are detailed below.

The regulatory domain of CaM kinase (amino acid 273-309) is capable of both inhibiting the kinase and binding calmodulin. It is subdivided into a region that interferes with ATP binding, a region that competes with peptide substrates, and finally the calmodulin-binding region (Smith et al., 1990). A peptide corresponding to amino acid 273-302 inhibits the kinase without binding calmodulin and inhibits LTP when injected into CA1 hippocampal neurons (Malinow et al., 1989). Autophosphorylation is an intersubunit reaction that occurs when two proximate subunits in a holoenzyme are bound by Ca^{2+}/calmodulin (Hanson et al., 1994). Calmodulin has a dual role—one molecule of calmodulin activates one subunit, whereas another molecule of calmodulin presents the autoinhibitory domain of a nearby subunit for autophosphorylation by the first subunit.

Autophosphorylation has two consequences that potentiate the active state of CaM kinase (Fig. 4–3). First, autophosphorylation of Thr^{286} dramatically reduces the off-rate of bound calmodulin with little change in its on-rate (Meyer et al., 1992). In the absence of autophosphorylation, calmodulin dissociates within 170 msec when Ca^{2+} is lowered, whereas this process takes at least 10 sec after autophosphorylation. At high Ca^{2+} levels, calmodulin does not dissociate from an autophosphorylated subunit for hundreds of seconds. Thus, autophosphorylation traps bound calmodulin. Importantly, the kinase is fully active while calmodulin is trapped, regardless of the Ca^{2+} concentration. The kinase is converted from an enzyme with one of the weakest affinities for calmodulin (45 nM) to an enzyme with one of the highest affinities (60 pM; Meyer et al., 1992). This change may allow CaM kinase to compete with other abundant calmodulin-binding proteins for a limited supply of calmodulin. The second consequence of autophosphorylation becomes apparent when calmodulin dissociates. The kinase does not fully deactivate, retaining 20 to 80% of its maximal Ca^{2+}-stimulated activity because the phosphorylated inhibitory domain is ineffective at inhibiting the kinase.

Ca^{2+}-stimulated autophosphorylation is the critical mechanism by which CaM kinase decodes Ca^{2+} signals. As a minimum, it allows the kinase to continue phosphorylating substrates after Ca^{2+} levels decline and until it is deactivated by dephosphorylation. An equally important effect may arise during repetitive stimuli or oscillations that do not fully activate the kinase, i.e. do not lead to binding of calmodulin to all subunits of a decameric enzyme (Meyer et al., 1992). Some calmodulin is retained during interspike intervals due to trapping. If a subsequent spike arrives before the dissociation of calmodulin, then autophosphorylation, which occurs between activated subunits, is enhanced. The kinase may then recruit more calmodulin with each successive spike in a positive feedforward manner. The autophosphorylation state of the kinase would therefore be affected by the frequency of stimulation. At high frequency the kinase becomes increasingly active as the fraction of calmodulin-bound subunits per decamer increases, whereas at low frequency dephosphorylation reduces the recruitment of calmodulin (Hanson et al., 1994).

Autophosphorylation in situ and Cross-Talk with PKC

Does the autophosphorylation of CaM kinase and its conversion to an autonomous enzyme occur *in situ*? In the pheochromocytoma cell line PC12, multiple signals stimulate a rise in intracellular Ca^{2+} and converge on CaM kinase (MacNicol and Schulman, 1992a,b). Ca^{2+}-independent activity of CaM kinase is enhanced by stimulation of PC12 cells with bradykinin (activating via the PI signaling pathway to release intracellular stores of Ca^{2+}), by nicotinic agents and ATP (stimulating Ca^{2+} entry via ligand-gated cation channels), as well as by membrane depolarization with high K^+ (leading to Ca^{2+} entry via voltage-sensitive Ca^{2+} channels). Such depolarization is also effective in stimulating the kinase in acute hippocampal slices, which can exhibit 10% to 20% autonomous CaM kinase activity even in the basal state (Molloy and Kennedy, 1991; Ocorr and Schulman, 1991). In cultured hippocampal neurons, glutamate elevates Ca^{2+} via NMDA receptors and increases the fraction of CaM kinase that is autonomous (Fukunaga et al., 1992). Glutamate stimulates phosphorylation of CaM kinase and two of its major neuronal substrates, synapsin I and MAP-2. Such activation of CaM kinase may occur during stimulation that elicits LTP. High-frequency stimulation in the stratum radiatum of the CA1 region of the hippocampus elicits a sustained increase in autonomous CaM kinase activity that is from 20% to 25% of maximal stimulated activity (Fukunaga et al., 1993).

Although calmodulin is a highly abundant protein, it serves a large number of calmodulin-binding proteins that together are equally abundant. Elevated Ca^{2+} stimulates binding of calmodulin to these proteins and is likely to markedly diminish the concentration of "free" calmodulin. For example, stimulation of fibroblasts leads to a shift of almost all calmodulin from a "free" to a bound state, suggesting that there is no excess calmodulin. Calmodulin-binding proteins with low affinity for calmodulin, such as CaM kinase, will therefore compete poorly for the limited calmodulin. Cells may regulate free calmodulin via "buffering" by such proteins as neuromodulin (GAP-43, F1, B-50), MARCKS, and neurogranin (Alexander et al., 1987; Klee, 1991; Skene, 1990). PKC may regulate the calmodulin-buffering capacity of cells by phosphorylating such proteins, reducing their binding to calmodulin, and increasing the availability of calmodulin for CaM kinase and other proteins. Thus, signals activating PKC may cross-talk with Ca^{2+}/calmodulin-dependent processes. In fact, pretreatment of PC12 cells with a phorbol ester to activate PKC enhances the submaximal stimulation of CaM kinase (MacNicol and Schulman, 1992a). PKC does not activate CaM kinase directly. Rather, it increases the extractable form of calmodulin by 60%, presumably by phosphorylation and disruption of the calmodulin-binding capability of such proteins as GAP-43. Thus, the excitability of neurons is not only dependent on the level of free Ca^{2+} but also on the availability of a mediator of Ca^{2+} action, calmodulin.

FUNCTIONAL STUDIES

Neuronal Excitability

CaM kinase has been implicated in numerous physiological processes in both neuronal and non-neuronal systems (Schulman, 1993; Schulman and Hanson, 1993). This section focuses on the general role of CaM kinase in neuronal excitability via phosphorylation of both presynaptic and postsynaptic elements. The best-characterized function of CaM kinase is modulation of synaptic release elicited by phosphorylation of the vesicle-associated protein synapsin I (Valtorta et al., 1992a). The current view is that synapsin embeds a reserve pool of synaptic vesicles in an actin meshwork because of its ability to bind to both synaptic vesicles and to actin. Dephosphorylated synapsin I, either free or vesicle bound, increases the rate of polymerization of actin filaments by reducing the critical monomer concentration and the lag time for polymerization (Benfenati et al., 1992a; Valtorta et al., 1992b). Phosphorylation of synapsin I by CaM kinase at two sites distinct from the site phosphorylated by PKA reduces its affinity for both actin and vesicles, thereby freeing the vesicles from the cytoskeleton and promoting their movement to the release sites. The rate of actin polymerization is also markedly reduced by phosphorylation of synapsin I by CaM kinase. A freeze-thaw technique for temporarily permeabilizing a preparation of rat brain nerve terminals or synaptosomes was used to test this model. Introduction of dephosphorylated synapsin I reduced the stimulated release of the transmitter glutamate from synaptosomes while having no effect on basal release (Nichols et al., 1990). Introduction of phospho-synapsin I (phosphorylated by multifunctional CaM kinase) or of control protein had no effect on release.

CaM kinase may also function as a binding protein for synapsin I (Benfenati et al., 1992b). Synapsin covers much of the surface of synaptic vesicles through the tight association of its N-terminal region with phospholipids and its C-terminal interaction with a protein that has now been identified as the α or major isoform of CaM kinase. Chemical cross-linking was used to demonstrate proximate interactions between a C-terminal fragment of synapsin and the α but not the β isoform of the kinase; both isoforms are present on purified vesicles. Co-localization of CaM kinase and synapsin I on synaptic vesicles and their direct interaction as a complex may add both specificity and speed to the phosphorylation that modulates synaptic release in response to neuronal stimuli that are typically brief.

Several actions of CaM kinase suggest that it can affect excitability by modulation of membrane depolarization and of intracellular Ca^{2+}. Cholinergic modulation of neuronal responsiveness seems to be mediated by CaM kinase (Muller et al., 1992). In CA1 and CA3 neurons, muscarinic agents block Ca^{2+}-activated K^{+} conductances, thereby suppressing the normal control of excitability based on adaptation of firing and slow afterhyperpolarization. Microinjection of a CaM kinase inhibitor representing amino acid 273-302 of the autoinhibitory domain (with alanine substituted for Thr^{286}) blocked the *g*K(Ca), thereby simulating the reduction of adaptation and slow afterhyperpolarization seen with muscarinic agents (Muller et al., 1992). CaM kinase phosphorylates and activates the brain and heart forms of the ryanodine receptor that control release of intracellular Ca^{2+} (Witcher et al., 1991) and N-type Ca^{2+} channels that control Ca^{2+} entry near release sites (Witcher et al., 1993). Finally, CaM kinase may be an important modulator of non-NMDA glutamate receptors (McGlade-McCulloh et al., 1993). In preparations of lysed synaptosomes and of postsynaptic densities, CaM kinase phosphorylates the GluR1-R4 receptors. Microinjection of an autonomous form of the kinase into CA1 neurons resulted in a three- to fourfold increase of kainate-induced current, suggesting that phosphorylation of the GluR1-R4 receptors or of a regulatory protein stimulates these receptor-linked channels, an effect that may relate to the role of CaM kinase in LTP (McGlade-McCulloh et al., 1993).

Gene Expression

Multifunctional CaM kinase may be responsible for some Ca^{2+}-induced gene expression (Bading et al., 1993; Sheng et al., 1991; Wegner et al., 1992). In a pituitary cell line (G/C) a DNA sequence corresponding to the binding site for C/EBPβ, a member of the bZip family of transcription factors, mediated the induction of a downstream reporter gene in response to elevated Ca^{2+}. C/EBPβ phosphorylated in situ was stimulated by Ca^{2+} influx and inhibited by KN-62, a cell-permeable inhibitor of CaM kinase. Co-transfection of a constitutive monomeric fragment of CaM kinase, α-CaM kinase(1-290), and C/EBPβ induced expression 60-fold of a CaM kinase response element-bearing reporter gene. Thus, promoters with binding sites for C/EBPβ, in addition to those with binding sites for CREB (Dash et al., 1991; Sheng et al., 1991), may confer Ca^{2+}-regulated transcription of genes in diverse cell systems. In hippocampal cultures a KN-62-sensitive process, presumably CaM kinase-mediated phosphorylation, was shown to be responsible for induction of c-fos in response to membrane depolarization (Bading et al., 1993). Interestingly, induction of c-fos elicited by Ca^{2+} influx via L-type Ca^{2+} channels was blocked by KN-62, whereas induction of c-fos by Ca^{2+} influx via NMDA receptors was insensitive and not likely to involve CaM kinase.

Gene Knock-Out and Memory

The most recent molecular technique to be applied to CNS studies is gene knock-out, a procedure by which the DNA encoding for a protein of interest is eliminated from the germ line by gene targeting of embryonic stem cells. Recently, knock-outs of the major neuronal isoform of CaM kinase and of the nonreceptor tyrosine kinase, *fyn*, have been used to demonstrate a role for these proteins in LTP and in spatial learning (Grant et al., 1992; Silva et al., 1992a,b). In most experiments, hippocampal slices from mutant animals displayed no LTP. The CaM kinase knock-out did not significantly affect normal transmission or the general development of brain morphology. Surprisingly, in two preparations LTP was essentially normal as measured by extracellular recording, suggesting that some stimuli can elicit LTP in a large population of neurons in mutant animals and that the lack of CaM kinase can be overcome. The finding that an increasing number of signaling molecules, such as CaM kinase, PKC, and *fyn*, are conditionally required for LTP suggests a redundant induction process.

It is not known whether the presence of a high concentration of an autophosphorylated CaM kinase *per se* is the essential change that participates in the induction of LTP or whether a key protein is phosphorylated by the kinase to facilitate induction of LTP. The high concentration of the kinase in PSD and its ability to be highly phosphorylated and to recruit calmodulin away from other proteins offer an alternative mechanism by which such an abundant and highly charged complex could participate in LTP.

The finding that mutant mice were defective in acquiring a form of memory measured in awake animals strengthens the notion that LTP is related to memory (Silva et al., 1992a). The mutant mice exhibited impaired learning in a Morris hidden platform task in which mice find the hidden platform submerged in a round pool of water that has been made opaque. Normal mice are able to use an effective strategy involving multiple spatial relationships between the hidden platform and distal visual cues outside the pool. Mutant mice perform less well because they are unable to perform true spatial learning and use other, less effective strategies to eventually find the escape platform. These studies greatly extend prior links made between LTP and learning and implicate α-CaM kinase in both.

Some caution should be applied in the interpretation of knock-out experiments. The elimination of a protein could have multiple indirect consequences. In the case of an abundant protein, such as CaM kinase, its absence may alter other components of the PSD, since it normally forms 20% of the structure. Genes that are necessary for appropriate development may cause minor or major dysfunctions that are unrelated to the function of the protein in mature animals. Knock-outs of some components of neuronal plasticity may yield false negative results. For example, the brain may compensate for a lack of an important signaling molecule by increasing the concentration of related isoforms or developing new strategies to overcome deficits. The specific manner by which LTP is induced *in vitro* involves a stronger, if not different, stimulus than its *in vivo* counterpart that animals may utilize in learning and memory. The *in vitro* induction paradigm for eliciting LTP may not require as many of the signal transduction elements as are essential for some forms of learning *in vivo*. Overall, however, the recent findings hold hope for our ability to link known molecular signaling elements to the standard *in vitro* preparations used to study neuronal function in the laboratory and to the actual behavior of animals.

5

Structure and function of ligand-gated channels

RAYMOND J. DINGLEDINE and JULIE A. BENNETT

The ligand-gated ion channel family is responsible for fast excitatory and inhibitory synaptic transmission throughout the brain. Inhibition is mediated by $GABA_A$ and glycine receptors, whereas excitation is mediated by nicotinic acetylcholine, serotonin, and glutamate receptors. Each of these receptors comprises a family of related subtypes. The glutamate receptor class has been further divided into NMDA receptors, AMPA receptors, and kainate receptors based on their pharmacological and genetic profiles. The ligand-gated ion channel receptors have several defining features. The receptor acts as both the ligand recognition site and the signal transducer, which in this case is an ion pore. Such an arrangement allows for the passage of ions across the membrane without the involvement of any other proteins and results in the transduction of a chemical signal to an electrical signal on a millisecond time scale. The ligand-gated ion channel receptors possess multiple ligand binding sites and consist of multiple subunits that are assembled in various combinations to impart different pharmacological and kinetic properties to the channel. These subunits share approximately 50% to 70% sequence homology within their individual receptor class and share approximately 15% to 40% homology across classes. Each subunit contains four distinct hydrophobic regions that may constitute transmembrane domains of these receptor subunits. The subunits typically contain a large extracellular amino terminus where the neurotransmitter binding site is thought to be located and a long cytoplasmic loop located between the third and fourth hydrophobic domains.

TRANSMEMBRANE STRUCTURE

The nicotinic acetylcholine receptor is the best-studied ligand-gated ion channel receptor and has been used as the model for all others. This receptor has five subunits that assemble in the lipid membrane to form a central conducting pore (Kubalek et al., 1987). Four membrane spanning segments were initially proposed for each subunit of the nicotinic receptor based on its hydrophobicity profile (Fig. 5–1). Mapping glycosylation sites in the amino terminal domain and the cellular orientation of discrete regions by the accessibility to proteolysis of a reporter fusion protein tag demonstrated that the four transmembrane domains assigned by hydropathy analyses were correct (Chavez and Hall, 1991, 1992).

The orientation of the transmembrane domains around the pore is important for understanding how ions permeate and block the channel. A plethora of site-directed mutagenesis experiments of the nicotinic receptor have shown that mutating select residues within the second transmembrane region affects ion conduction through the pore, and it was concluded that the second transmembrane domain (TM2) lines the channel pore (Akabas et al., 1992; Charnet et al., 1990; Imoto et al., 1988; Leonard et al., 1988; Villarroel and Sakmann, 1992). Furthermore, the use of photoaffinity labeling by desensitizing noncompetetive antagonists pre-

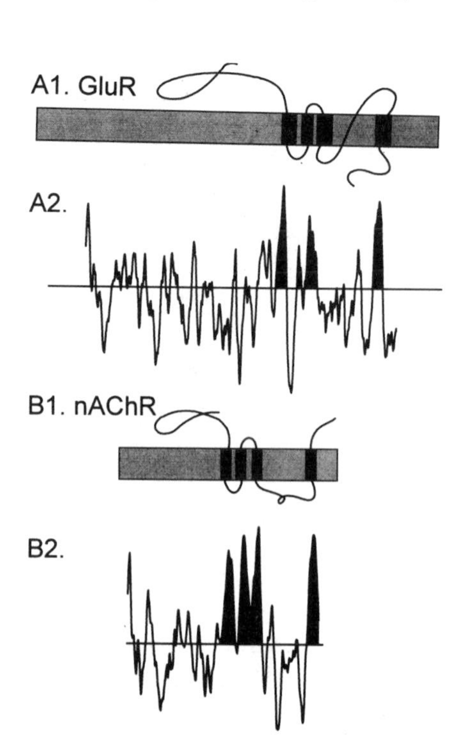

FIGURE 5–1. Possible transmembrane topologies of the glutamate receptors and the nicotinic receptors in alignment with their hydophobicity profiles. A2 and B2 represent the hydrophobicity profiles as predicted by the method of Kyte and Dolittle where areas above the x-axis are hydrophobic and areas below the axis are hydrophilic. The solid areas in the schematic representations and the hydropathy plots correspond to the proposed transmembrane domains of the receptor subunit.

dominantly labeled residues in TM2 (Giraudat et al., 1987, 1989; Hucho et al., 1986; Leonard et al., 1988; Pedersen et al., 1992; Revah et al., 1990). However, a photoreactive open channel blocker of the nicotinic receptor labeled a peptide fragment containing TM1, suggesting that TM1 also contributes some amino acid residues to the open channel (DiPaola et al., 1990). The secondary structures of the transmembrane regions have been assumed to be α-helices, in part because site-directed mutagenesis studies of TM2 have indicated that residues 6 and 10, which could be one helical turn apart, interact with an open-channel blocker of the nicotinic receptor (Charnet et al., 1990). However, probing cysteine-substitution mutants with a sulfhydryl reagent that acts as a channel blocker indicated that every other residue has access to the pore, suggesting that the second transmembrane region contains at least some β-sheet secondary structure (Akabas et al., 1992). Averaging of electron microscopy images at 9 Å resolution showed that only a single region, assumed in these studies to be TM2, forms an α-helix that kinks and tilts radially outward on either side of the membrane. This region was flanked by a continuous rim of density, which was interpreted to mean that the remaining transmembrane domains are comprised entirely of β-sheet (Unwin, 1993). The controversy surrounding the structure of the channel-lining transmembrane domain for even the very extensively studied nicotinic acetylcholine receptor suggests caution in making assumptions regarding structural models for the other, less well-studied ligand-gated ion channels.

The other ligand-gated ion channel receptors—$GABA_A$, $5\text{-}HT_3$, glycine, and glutamate receptors—were initially assumed to possess the same structural elements as the nicotinic acetylcholine receptor based on similarities of their hydrophobicity profiles and function (Fig. 5–1). However, recent studies of the glutamate receptors have shown that this is likely not a valid assumption. A protein kinase A site located in the proposed large intracellular loop between the third and fourth hydrophobic domains in the kainate receptor subunit GluR6 was shown to be phosphorylated upon the intracellular application of the catalytic subunit of protein kinase A, supporting the intracellular location of this region (LA Raymond et al., 1993; Wang et al., 1993). However, a N-linked glycosylation site in GluR6 was found 36 amino acids downstream of the phosphorylation site, but still within the region between the third and fourth hydrophobic regions. This site was glycosylated in an *in vitro* assay in the presence of microsomal membranes, indicating that it should be extracellular (Roche et al., 1994; Taverna et al., 1994). Similar results have recently been obtained with mutant GluR3 receptors (Bennett and Dingledine, 1995), supporting the generality of these conclusions. Consistent with this being a fifth transmembrane domain in the glutamate receptors, four serines were identified in the carboxyl terminal of the NMDA receptor subunit NR1 following the final hydrophobic domains that were phosphorylated by the intracellular application of PKC (Tingley et al., 1993). This finding places the C-terminus of the protein on the intracellular side of the membrane, which is inconsistent with the nicotinic receptor profile but would follow from the presence of an additional transmembrane domain (Fig. 5–1A), or if one of the membrane domains was re-entrant rather than membrane crossing (Bennett and Dingledine, 1995). The discrepancy in the transmembrane topologies between the nicotinic and the glutamate receptors brings into question the assumption that all ligand-gated ion channels belong to a structurally similar superfamily of receptors and challenges the validity of generalizing across receptor classes. The nicotinic acetlycholine receptors and the glutamate receptors may have arisen from a common ancestor, but the two receptor classes must have undergone structural changes through the years, resulting in different transmembrane topologies.

GLUTAMATE RECEPTOR FAMILIES

Cortical neurons, like all other central neurons and some glia, express glutamate receptors. We have focused our attention on the glutamate receptors as they are the primary players in excitatory synaptic transmission throughout the brain. Since December 1989, cDNAs encoding 23 glutamate receptor subunits have been cloned from rat brain. These include 7 metabotropic receptors, 14 functional ionotropic receptor subunits (Fig. 5–2), and 2 orphan receptor subunits, δ-1 and δ-2, for which no function has yet been identified. The distribution of metabotropic receptors and ionotropic receptor subunits is quite varied, as judged by *in situ* hybridization of subunit mRNA and immunocytochemical localization of the subunit proteins. This variability is consistent with the notion that the repertoire of glutamate receptor subtypes expressed by neurons is large and heterogeneous. The metabotropic receptors are now known to serve traditional roles in synaptic transmission, both presynaptically as autoregulators of transmitter release and postsynaptically as transducers of electrical signals (Conn and Patel, 1994). Roles for metabotropic receptors in plasticity and even excitotoxicity are now emerging. However, our focus here is on the ionotropic glutamate receptors.

The grouping of ionotropic glutamate receptor subunits into five structural families by sequence homology (Fig. 5–2) also separates them functionally as determined by studying the receptors in heterologous expression systems, such as *Xenopus* oocytes (Fig. 5–3) and mammalian cell

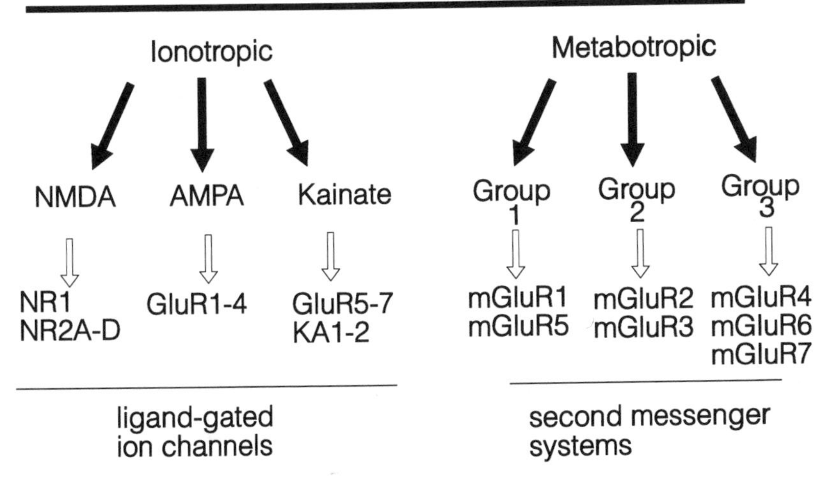

FIGURE 5–2. Twenty-one genes encoding functional glutamate receptor subunits. The solid arrows lead from the class of receptor (ionotropic, metabotropic) to pharmacologically defined receptor types (NMDA, AMPA, kainate, Group 1-3). The open arrows lead to individual genes encoding receptor subunits.

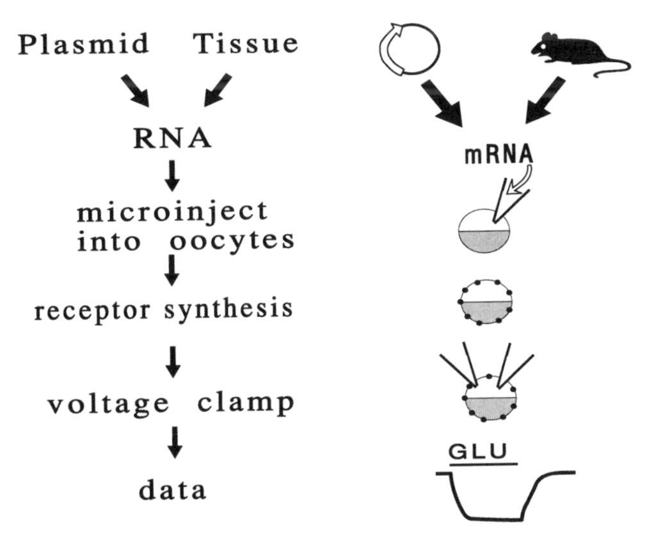

FIGURE 5–3. Schematic depiction of *Xenopus* oocyte expression system. Messenger RNA isolated from brain or transcribed from plasmids is injected into individual oocytes, which are cultured for 2 to 5 days to allow receptor synthesis, assembly, and insertion into the plasma membrane. Two electrode voltage-clamp techniques are then used to measure ionic current flowing through glutamate receptor channels in response to agonists.

lines. Importantly, NMDA receptors assembled from various combinations of NR1 and NR2 subunits exhibit the general pharmacological profile of native NMDA receptors studied in neurons; namely, the requirement for both glycine and glutamate as co-agonists, the voltage-dependent block by magnesium and phencyclidine analogs, the competitive block by D-AP5, and the potentiation by spermine and similar polyamines. AMPA receptors, assembled from GluR1-4, and kainate receptors, combinations of GluR5-7 and KA1-2, are insensitive to NMDA, but are activated by kainate and, for some subunit combinations of both types, by AMPA. Because of the overlapping agonist profile an alternate means of distinguishing these two classes of receptors is needed. AMPA receptors exhibit rapidly desensitizing responses to AMPA and to kainate that can be blocked by cyclothiazide and similar diuretic thiazides. Kainate receptors, in contrast, exhibit strong desensitization when kainate is the agonist, and this desensitization can be nearly eliminated by concanavalin A (Partin et al., 1993). Since cyclothiazide does not appear to act on kainate receptors and concanvalin A has little or no effect on AMPA receptors, the pharmacology of desensitization seems to be diagnostic for distinguishing AMPA from kainate receptors.

STRUCTURAL DETERMINANTS OF PERMEATION AND CHANNEL BLOCK

Ligand-gated ion channels can be considered to have four elementary functions: ligand recognition, channel gating, permeation, and desensitization. One anticipates that each elementary function may be carried out by a separate, somewhat independently acting structural domain since many proteins are built with a modular design. The isolation and identification of cDNAs encoding glutamate receptor subunits open the way to identify the structural domains responsible for the various elementary functions of these receptors. The domains mediating ion permeation are the most immediately accessible and are of great functional importance for glutamate receptors. Ionic selectivity holds special significance since calcium entry through glutamate receptor channels plays a key role in such diverse events as long-term potentiation, activity-dependent synapse formation, and excitotoxic neuron damage following ischemia or severe seizures. All glutamate receptors exhibit measurable calcium permeability, but the *degree* of calcium permeability varies for different subunit combinations within a family, especially for the AMPA receptors. Notably, recombinant AMPA receptors lacking the GluR2 subunit exhibit about the same high degree of calcium permeability as NMDA receptors (Burnashev et al., 1992;

Hume et al., 1991), whereas receptors containing the GluR2 subunit exhibit very low calcium permeability. Native AMPA receptors that exhibit high calcium permeability and seem to lack the GluR2 subunit are expressed by some neurons (Bochet et al., 1994). Most AMPA receptors expressed by neurons do seem to contain GluR2, which is consistent with the very wide distribution of this subunit in *in situ* hybridization assays.

Overactivation of glutamate receptors on cortical neurons can cause epileptic seizures and, with ischemia, excitotoxic neuron damage. Although neuronal injury that follows transient, focal ischemia is mediated in part by NMDA receptors (Choi and Rothman, 1990), it is now recognized that global ischemia produces cortical cell death that is insensitive to NMDA receptor blockers, but can be ameliorated by specific AMPA receptor antagonists (Sheardown et al., 1990). Calcium entry through GluR2-lacking AMPA receptors is therefore likely to mediate some of the excitotoxic death that accompanies global ischemia. Investigations into the structural basis of calcium permeation through glutamate receptor channels have begun to provide some insight that could be helpful in designing specific blockers of the calcium-permeable glutamate receptors.

Site-directed mutagenesis has been used to study the permeation properties of mutant glutamate receptors. In these experiments single codon mutations are introduced by now-conventional techniques into the cDNA sequence encoding receptor subunits, and the resulting subunit mutants are expressed in oocytes or other cells for functional study. One of the first findings from this approach was that the divalent permeation properties of recombinant AMPA receptors are specified by a single amino acid in a putative transmembrane region. This site, now commonly termed the Q/R/N site since all known glutamate receptor subunits contain either a glutamine, arginine, or asparagine at this position, is identified by the boxed position in Fig. 5–4, which depicts the amino acid sequence of all known functional glutamate receptor subunits in the rat. In this region the AMPA receptor subunits—GluR1-GluR4—differ *only* in this site: an arginine is present in GluR2, whereas a glutamine occupies this position in the other subunits, although all of the genomic sequences encode for a glutamine. It has been postulated that subunit-specific RNA editing occurs that results in the switch from glutamine to arginine codon in the mRNA of GluR2 (Higuchi et al., 1993). All known NMDA receptor subunits contain an asparagine at this position. One of the most important conclusions drawn from the mutagenesis experiments is that the Q/R/N site is a major structural determinant of ion permeation in all three classes of glutamate receptor channels. NMDA and AMPA receptors that contain one or more subunits with an arginine in this position exhibit very low calcium permeability, whereas receptors that contain only glutamine or asparagine in this site exhibit comparably high calcium permeability (Burnashev et al., 1992; Dingledine et al., 1992; Hume et al., 1991; Kawajiri and Dingledine, 1993; Mori et al., 1992; Sakurada et al., 1992). The situation is similar in kainate receptors, but the effect of switching arginine and glutamine seems to depend on the cell type used for expression (Egebjerg and Heinemann, 1993; Kohler et al., 1993). The same site controls voltage-dependent magnesium block in NMDA receptors as shown in Fig. 5–5.

The mutagenesis results suggest that the ion channel domains of all glutamate receptors share much structural similarity. A most interesting observation that supports this suggestion is that certain polyamine venoms from spiders and wasps can exert a voltage- and subunit-dependent block of AMPA and NMDA channels, depending on the amino acid residing in the Q/R/N site. If one or more subunits contain an arginine in this position there is no block, but the calcium-permeable glutamate receptor channels are all potently blocked by Joro spider toxin and similar polyamine compounds (Blashke et al., 1993; Herlitze et al., 1993). These spider toxins should thus be useful experimental tools for investigating the roles of calcium-permeable AMPA receptors in cortical circuits and might also find important uses in the treatment of stroke-induced brain damage.

transmembrane — Q/R/N site — *

GluR1	S N E F G I F N S - L W F S L G A F M Q Q G C - D I
GluR2	T N E F G I F N S - L W F S L G A F M R Q G C - D I
GluR3	P N E F G I F N S - L W F S L G A F M Q Q G C - D I
GluR4	P N E F G I F N S - L W F S L G A F M Q Q G C - D I
NR1A	E E D A L T L S S A M W F S V G V L L N S G I G E G
NR2A	H G P S F T I G K A I W L L W G L V F N N S V P V Q
NR2B	G G P S F T I G K A I W L L W G L V F N N S V P V Q
NR2C	G G P S F T I G K S V W L L W A L V F N N S V P I E
NR2D	G G S T F T I G K S I W L L W A L V F N N S V P V E
GluR5	V E N N F T L L N S F W F G V G A L M Q Q G S - E L
GluR6	V E N N F T L L N S F W F G V G A L M R Q G S - E L
GluR7	V E N N F T L L N S F W F G M G S L M Q Q G S - E L
KA-1	L V N Q Y S L G N S L W F P V G G F M Q Q G S T I A
KA-2	L E N Q Y T L G N S L W F P V G G F M Q Q G S E I M

FIGURE 5–4. Alignment of all ionotropic glutamate receptor sequences in the channel-lining membrane domain. Residues that are identical in at least two receptor types are shaded. Residues in the Q/R/N site that play a major role in ionic selectivity of the open channel are boxed. The aspartate residue four amino acids downstream of the Q/R/N site, indicated by an asterisk, plays a role in determination of the shape of the current voltage curve of AMPA receptors (From Dingledine et al., 1992).

STRUCTURAL DETERMINANTS OF MODULATION

Several regions in glutamate receptors have been identified that are involved in modulation of the ion channel. The response of glutamate receptors to glutamate or kainate in cultured neurons is potentiated by protein kinase A (Greengard et al., 1991). It appears that this potentiation is a result of the phosphorylation of a single serine residue, ser 684, in the case of the kainate receptor subunit GluR6 (Raymond et al., 1993; L Wang et al., 1993). Phosphorylation of AMPA receptors by calcium/calmodulin protein kinase II also results in a potentiation of agonist-induced currents (McGlade-McCulloh et al., 1993). Interestingly, the calcium increase resulting from the activation of NMDA receptors engages the calcium/calmodulin protein kinase II system that can phosphorylate AMPA receptors (Fukunaga et al., 1992; Tan et al., 1994). A serine residue, ser 627, is embedded in a consensus phosphorylation site for calcium/calmodulin protein kinase II that is conserved across the AMPA receptor family and may be important in this modulation pathway.

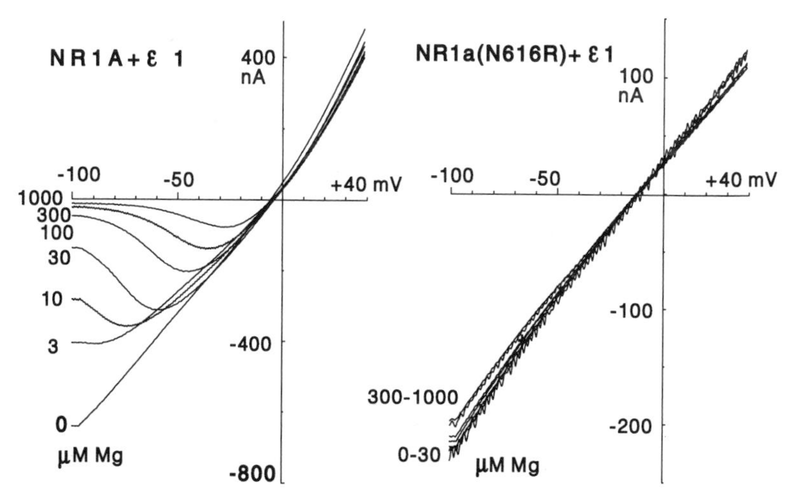

FIGURE 5–5. Role for the asparagine at position 616 in voltage-dependent Mg^{2+} block in recombinant NMDA receptors. The left panel shows current-voltage curves for heteromeric NMDA receptors expressed in oocytes injected with mRNA encoding NR1A and ϵ1 (the mouse equivalent of NR2A). This oocyte was perfused with 100 μM NMDA and 10 μM glycine plus the indicated concentrations of Mg^{2+}. The right panel demonstrates that the voltage-dependent block by Mg^{2+} is not present in mutant NMDA receptors in which arginine is substituted for asparagine at N616 in the NR1A subunit. (Adapted from Kawajiri and Dingledine, 1993).

Other potential modulatory regions have been identified in the glutamate receptor family by analyzing the functional differences of naturally occurring alternative splice variants. A 38 amino acid stretch designated the flip/flop region was discovered in the AMPA receptors where one of two amino acid modules are alternatively spliced into the receptor (Fig. 5–6A). Receptors made from either flip or flop show a rapid desensitization to glutamate followed by a steady-state plateau. Receptors containing the flip cassette show a four to five times greater response to L-glutamate, either alone or in the presence of cyclothiazide, than those containing flop. The prediction was made that flip could cause an enhanced synaptic response, which might underlie at least part of the AMPA receptor requirement for the maintenance of LTP (Partin et al., 1993; Sommer et al., 1990). It is clear that the flip/flop module is intimately involved in AMPA receptor desensitization; however, the mechanism is not yet known.

In the NMDA receptor, spermine can potentiate the glutamate response in two potentially distinct manners: glycine dependent and independent modes. Alternative splicing of three exons generates eight NMDA receptor subunit variants of NR1 (Fig. 5–6B). Two classes of NR1 splice variants have been identified in which the fifth exon is either spliced in or out to generate distinct subunits. Receptors lacking the fifth exon exhibit potentiation by spermine at saturating glycine concentrations. However, receptors containing the exon exhibit little spermine effects under the same conditions, showing that the region encoded by the fifth exon of NR1 is involved in the glycine-independent effects of spermine. Within the fifth exon there are six positively charged residues that seem to be responsible for these effects. This region and the same six amino acids seem to control potentiation of the NR1 subunit by Zn^{2+} as well (Durand et al., 1993;

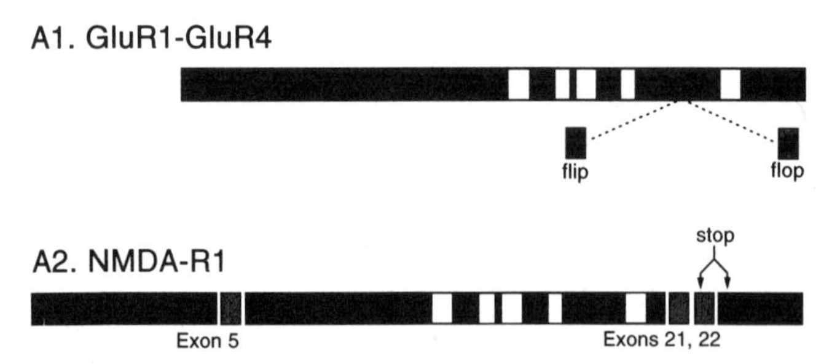

FIGURE 5–6. Representation of the alternative splicing patterns for the AMPA receptor subunits and the NMDA-R1 subunit. The white boxes indicate the proposed transmembrane domains, and the stippled boxes represent the regions that are alternatively spliced. Breaks surrounding the alternative splice sites depict regions that can be either inserted or deleted in the subunit.

Hollmann et al., 1993; Zheng et al., 1994). The consequences of alternative splicing and phosphorylation of the glutamate receptors have provided some initial insights into the structure-function correlates in these receptors.

FUNCTIONAL SIGNIFICANCE OF MOLECULAR DIVERSITY IN GLUTAMATE RECEPTORS

One of the hallmark conclusions of the first 4 years of studying recombinant glutamate receptors is that the high degree of molecular diversity achieved by subunit selection, alternative splicing, and RNA editing is paralleled by an equally high degree of functional diversity. As already noted, the permeability properties of recombinant AMPA receptors are influenced greatly by the presence or absence of the (edited) GluR2 subunit. An array of regulatory opportunities has also been identified for NMDA receptors by studying receptors with different subunit compositions. The ability of such polyamines as spermine to potentiate NMDA receptor activation depends critically on whether the fifth exon is present in the NR1 subunit. Several additional examples can be listed that are expected to have physiological consequences for synaptic transmission mediated by glutamate receptors. The potency of glycine as a co-agonist is different for heterotrimeric receptors assembled from NR1a/NR2A/NR2C compared with either NR1a/NR2A or NR1a/NR2C receptors (Wafford et al., 1993). Magnesium has less than one-tenth the potency as a blocker of NR1a/NR2C receptors than either NR1a/NR2A or NR1a/NR2B receptors (Ishii et al., 1993; Monyer et al., 1992). Finally, responses to the rapid perfusion of glutamate and glycine decay more slowly for NR1a/NR2C and especially NR1a/NR2D receptors when compared with NR1a/NR2A receptors (Monyer et al., 1992, 1994). None of these subunit-specific properties is understood yet in mechanistic terms.

In most cases the correlation between structure and function has been achieved only for recombinant receptors of defined composition. However, attempts are beginning to be made to identify the subunit composition of native glutamate receptors at known excitatory synapses. The approach combines electrophysiology and molecular biology on the same neuron. As schematized in Fig. 5–7, a functional diagnostic test designed to predict which glutamate receptors are expressed by a neuron is first carried out in whole-cell patch clamp, and then the cytoplasmic contents of the cell are drawn into the pipette by suction. The cellular mRNA harvested in this manner is used as template for reverse transcriptase, and the resulting single strand cDNA is subsequently used as template for PCR amplification of selected transcripts. The PCR products can be separated by electrophoresis on agarose or polyacrylamide gels and visualized by ethidium bromide staining or the incorporation of radioactive nucleotides. This dual approach has been used by Bochet et al. (1994) to identify the AMPA receptor subunits expressed by two populations of hippocampal neurons.

Single-cell PCR is admirably suited for several applications; for example, genotyping an individual sperm for which a simple yes or no answer for a given gene is the only result sought. When applied to correlating the structure and function of individual cortical neurons, however, this technique has two major conceptual drawbacks. First, the PCR analysis, like *in situ* hybridization, reveals which mRNAs rather than which proteins are expressed by the neuron, whereas the electrophysiological analysis reflects protein expression. A growing number of examples of mismatches between mRNA and protein expression (e.g., Sucher et al., 1993) indicate significant translational regulation of receptor expression and caution against the simple interpretation of the PCR results. Second, in contrast to genotyping sperm, a present or not-present answer for a given mRNA

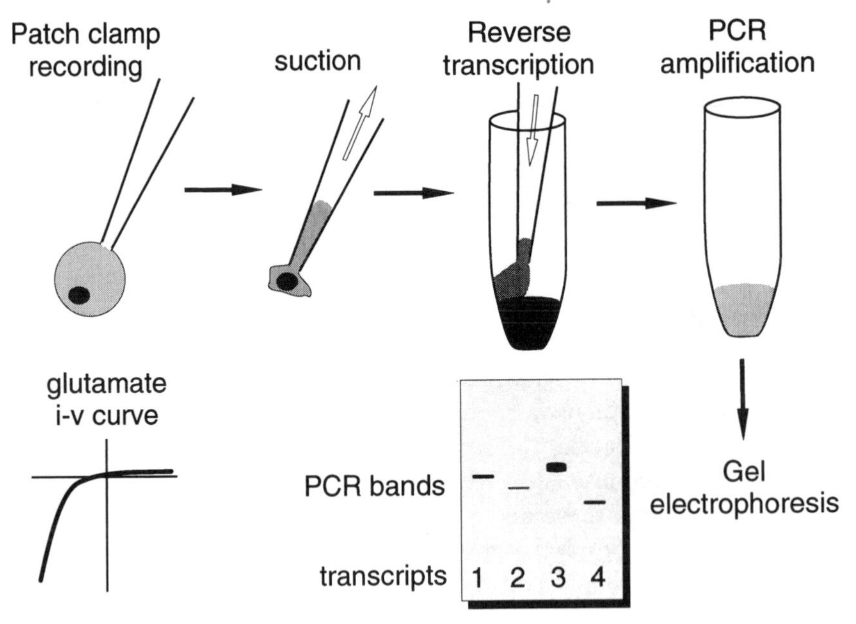

FIGURE 5–7. Schematic depiction of single-cell PCR technique. After electrophysiological phenotyping by whole-cell recording, the cytoplasm is withdrawn into the patch pipette and then expeled into a siliconized tube containing reverse transcriptase and its buffers. The resulting single-stranded cDNA is amplified by one or more rounds of PCR with subunit-specific oligonucleotide PCR primers. The PCR products can then be sized by agarose gel electrophoresis as shown.

is insufficient to explain the subunit composition of ligand-gated ion channels. This is the case because both subunit stoichiometry and subunit identity are probably functionally significant and, more importantly, because neurons may express a mosaic of several receptor subtypes, some of which contain a particular subunit and others of which do not. Even quantitative PCR will not address the latter issue, and it seems that the most secure approach to identification of receptor composition remains the careful comparison of functional properties of native receptors with those of recombinant receptors of known composition.

SUMMARY

The past 5 years have witnessed the cloning and initial characterization of numerous glutamate receptor genes with their associated splice and editing variants. Although additional molecular species of glutamate receptor are likely to be identified in the near future, the next phase of research must take advantage of the wealth of molecular detail amassed and begin to address the roles of glutamate receptor subtypes in important synaptic pathways in the brain. It is no longer satisfying to identify an excitatory synapse as mediated by an "NMDA" or "non-NMDA" receptor. The challenge for synaptic physiologists who study cortical circuitry is to incorporate into our own experimental questions the myriad of new insights into receptor structure-function correlates provided by the molecular biologists. For example, developing a mechanistic understanding of how subunit composition influences functional properties of glutamate receptors is almost certain to provide new opportunities for modulation of these receptors, in some cases to therapeutic advantage. Several other questions can now be addressed that will undoubtedly be rewarding for understanding cortical circuitry: How does an individual neuron decide which genes to transcribe, whether to edit a primary transcript, which exons to include in the mRNA, which subunits to assemble together, and where to route the resulting functional receptor? These questions have traditionally been in the realm of the developmental biologist, but it is becoming clear that even in adult neurons experience-driven alterations of receptor expression may contribute to synaptic plasticity. Answers to these questions will require a concerted multidisciplinary collaboration among molecular biologists and cellular electrophysiologists, which should lead to rewarding new insights about the cellular and molecular underpinnings of cortical function.

6

Synaptic actions of amino acid neurotransmitters

SCOTT M. THOMPSON, DANIEL V. MADISON, and ISTVAN MODY

SYNAPTIC ACTIONS OF GLUTAMATE

Fast synaptic excitation in the neocortex is mediated by the release of glutamate from the axon terminals of thalamocortical projection cells and of locally and intracortically projecting pyramidal cells (Streit, 1985; Tsumoto, 1990). The process of synaptic excitation consists of coupling the invasion of the action protential to the vesicular release of glutamate, which then diffuses across the synaptic cleft to bind with receptors and activate their associated ion channels in the postsynaptic membrane. The strength of synaptic excitation depends on numerous processes, such as the probability that a presynaptic action potential triggers glutamate release, the amount of transmitter released, the number and types of receptors on the postsynaptic membrane, and the conductance and kinetics of the ion channels gated by these receptors. These steps are considered in sequence to provide as coherent a picture of an excitatory cortical synapse as possible. Clearly, more details are availabe for some of these processes than for others. Where necessary, gaps are filled with data from more numerous hippocampal studies.

Release of Glutamate in the Neocortex

As discussed in Chapter 2, the quantitative description of transmitter release at individual synapses in the CNS presents numerous technical challenges that are only now beginning to be surmounted. Nonetheless, existing data permit a valuable qualitative description. Studies of excitation at synapses formed between pyramidal cell pairs (Thomson and West, 1993) and at synapses between pyramidal cells and interneurons (Stern et al., 1992; Thomson et al., 1993b) indicate that a single presynaptic action potential is very likely to produce a response in the postsynaptic cell. The amplitudes of individual synaptic responses exhibit considerable fluctuation, however, and the number of release sites that make up these synapses has not been defined. Therefore, it has not yet been possible to determine the release probability at an individual release site. The number of release sites is most likely small, suggesting that the probability of release from each site must be relatively high if few failures of transmission are to occur in paired recordings.

The probability of evoked glutamate release in the neocortex seems to be modifiable by several important coditions: upon activation of presynaptic inhibitory receptors, in response to high frequencies of presynaptic discharge (Thomson et al., 1993a,b), and after activation of presynaptic protein kinases.

Presynaptic receptors whose activation results in a decrease in synaptic excitation have been identified for acetylcholine (Vidal and Changeux, 1993), adenosine (Phillis et al., 1979), γ-aminobutyric acid (GABA; Howe et al., 1987b), glutamate (Sladeczek et al., 1993), and norepinephrine (Dodt et al., 1991). These effects are mediated by muscarinic M_2, purinergic A_1, $GABA_B$, metabotropic, and α1 receptors, respectively, which are all probably coupled to their effectors via heterotrimeric GTP-binding proteins. The mechanisms of action of these receptors have not yet been characterized in great detail, but are probably similar to those described in the hippocampus (Thompson et al., 1993a,b). Essentially, it is likely that their activation either (1) inhibits voltage-dependent Ca^{2+} influx into the axon terminal upon invasion of an action potential; (2) directly inhibits the exocytotic process at some point subsequent to Ca^{2+} influx, as shown by the ability of agonists at these receptors decrease the frequency of miniature excitatory synaptic currents; or (3) exhibits some combination of both processes. Functionally, these receptors allow for global regulation of synaptic strength by extrinsic neuromodulatory systems (acetylcholine and norepinephrine), as well as activity-dependent, feedback control of local excitability (adenosine, GABA, glutamate).

In contrast to receptor-mediated presynaptic inhibition of release, several protein kinases are localized in presynaptic terminals and are capable of facilitating glutamate release in the cerebral cortex (e.g., Nichols et al., 1987). There is evidence, at least in hippocampus, that the activation of presynaptic protein kinases and subsequent facilitation of their release may underlie the presynaptic expression of some types of synaptic plasticity (see Chapter 7). By analogy to the hippocampus, both the amplitude of stimulus-evoked glutamatergic synaptic responses and the frequency of action potential-independent, spontaneous glutamatergic synaptic responses are increased after application of activators of protein kinase C, such as phorbol esters (Finch and Jackson, 1990; Malenka et al., 1987) or of protein kinase A, such as forskolin (Chavez-Noriega and Stevens, 1994). Quantal analysis of putative unitary EPSPs suggests that these increases result from an increased probability of release from the presynaptic terminal (Chavez-Noriega and Stevens, 1994; Yamamoto et al., 1987). The mechanism responsible for at least part of this increased release probability occurs at some point after voltage-dependent Ca^{2+} influx, as considerable phorbol ester-induced potentiation of glutamate release can occur in the absence of extracellular Ca^{2+} (Malenka et al., 1987) or after the block of Ca^{2+} influx (Parfitt and Madison, 1993). These Ca^{2+}-independent effects of protein kinase stimulation and the Ca^{2+}-independent effects of agonists at presynaptic inhibitory receptors on release probability may result from opposing actions on common substrates in the presynaptic terminal.

Interestingly, the levels of protein kinase expressed by cortical neurons may be regulated both as a consequence of changes in afferent activity and during developmental, reaching their highest levels during the critical period before declining to adult levels (Jia et al., 1990). The relationship between changes in PKC levels and the critical period for synaptic plasticity is not yet fully defined. Deafferentation of the cortex early in development was first shown to prevent the subsequent maturational decrease in PKC expression (Jia et al., 1990). More recently, however, dark rearing, which is known to prolong the critical period, was shown to result in even larger decreases in PKC immunoreactivity than that seen during normal development (Elkabes et al., 1993).

Physiology of Synaptic Glutamate Receptors

The development of selective agonists and antagonists for glutamate receptor subtypes and of in vitro cortical slice preparations over the last 15 years has led to numerous descriptions of excitatory postsynaptic potentials and currents (EPSP/Cs) in the neocortex. As previously observed in the hippocampus, extracellular activation of many excitatory afferent fibers to both pyramidal cells and interneurons results in a dual component EPSP (e.g. Burgard and Hablitz, 1993; Hwa and Avoli, 1992a,b; Jones and Baughman, 1988; LoTurco et al., 1990; Stern et al., 1992; Thomson, 1986; Thomson et al., 1985, 1989a,b). The early, fast component becomes decreased in amplitude with depolarization from the resting membrane potential and is blocked by such substances as quinoxalinediones (CNQX and DNQX), indicating that it is mediated by the AMPA/kainate classes of glutamate receptor. The late, slow component becomes increased in amplitude with depolarization from the resting potential, is sensitive to glycine and to changes in the extracellular concentration of Mg^{2+}, and is blocked by such substances as D-2-amino-5-phosphovalerate (AP5) and ketamine, indicating that it is mediated by the NMDA class of glutamate receptor.

Descriptions of excitatory synaptic connections between pairs of intracellularly recorded pyramidal cells within cortical columns in slices of visual (Mason et al., 1991) or sensorimotor (Thomson and West, 1993; Thomson et al., 1993a) cortex indicate that unitary EPSPs range in amplitude from about 0.5 to 3 mV and have a duration of roughly 15 to 60 msec. Similar properties are reported for unitary excitatory synapses between pyramidal cells and inhibitory interneurons (Thomson et al., 1993b). These EPSPs seem to decay with a time constant corresponding to the membrane time constant, at least when measured at hyperpolarized membrane potentials.

In primary cultures of cortical neurons, "hot spots" containing high densities of both NMDA and non-NMDA receptors can be found at sites that are opposed to presumptive excitatory synapses (Jones and Baughman, 1991). Does this co-localization also hold true for cortical neurons in more intact tissue? If the two receptor types are co-localized, then action potentials in single cells should elicit dual component EPSPs in the postsynaptic cells. Unfortunately, this possibility has not yet been fully addressed in the studies to date of EPSPs between pairs of cortical neurons, although there is evidence of increases in unitary EPSP amplitudes upon depolarization from the resting membrane potential (Mason et al., 1991; Thomson et al., 1993a). The evidence for co-localization from the study of miniature excitatory synaptic currents (mEPSCs), recorded in the presence of TTX to block action potentials, is conflicting. One study did report that mEPSCs in neocortical slices were fully blocked by CNQX and displayed a time course of decay that could well fit with a single exponential, suggesting that they were entirely mediated by AMPA/kainate receptors (LoTurco et al., 1990). Pure NMDA receptor-mediated miniature EPSCs have been found in other more recent studies, at least when sucrose is focally applied to increase the probability of release (Carmignoto and Vicini, 1992). It thus seems likely that at least some synapses express both NMDA and non-NMDA receptors.

How, and under what conditions, do these distinct forms of synaptic excitation contribute to information processing? It is usually assumed that NMDA receptors do not contribute to normal synaptic transmission because the membrane potential is not sufficiently depolarized to relieve the block of the ion channel by Mg^{2+} ions and because of temporal overlap with the inhibitory synaptic potential. The NMDA receptor-mediated component of EPSPs is therefore typically studied either with the postsynaptic membrane strongly depolarized, in the absence of extracellular Mg^{2+}, or after blocking synaptic inhibition. Nevertheless, NMDA antagonists have been shown to block components of synaptic excitation in visual and sensorimotor cortices elicited with relatively natural stimuli in vivo.

In the adult cat visual cortex (Fox et al., 1990), the involvement of NMDA receptors seems somewhat layer specific: AP5 affects responses to stimuli of varying contrast more strongly in layers II/III than in layers IV to VI. Importantly, it was shown that the proportion of the evoked excitation in the superficial cortical neurons that was sensitive to AP5 was the same, regardless of the intensity of the visual drive, i.e., the amount of contrast in the stimulus. Similarly, in the barrel fields of rat cortex (Armstrong-James et al., 1993), neurons in layer IV are excited by a 3-msec deflection of the principal whisker and fire action potentials anywhere from 5 to 100 msec after the stimulus. Local iontophoresis of AP5 largely abolishes action potentials occurring later than 10 msec after the stimulus, but leaves the initial response unaffected. Iontophoresis of the AMPA/kainate receptor antagonist DNQX, in contrast, blocks the initial response. Interestingly, layer II cells only respond to the stimulus with discharges that are delayed more than 10 msec, and these are fully blocked by AP5.

It thus seems that the commonly held idea that NMDA receptors only contribute to synaptic excitation after reduction of inhibition or with high frequencies of stimulation, which was derived from experiments in hippocampal slices (e.g. Collingridge et al., 1988a,b) is not necessarily true in vivo, at elast for the neocrotex. This is a particularly important conclusion when considering the various conditions likely to trigger synaptic plasticity (see Chapter 7). These experiments do, however, support previous suggestions from studies of cortical slices that NMDA receptors play a more important role in the superficial layers (Thomson, 1986). In addition, these results confirm earlier binding studies indicating that there are more NMDA receptors in the upper cortical layers than in middle and deeper layers, e.g., Monaghan and Cotman (1985).

Kinetics of Excitatory Synaptic Events

The AMPA/kainate receptor-mediated component of the EPSC in adult cortical neurons decays with a time constant of roughly 3 msec (Hestrin, 1992), whereas the NMDA receptor-mediated component has a time constant of roughly 40 msec (Carmignoto and Vicini, 1992). Why is the NMDA receptor-mediated component of the EPSP so much slower than the AMPA/kainate receptor-mediated component? The affinity of glutamate at the NMDA receptor is about eightfold higher than at AMPA/kainate receptors on cultured hippocampal neurons (Patneau and Mayer, 1990), although this will certainly vary *in vivo* depending on the exact subunit composition. Does synaptically released glutamate bind only once to the AMPA/kainate receptors and several times to the NMDA receptor, or are the kinetics of the NMDA receptor-gated channels much slower than those gated by AMPA/kainate receptors? Considerable progress has been made in the past several years in creating detailed kinetic schemes that can describe the molecular events responsible for the time course of dual component EPSP/Cs. Many of these studies have been performed in the hippocampus, but the conclusions probably apply to the neocortex as well.

Using cultured hippocampal neurons, Lester et al. (1990) established elegantly and conclusively that the slow time course of the NMDA receptor-mediated component of the EPSP results from a slow rate of channel closing and not from rebinding of glutamate to the receptor. Rapid application of AP5, which prevents the rebinding of glutamate to the NMDA receptor, immediately after evoking unitary NMDA receptor-mediated EPSCs between pairs of cells in Mg^{2+}-free saline has no effect on the synaptic current. In contrast, rapid application of Mg^{2+}, which occludes open channels but has no effect on the binding of glutamate to the NMDA receptor, results in the immediate block of the current. Using excised patches from these cells and recording in the presence of CNQX, it could then be shown that application of glutamate followed by AP5 results in prolonged single channel openings, despite block of the receptor, and that very brief (less than 5 msec) pulses of glutamate induce currents in excised patches with the same time course as the NMDA receptor-mediated synaptic current. Long-lasting NMDA receptor-mediated EPSPs thus result from a single binding and slow unbinding of glutamate to the receptor.

The rapid time course of the AMPA/kainate receptor-mediated EPSP/C is also apparently determined by the closing rate of the channels, although this process occurs much more quickly than for NMDA receptor-gated channels. Initial studies in hippocampal neurons had suggested that receptor desensitization might play the critical role in limiting AMPA/kainate receptor-mediated EPSCs (Tang et al., 1991; Trussel and Fischbach, 1989). Studies on neocortical neurons, however, indicate that AMPA/kainate receptor-mediated EPSC, as well as responses to brief applications of glutamate in the presence of NMDA receptor antagonists, decay more quickly (time constant = 3 msec) than the rate at which the receptors desensitize (time constant = 8 msec; Hestrin, 1992b). It is thus likely that the decay of the fast EPSP reflects the closing of AMPA/kainate receptor-gated ion channels upon unbinding of glutamate. The rate at which unbinding occurs is presumably faster at these receptors than at NMDA receptors because of the difference in their affinity for glutamate. Indeed, NMDA receptor-mediated currents in response to the brief application of agonist are faster when exogenous agonists having a lower affinity than glutamate are used. Only glutamate has an affinity that is consistent with the time course of the slow excitatory synaptic currents (Lester and Jahr, 1992).

As a corollary to these observations, it can be concluded that the concentration of glutamate in the vicinity of the synaptic receptors reaches a high concentration, probably about 1 mM (Clements et al., 1992), and then decreases precipitously. If not, the receptors would be activated more than once, resulting in EPSC decay time constants that are slower than the single channel kinetics. The rapid decline in the concentration of glutamate seems to be determined solely by diffusion under normal conditions, as inhibitors of glutamate uptake have no effect on the decay of EPSCs, at least in the hippocampus (Hestrin et al., 1990).

SYNAPTIC ACTIONS OF GABA

Inhibition in the mammalian brain is mainly caused by the binding of a small amino acid, GABA, to its receptors (Biggio et al., 1992; Roberts, 1986; Sivilotti and Nistri, 1991). Approximately 17% of the synapses in the mammalian cerebral cortex have been estimated to be GABAergic (Halasy and Somogyi, 1993). The inhibition mediated by GABA in the cerebral cortex (Krnjevic and Phillis, 1963) is critically involved in both short- and long-term regulation of neuronal excitability in virtually all modes of functioning and malfunctioning of the mammalian brain, including that of humans (McCormick, 1989b).

At physiological pH, GABA is an electroneutral zwitterion (Roberts and Sherman, 1993; pI = 7.3). This property, unique among neurotransmitters, allows GABA to traverse the narrow synaptic cleft virtually unimpaired by electrostatic interactions (Roberts and Sherman, 1993). Once released from the presynaptic terminal, GABA acts at distinct receptors, according to the principle of divergence in neurotransmitter action (Nicoll, 1988). Three different types of GABA receptor—A, B, and C—are known to be activated by GABA in the central nervous system (Sivilotti and Nistri, 1991). At most inhibitory synapses in the brain, GABA predominantly activates two qualitatively different inhibitory mechanisms via $GABA_A$ and $GABA_B$ receptors (Bormann, 1988; Bowery, 1993; Sivilotti and Nistri, 1991).

The $GABA_A$ receptors are members of the ligand-gated ion channel superfamily (Schofield et al., 1987). Most "fast" synaptic inhibition in the cortex is accomplished through activation of these receptors (see Chapter 16). There are 16 known $GABA_A$ receptor subunits that may assemble in various combinations of five (pentamers) to form functional Cl^- channels (Nayeem et al., 1994). The total number of possible configurations is over half a million, but certain combinations are clearly preferred during assembly (Angelotti and Macdonald, 1993; Angelotti et al., 1993; Backus et al., 1993). The subunit composition of the channels can influence their main conductance state, their kinetics, and their sensitivity to modulators (Puia et al., 1991; Sigel et al., 1990; Verdoorn et al., 1990). However, regardless of subunit composition, all $GABA_A$ channels are permeable to Cl^-. Therefore, activation of $GABA_A$ channels in an intact cortical neuron during an inhibitory postsynaptic potential (IPSP) can lead to either a hyperpolarization, little or no change in membrane potential, or to a depolarization, solely depending on the relationship between the Cl^- reversal potential (E_{Cl}) and the resting membrane potential of the neuron. As simple as it sounds, this critical relationship is not straightforward to determine. The differences between the unperturbed E_{Cl} have recently been measured in granule cells and hilar neurons of the dentate gyrus (Soltesz and Mody, 1994) with a noninvasive technique that uses single K^+ channels as transmembrane voltage sensors (Zhang and Jackson, 1993). No matter in which direction the membrane potential deviates during an IPSP, activation of $GABA_A$ receptors always results in large increases in membrane conductance and strong shunting of excitatory current (Staley and Mody, 1992).

The second type of GABA receptor ($GABA_B$), although not yet cloned, seems to be part of the seven transmembrane spanning, G-protein-coupled receptor superfamily (Bowery, 1993). Several subtypes of $GABA_B$ receptors may exist analogous to the large diversity of cloned G-protein-coupled amino acid receptors, such as the metabotropic glutamate receptors (mGluR; Tanabe et al., 1992). The localization of $GABA_B$ receptors is a key component of GABAergic synapses. Depending on their synaptic location, whether pre- or postsynaptic, activation of $GABA_B$ receptors may produce inhibition or disinhibition. Postsynaptically, there is a well-characterized relatively long increase in a K^+ conductance that is responsible for generating "slow" inhibitory events in CNS neurons (Dutar and Nicoll, 1988; Otis et al., 1993). On the presynaptic side, activation of K^+ channels or inhibition of Ca^{2+} conductances through $GABA_B$ autoreceptors may diminish GABA release and thereby reduce the overall level of GABA-mediated inhibition (Thompson et al., 1993a). Such mechanisms may operate to reduce the amount of inhibition when plastic changes in excitability are induced (Davies et al., 1991). As $GABA_B$ receptors are also present on excitatory terminals, diffusion of GABA to neighboring glutamatergic terminals may curtail the relase of excitatory amino acids (Isaacson et al., 1993; Thompson, et al., 1993a).

Properties of GABAergic Inhibition

It is possible to learn a great deal about the properties of neural inhibitory circuitry and the modulation of that circuitry by recording in the target neuron of that inhibition. As most such recordings have been done in the CA1 and dentate gyrus regions of the hippocampal formation, the principal pyramidal cells and granule cells of these regions serve as "inhibitory targets" for the purpose of this review. Using sharp microelectrodes or whole-cell pipettes, four types of synaptic inhibition can be used to assess the properties of GABAergic inhibition. Two of these types are evoked by electrical stimulation through an electrode. The other two types of inhibition arise spontaneously from the GABA-containing neurons themselves. The four measures, shown in Figure 6–1 are (1) polysynaptic-evoked inhibitory postsynaptic potentials (IPSPs), (2) monsynaptic-evoked IPSPs, (3) spontaneous action potential-dependent inhibitory postsynaptic currents (IPSCs), and (4) spontaneous miniature IPSCs.

Polysynaptic IPSPs. Polysynaptic-evoked IPSPs are the most frequently used measures in studies of synaptic inhibition. In the CA1 region of the hippocampus, they consist of hyperpolarizations evoked by brief electrical shocks delivered to the excitatory Schaffer collateral fiber pathway. Stimulation of Schaffer collaterals causes the activation of inhibitory interneurons and the synaptic inhibition of the principal pyramidal neurons by two different paths. First, Schaffer collaterals are thought to contact directly all known subtypes of inhibitory interneurons of area CA1. These interneurons then release GABA onto the pyramidal neurons. This is known as feedforward inhibition. Second, Schaffer collaterals directly excite CA1 pyramidal cells, which in turn form excitatory connections onto a subset of inhibitory interneurons. The interneurons in turn form feedback inhibitory synapses onto the pyramidal cells. These components involve the serial activation of at least two (for feedforward) or three (feedback) synapses; thus, the name polysynaptic. The IPSP that arises in pyramidal neurons from Schaffer collateral stimulation is a mixture of both feedforward and feedback components.

There are at least three types of inhibitory interneuron found in area CA1 of the rat hippocampus. Two of these types, the basket cell and the oriens-alveus interneuron (named for its location at the stratum oriens-alveus border), receive the same inputs and express similar outputs, as far as is known, so they can be considered together as one type in this context.

These two types of interneuron receive both direct feedforward input from the Schaffer collaterals and feedback input from the CA1 pyramidal cells. Their terminals are primarily on the somata of pyramidal cells, where they form inhibitory synapses. The dominant postsynaptic receptors at these particular synapses are of the $GABA_A$ subtype. The third type of interneuron, lacanosum-moleculare interneurons, is also named after its anatomical location. This interneuron is thought to receive feedforward input from the Schaffer collaterals almost exclusively. It terminates primarily onto the

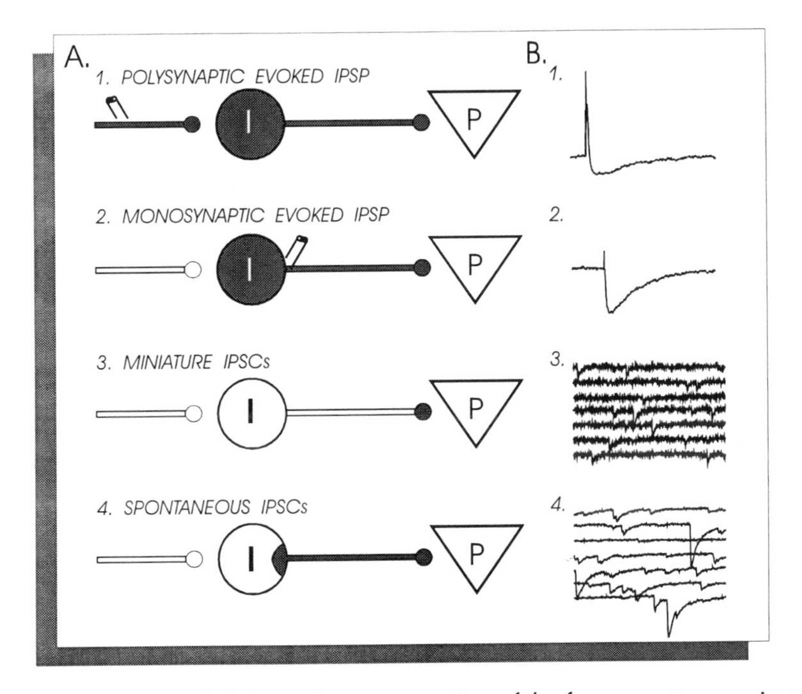

FIGURE 6–1. **A.** Schematic representation of the four ways to experimentally evaluate the efficacy of inhibition in brain slices. in each panel, the shaded areas represent the synaptic mechanisms responsible for the control of inhibition. I is the inhibitory interneuron, and P denotes the principal cell. (1) A polysynaptic inhibition can be evoked with a stimulating electrode that activates the excitatory afferents onto the inhibitory cell. (2) In the presence of excitatory receptor antagonists, monosynaptic inhibitory responses can be evoked by directly stimulating the inhibitory terminals. (3) Miniature inhibitory events can be recorded when excitation is blocked and action potentials of presynaptic calcium entry are also inhibited. (4) In an experimental paradigm similar to (2), but with no exogenous stimulation of the inhibitory terminals, spontaneous release of GABA will also depend on the frequency of action potentials reaching the terminals. B. Sample recordings corresponding to the experimental approaches illustrated in A1–A4. The two top panels (B1 and B2) are current-clamp recordings, whereas the two bottom panels (B3 and B4) were obtained in whole-cell patch configuration.

dendrites of pyramidal neurons. The synapses formed by the terminals of this type of interneuron use the $GABA_B$ subtype of GABA receptor.

Because of the connections formed among Schaffer collaterals, pyramidal cells, and interneurons, the polysynaptic IPSP has two phases. The first phase, called the fast or early IPSP results primarily from activation of the basket/oriens-alveus interneuron (both feedforward and feedback). Since synapses from these interneurons use the $GABA_A$ receptor, the early IPSP is mostly mediated by an increase in postsynaptic Cl^- conductance. The second phase of the IPSP, the slow or late IPSP, may arise from activation of the stratum radiatum interneurons. The synapses between these interneurons and pyramidal cells seem to utilize primarily the $GABA_B$ receptor subtype, and this IPSP is mediated primarily by an increase in potassium conductance (Samulack and Lacaille, 1993). Even though the late IPSP is almost exclusively disynaptic (it has no known feedback component), its onset is lower thant the early IPSP presumably because of the slow activation kinetics of the $GABA_B$-activated channel, which requries the intercession of a second messenger.

Monosynaptic Evoked IPSPs. The second type of inhibitory transmission can only be seen under experimental conditions. The monosynaptic IPSP can be recorded in any CA1 pyramidal cell and is evoked by directly activating interneurons with electrical shocks delivered through a stimulating electrode. During an experiment, this is accomplished by performing two maneuvers. First, a stimulating electrode must be placed close to the interneurons. Since interneurons are scattered about the hippocampus, this is done in practice by placing the stimulating electrode close to the location of a pyramidal cell where intracellular recordings are being made. This approach takes advantage of the divergent efferents of interneurons. An interneuron in the vicinity of the stimulating electrode is almost certain to contact the recorded pyramidal cell. The second manipulation is to bathe the preparation, usually a slice of hippocampus, in pharmacological blockers of glutamate receptors. This step is necessary because, even when stimulating close to the interneuron, glutamatergic axons are also stimulated, evoking polysynaptic IPSPs. In the presence of glutamate antagonists, stimulation is only able to activate inhibitory synapses from the interneurons in that local area. This is apparent by two measures. First, if the electrode is placed more than 1 mm away from the recording electrode, no monosynaptic IPSP can be evoked. Second, stimulation in the stratum radiatum evokes both an early and late ($GABA_A$ and $GABA_B$) monosynaptic IPSP, whereas stimulation a short distance away in the stratum oriens produces only an early IPSP. This is presumably because the electrode in the stratum oriens spreads to oriens/alveus and perhaps basket interneurons, which only make $GABA_A$ synapses.

Measures of Spontaneous Inhibition. The two remaining measures of inhibition both occur spontaneously under physiological conditions. We refer to these two types of events as miniature IPSCs (mIPSP), and spontaneous IPSCs (sIPSP). Both of these types of inhibitory currents arise from the spontaneous release of small amounts of GABA which causes the opening of postsynaptic $GABA_A$-associated chloride channels. Because of the small size of these events, they are generally recorded with chloride-containing electrodes, to reverse the cell's transmembrane chloride gradient and to increase the difference between E_m and E_{Cl}. Doing so has the result of reversing these chloride-mediated inhibitory events to the depolarizing direction, or making them inward currents in voltage clamp. It also has the salutary effect of making these events larger and easier to detect and measure.

Miniature IPSCs are the smaller and less frequent of the spontaneous inhibitory events. These events persist in the presence of the sodium channel blocker tetrodotoxin (TTX) and therefore do not depend on action potentials (Edwards et al., 1990; Otis and Mody, 1992b; Otis et al., 1991; Ropert et al., 1990). They likely represent the quantal release of GABA from interneuron terminals. In a hippocampal slice maintained at 22°C in 2.5 mM external potassium, these events are detected in CA1 pyramidal cells with an average frequency of between 1 to 2 Hz. With 100 mM KCl in the whole-cell recording electrode, they average 25 pA in amplitude.

Spontaneous IPSCs only occur in the absence of TTX, and so they likely arise from the ongoing action potential discharge of inhibitory interneurons. Spontaneous IPSCs are generally larger in amplitude than miniature IPSCs, usually at 2 to 3 Hz under the conditions described above. They also have a higher average amplitude than mIPSCs, perhaps reflecting an ability of action potentials to release multiple transmitter quanta. The average size of sIPSCs is approximately 50 pA.

Since each of these measures of inhibition arises from different parts of the inhibitory circuitry, the study of each gives a window to its part of the circuit.

Activation and Saturation of Synaptic $GABA_A$ Receptors

Based on the low variance at the peak of synaptic responses and their small amplitudes, a fair amount of receptor saturation has been proposed to exist for some time at excitatory synapses (Jack et al., 1981; Redman, 1990). More recently, detailed electrophysiological studies (Clements et al., 1992; Tong and Jahr, 1994) have reinforced the notion of the pharmacological saturation of glutamate receptors. Similarly, at inhibitory synapses, the low quantal variance of evoked IPSCs, the skewed amplitude distribution and the rapid rise times of mIPSCs, the postulated low number of

GABA receptors activated during mIPSCs (Edwards et al., 1990), and elegant modeling studies (Busch and Sakmann, 1990; Faber et al., 1992) are all consistent with a saturation of synaptic $GABA_A$ receptors. In other words, the concentration of GABA released from presynaptic vesicle(s) into the synaptic cleft is overwhelming in comparison to the low number of postsynaptic $GABA_A$ receptor channels. In fact, a recent report demonstrated the similarity in the time course of mIPSCs and that of responses elicited by rapid (less than 1 msec) applications of large concentrations of GABA onto excised patches containing a few $GABA_A$ receptors (Maconochie et al., 1994).

In light of these observations, the kinetics of synaptic $GABA_A$ channels after their brief activation by excessive agonist concentrations must be solely responsible for the time course of synaptic currents. At the synapse, the amount of GABA released into the cleft diffuses away so rapidly that GABA clearance by cellular uptake has little to do with the decay phase of spontaneous IPSCs, as demonstrated by the ineffectiveness of GABA uptake blockers on mIPSC kinetics (Otis and Mody, 1992b; Thompson and Gähwiler, 1992). Thus, drug effects on the kinetics of synaptic $GABA_A$ receptor channels can be studied simply by examining changes in the time course (rise and decay time constants) of mIPSCs. For example, the benzodiazepine agonists that alter $GABA_A$ channel kinetics have been shown to prolong the decay time constant of TTX-resistant mIPSCs (De Koninck and Mody, 1994; Otis and Mody, 1992b; Soltesz and Mody, 1994). The unaltered amplitude distribution of mIPSCs in the presence of benzodiazepines, drugs that increase the on-rate of GABA binding (Rogers et al., 1994), cannot be explained unless it is assumed that the synaptic receptors are already saturated by the GABA concentration in the cleft. Is it then at all possible for such drugs to increase the amplitude of synaptic events resulting from the summation of several mIPSCs? Given the smallest amount of jitter in the alignment of synaptic currents that may originate from several synapses, the prolongation of individual mIPSCs will be reflected in the increased amplitude of the compound inhibitory response.

An established technique, originally developed by Sigworth (1980) for the study of Na^+ channels, made it possible to examine more closely the behavior of channels activated by GABA in the postsynaptic membrane during mIPSCs. The nonstationary fluctuation analysis allowed us to resolve the single channel conductance and number of postsynaptic $GABA_A$ receptors opened by a packet of GABA released at a central mammalian synapse (De Koninck and Mody, 1994). The conclusions from this study and computer simulations fully support the hypothesis that, there is a near physiological saturation (i.e. nearly all channels open) at the peak of synaptic responses.

Pre- and Postsynaptic Control of GABAergic Inhibition

The fundamental consequence of receptor saturation is that there are only certain ways in which the efficacy of GABAergic inhibition may be enhanced in the mammalian brain. Modulation of the frequency, time course, amplitude, or a combination of these three parameters can all result in significant changes in the effectiveness of inhibition. A change in the frequency of spontaneous IPSCs that continuously bombard most neurons in the mammalian CNS (Otis et al., 1991) may be produced by purely presynaptic mechanisms without a change in the amplitude or time course of individual synaptic responses. This change seems to be a fundamental mechanism whereby opiates modify the efficacy of inhibition in the brain (Capogna et al., 1993; Cohen et al., 1992; Thompson et al., 1993a).

Even with saturated postsynaptic receptors, several mechanisms can change the time course or amplitude of inhibitory events. One possibility is to change the kinetics of $GABA_A$ channel openings and consequently to prolong the duration (but not the amplitude) of spontaneous IPSCs. At clinically relevant concentrations of certain GABA-modulatory drugs, such as barbiturates, this prolongation may be sufficient to merge individual synaptic events into each other (Otis and Mody, 1992b), resulting in the elevation of the "tonic" Cl^- shunting produced by $GABA_A$ receptor activation. Similar mechanisms may underlie the effect of benzodiazepines (Otis and Mody, 1992b) and volatile anesthetics (Mody et al., 1991; Tanelian et al., 1993).

Perhaps more relevant to the physiological control of transmission at a GABA synapse is another mechanism for the enhancement of inhibition through synchronization of the release from the approximately eight to ten terminals that all converge onto a principal cell and originate from a single inhibitory neuron (Buhl et al,. 1994). A short current pulse delivered through an extracellular stimulating electrode in a slice preparation may produce asynchrony of release from such terminals. An imperfectly synchronized release may be responsible for the increased rise-time and decay time constant of stimulus-evoked IPSCs. Even the use of "minimal" stimulation that elicits a single fiber IPSCs, about eight to ten times the size of average mIPSCs (Lambert and Wilson, 1993), may cause relatively slowly rising and slowly decaying IPSCs. In paired-pulse stimulation experiments, as action potentials in the axon probably fail to invade certain release sites during the second of the two pulses, only the time course of the IPSC evoked by the second stimulus approaches that of a spontaneous event. A transmission failure secondary to the lack of action potential invasion of branched axons (Manor et al., 1991) seems to be a prominent feature in controlling the efficacy of GABAergic synapses. The control of action potential propagation at branchpoints (Manor et al., 1991) may provide an effective means of controlling the efficacy of inhibition by a presynaptic mechanism when increased neurotransmitter release at a given release site may be unable to do the same because of the pharmacological saturation of synaptic receptors.

The different time course of IPSCs elicited by conditioning pulses can most likely be attributed to an imperfect synchrony of release and is a hallmark of most events evoked by extracellular stimulating electrodes. Thus, in pharamcological studies of IPSP(C)s evoked by stimulating electrodes it is difficult to distinguish an effect on synchrony from a true effect on time course. Such experiments will need to be supplemented with data on spontaneous events resulting from action potentials normally generated by interneurons and naturally propagating down the axonal tree. These latter sIPSCs produce events that are indistinguishable in time course from mIPSCs (Otis and Mody, 1992). An extracellular current spread during stimulation, resembling the imperfect synchronization during repetitive bursting of interneurons (Otis and Mody, 1992a) may be responsible for the lack of synchrony in stimulus-evoked IPSP(C)s.

If presynaptic factors that control GABA release at a single bouton have marginal effect on the size of mIPSCs, enhancing individual mIPSC amplitude might still be possible through a change in the single channel conductance or an alteration in the number of postsynaptic $GABA_A$ channels opposite the release site. As such enhancement does not depend on presynaptic synchronization, the amplitude of action potential-resistant m-IPSCs should also change size accordingly. An increase in the number of functional $GABA_A$ receptors actually takes place during the long-term increase in mIPSC size associated with kindling-induced epilepsy (Otis et al., 1994). Kindling is one of the most widely used models of experimental epilepsy (McNamara, 1992). Increased inhibition after kindling is consistent with the observed increase in the number of muscimol binding sites (Shin et al., 1985b; Valdes et al., 1982), but an increase in mIPSC size was a counterintuitive finding in a chronically epileptic tissue. Although the conductances of mIPSCs after kindling were larger, their rise and decay kinetics were comparable to controls. Although there might be some synchrony of release between adjacent release sites even in the absence of action potentials (Korn et al., 1993), the findings could be best explained by a nearly doubling of the number of postsynaptic $GABA_A$ receptor channels (Otis et al., 1994).

How can such a significant increase in the number of postsynaptic receptors take place? Certainly, the postsynaptic area is sufficiently large to accommodate more than the 70 to 80 channels that are open at the peak of the largest IPSCs (Otis et al., 1994). Are there somewhere in the postsynaptic cell any spare, nonfunctional subunits waiting to assemble into functional receptors? It has been speculated (Otis et al., 1994) that $GABA_A$ receptor channels might be inserted in the membrane as a "bunch" containing tens of receptors. For glycine receptors, a specific protein (gephyrin) has recently been shown to be responsible for their clustering and aggregation in cultured spinal cord neurons (Kirsch et al., 1993). Clearly, we will need to learn more about both the physical and biochemical factors governing functional receptor assembly in the postsynaptic membrane. Some recent studies already point in the direction of phosphorylation-dependent mechanisms in the aggregation and assembly of ligand-gated channels (Green et al., 1991).

Modulation of Inhibition

Modulation of inhibition in the mammalian brain has profound implications for both normal physiological processes (*e.g.* learning and memory) and pathological states of brain function, *e.g.* epilepsy; Alzheimer's, Huntington's, or Parkinson's disease; and psychiatric disorders. Several therapeutical approaches involve enhancing the effectiveness of GABAergic inhibition (Biggio et al., 1992). Yet, the physiology and topography of GABAergic synapses put several major constraints on how the function of GABA synapses can be enhanced. An immediate concern for pharmacologists is the high GABA concentration in the synaptic cleft. Because drugs may act differently at receptors in the presence of saturating concentrations of agonists, there should be a thorough re-evaluation of experiments using low GABA concentrations at extrasynaptic sites or in $GABA_A$ receptor expression systems. One may justifiably ask which of the several kinds of inhibitory responses should be used to evaluate the effect of a drug or neuromodulator on inhibition. For example, a disinhibitory neurotransmitter can change the properties of a parpticular neural circuitry by reducing synaptic inhibition. This increases the effectiveness of excitatory transmission in bringing principal cells to threshold, since the counteracting inhibitory transmission is reduced. In the hippocampus, several neurotransmitters—norepinephrine, enkephalin, acetylcholine, and GABA itself—share the property of being disinhibitory. To understand how each of these transmitters actually disrupts the inhibitory circuitry requires an understanding of how each of them acts on the individual parts of the circuit. By studying the effects of a particular modulator on each of the four measures of inhibition, it is possible to form conclusions about what parts of the circuit are affected by the modulator and to make predictions about what kinds of actions might mediate this modulation.

The usefulness of the four measures of inhibition discussed above can best be demonstrated by comparing the actions of two different modulators on inhibitory transmission. Adrenergic agonists and opioid peptides are both disinhibitory in the hippocampus. If one examines their actions on polysynaptic IPSPs, they both seem to have virtually identical actions. However, when the actions of these two modulators are studied on the other measures of inhibition, their actions diverge. The opioid peptide strongly decreases the amplitude of monosynaptic IPSPs, whereas, epinephrine has no effect on this monosynaptic inhibitory transmission. Thus, opioids exert an action somewhere in the portion of the inhibitory circuitry that supports the monosynaptic IPSP, whereas adrenergic actions must act somewhere else.

As described above, the monosynaptic IPSP results from direct stimulation of interneurons. Thus, the action of opioids arises either from an action on the interneuron itself or on the postsynaptic pyramidal cell. A more exact localization of opioid action can be determined by examining the effects of enkephalin on the measures of spontaneous inhibition. Opioids exert two effects on spontaneous, action-potential-dependent IPSCs (Cohen et al., 1992). First, they decrease their frequency. Assuming that such IPSCs arise from all-or-none action potentials, the simplest explanation of this effect is that opioids cause a decrease in the spontaneous action potential discharge of interneurons. Such an effect would likely arise from the known effect of opioids to hyperpolarize interneurons. Adrenergic agonists, on the other hand, strongly increase the frequency of spontaneous IPSCs, showing that far from inhibiting the discharge of interneurons, these agonists actually stimulate them. Second, opioids decrease the amplitude of spontaneous IPSCs. This property indicates that opioids may be having an action directly at the interneuron terminal to decrease GABA release for each action potential or may be decreasing the postsynaptic GABA responsiveness. Adrenergic agonists have no effect on the amplitude of spontaneous IPSCs, confirming the conclusion that they have no effect on this part of the inhibitory circuit (Doze et al., 1991).

The contrasting actions of opioids and adrenergic agonists extend to their effects on miniature IPSCs. Opioid peptides cause a marked decrease in the frequency of mIPSCs, but cause no change in their amplitude. Thus, in agreement with previous findings, opioids do not diminish postsynaptic GABA sensitivity. Moreover, it is likely that this effect reflects an inhibition of postsynaptic GABA release. Interestingly, the effect of opioids is not sensitive to inorganic calcium channel antagonists, suggesting that it does not result from an inhibition of calcium influx in to the interneuron terminal. Based on these results, it is likely that opioids are acting directly on the release process to decrease the spontaneous release of GABA. This action may well account for the decrease in action-potential-dependent GABA release observed with spontaneous IPSCs. In contrast, adrenergic agonists have no effect on mIPSCs.

SUMMARY

In summary, excellent characterizations of the physiology and molecular biology of glutamate-mediated synaptic excitation in the neocortex are currently available. Significant progress is being made in elucidating the cellular basis for short- and long-term modifications in the strength of excitation, as well as the role played by the different receptor types in cortical information processing. On the inhibitory side, four measures of inhibition can be used effectively to localize and study the mechanisms of disinhibition by various neurotransmitters. Such studies have revealed that different disinhibitory neurotransmitters work by completely different mechanisms. This mechanistic divergence may prove to be highly significant in a functional sense, allowing each transmitter to modulate the inhibitory circuitry in a unique manner.

Detailed information about the functioning of excitatory and inhibitory amino acid receptors, including their pharmacological saturation and molecular diversity, may open a new era in the design of receptor-specific pharmacological tools for the fine tuning of cortical excitation and inhibition.

7

Synaptic plasticity in hippocampus and neocortex: a comparison

ROBERT C. MALENKA

Long-lasting activity-dependent changes in the strength of synaptic transmission are of fundamental importance for the development of neural circuits and for information storage in the nervous system. It has long been assumed that the storage of memories occurs primarily in the cerebral cortex (Hebb, 1949), and thus it is not surprising that some of the first attempts to elicit experimentally such changes in synaptic efficacy used preparations of cerebral cortex (Bliss et al., 1968). However, these initial attempts yielded confusing results, and attention was then turned to the hippocampus, another structure thought to play a critical role in information storage. Here, *in vivo* experiments met with much greater success, resulting in the landmark first description of long-term potentiation (LTP), a sustained increase in synaptic strength normally elicited by brief trains of high-frequency afferent stimulation (Bliss and Lomo, 1973). Over the last decade LTP in the hippocampus has been the most intensively studied model of activity-dependent synaptic plasticity in the mammalian brain primarily because it can be elicited reliably in isolated slices of hippocampus, a preparation that offers many experimental advantages. As a result of this research effort, we now know a great deal about the cellular mechanisms responsible for hippocampal LTP, particularly at the synapses between Schaffer collateral/commissural afferents and CA1 pyramidal cells.

The examination of the mechanisms responsible for LTP in the cerebral cortex has proved to be a much more challenging and daunting task. In large part, this is due to the much greater complexity of cortical circuitry, which does not readily lend itself to the stimulation of defined inputs and the recording of defined monosynaptic responses. Nevertheless, reasonable progress has been made over the last few years in the examination of the properties of LTP in various areas of neocortex. In this chapter I begin by briefly reviewing some of the salient aspects of the LTP elicited in CA1 cells in the hippocampus, which will serve as a model to which the various forms of LTP observed in cortical preparations can then be compared.

LTP IN THE HIPPOCAMPUS

Although LTP was first described in the dentate gyrus, the vast majority of experiments aimed at elucidating underlying cellular and biochemical mechanisms have been performed in the CA1 region of rodent hippocampal slices. The advantages of examining synaptic plasticity in this preparation include the following: (1) LTP can be elicited with high probability (70% to 90% of attempts in most laboratories) by simple manipulations, such as brief tetanic stimulation, (2) well-defined monosynaptic responses can be recorded in response to stimulation of known afferents without contamination by antidromic action potentials, and (3) excitatory postsynaptic potentials (EPSPs) can be measured using either extracellular field potential recording or by recording directly from individual cells.

The properties and mechanisms of LTP in the hippocampus have been reviewed extensively over the last few years (Bliss and Collingridge, 1993; Brown et al., 1990; Madison et al., 1991; McNaughton et al., 1993; Nicoll et al., 1988). This chapter focuses on those properties of hippocampal LTP that may prove useful for comparison with the results in neocortex. It is important to note that, even within the hippocampus, forms of LTP with strikingly different properties can be found (Harris and Cotman, 1986; Johnston et al., 1992; Zalutsky and Nicoll, 1990). This brief review, only discusses the form of LTP observed in CA1 cells.

The Role of NMDA Receptors and Ca^{2+}

Despite the large amount of work on hippocampal LTP, only a few of its properties are universally accepted. First, LTP exhibits input specificity in that only those inputs that are activated by the LTP-inducing stimulation exhibit LTP (Andersen et al., 1977; Lynch et al., 1977). Second, under most conditions, LTP induction requires activation of postsynaptic NMDA receptors during concomitant postsynaptic depolarization. The evidence that supports this statement is compelling and includes the findings that (1) AP5, a competitive NMDA receptor antagonist, reversibly blocks LTP (Collingridge et al., 1983); (2) strong hyperpolarization during application of a LTP-inducing tetanus also blocks LTP (Kelso et al., 1986; Malinow and Miller; 1986); and (3) LTP can be elicited by "pairing" low-frequency afferent stimulation with postsynaptic depolarization (Kelso et al., 1986; Sastry et al., 1986; Wigström et al., 1986). In addition, the generation of LTP requires a rise in postsynaptic Ca^{2+} concentration ($[Ca^{2+}]_i$). This conclusion seems warranted because injection of Ca^{2+} chelators into CA1 cells blocks LTP (Lynch et al., 1983; Malenka et al., 1988, 1992), and directly raising $[Ca^{2+}]_i$ within the postsynaptic cell can cause an enhancement of synaptic transmission (Malenka et al., 1988), although the exact relationship of this form of potentiation to LTP has not been determined.

Several recent findings indicate that the exact role and source of the requisite increase in postsynaptic $[Ca^{2+}]_i$ may be more complex than originally postulated. Independent of NMDA receptor activation, strong and prolonged activation of L-type voltage-dependent Ca^{2+} channels (VDCCs) can cause a long-lasting increase in synaptic strength (Aniksztejn and Ben-Ari, 1991; Grover and Teyler, 1990; Huang and Malenka, 1993), even though normally LTP does not require activation of L-type VDCCs (Huang and Malenka, 1993; Kullman et al., 1992). One possible explanation for these observations is that the large rise in postsynaptic $[Ca^{2+}]_i$ elicited by repetitive activation of L-type VDCCs overcomes endogenous buffering mechanisms and activates processes that are normally activated by entry of the Ca^{2+} into the dendritic spine via NMDA receptors (Huang and Malenka, 1993). The possibility remains, however, that the entry of Ca^{2+} via VDCCs may also activate intracellular processes that are distinct from those activated by Ca^{2+} entry via NMDA receptors (Bading et al., 1993).

Although the entry of Ca^{2+} across the cell membrane is necessary for LTP induction, the release of Ca^{2+} from intracellular stores may also be important (Bortolotto and Collingridge, 1993; Obenaus et al., 1989). Moreover recent experiments suggest that a large rise in postsynaptic $[Ca^{2+}]_i$, in and of itself, may not be sufficient to generate stable, nondecrementing LTP. Loading cells with Ca^{2+} via activation of VDCCs with repetitive long depolarizing voltage steps caused an increase in synaptic efficacy that was always transient (20 or so minutes), even though the initial potentiation could be quite large (300% to 400%). However, if these depolarizing pulses were paired with low-frequency afferent stimulation (in the presence of AP5), a stable potentiation was elicited (Kullman et al., 1992). This finding, along with the observation that direct iontophoretic application of NMDA also causes only a transient synaptic enhancement (Collingridge et al., 1983; Kauer et al., 1988), suggests that an additional factor associated with synaptic transmission may be required for LTP. A prime candidate for this additional factor is activation of the metabotropic glutamate receptor since a specific metabotropic glutamate receptor antagonist blocks LTP (Bashir et al., 1993) and application of a specific agonist may cause a slowly developing but long-lasting synaptic enhancement (Bortolotto and Collingridge, 1993).

The relationship between the transient synaptic enhancement (often termed short-term potentiation or STP) and stable LTP remains enigmatic. It has been suggested that the magnitude of the increase in postsynaptic $[Ca^{2+}]_i$ may determine whether STP or LTP is generated (Malenka, 1991). Specifically, stable LTP may require a critical "threshold" level of postsynaptic $[Ca^{2+}]_i$, and if this threshold is not reached, STP results. This hypothesis can be considered consistent with all of the aforementioned results if part of this increase in $[Ca^{2+}]_i$ is due to release from intracellular stores that are influenced by phosphoinositide turnover. Alternatively, LTP may require both a threshold level of $[Ca^{2+}]_i$ and activation of additional, Ca^{2+}-independent biochemical processes.

The Locus of Expression of LTP

Perhaps no subject has engendered more controversy than whether the expression of LTP involves pre- or postsynaptic modifications. Despite the obvious importance of this issue, it has not yet been examined in neocortex, and thus I only discuss it briefly. Readers interested in a more thorough treatment may refer to any number of recent reviews (Bliss and Collingridge, 1993: Lisman and Harris, 1993; McNaughton, 1993). Suffice it to say that at this juncture, there is evidence that LTP involves both pre- and postsynaptic modifications. For example, on the postsynaptic side it has been found that the amplitude of miniature EPSCs increases after application of NMDA (Manabe et al., 1992) and that the responses to exogenous application of AMPA (a glutamate receptor agonist) slowly grow after LTP induction (Davies et al., 1989). In support of a presynaptic modification is the reported increase in the overflow of radiolabeled or endogenous L-glutamate after the induction of LTP (Bliss et al., 1986; Dolphin et al., 1982).

Recently, quantal analysis, which at the neuromuscular junction is a valid and unambiguous technique for determining the locus of synaptic change, has been used by several laboratories to examine the changes that occur with LTP. It is clear that the validity and usefulness of this technique in the mammalian CNS depend on the validity of the assumptions that underlie the chosen statistical approach (Korn and Faber, 1991). Furthermore, any measured changes in quantal parameters can only put constraints on possible pre- or postsynaptic mechanisms and cannot definitively localize the change. Nevertheless, despite these caveats, the present consensus is that changes in both quantal size (q; classically a postsynaptic variable) and quantal content (n; a presynaptic variable) can occur during LTP (Kullman and Nicoll, 1992; Larkman et al., 1992; Liao et al., 1992) and that the relative contribution of the change in each presynaptic or postsynaptic variable may depend on the initial "state" of the synapse, specifically whether the initial probability of release is relatively high or low (Larkman et al., 1992; Liao et al., 1992).

Biochemical Mechanisms in LTP

The most common experimental approach for examining the biochemical processes involved in LTP has been to apply, either extracellularly or intracellularly, agents that specifically inhibit the activity of one or more enzymes and determine whether LTP can be generated. Such studies have implicated a myriad of second-messenger systems (Bliss and Collingridge, 1993; Madison et al., 1991), which have yet to be examined in models of cortical synaptic plasticity. It is fair to state that for hippocampal LTP the most compelling results, both in terms of volume and reliability, indicate that a network of protein kinases (including protein kinase C, Ca/calmodulin-dependent protein kinase II, and an as yet unspecificed tyrosine kinase) must be active to generate LTP. Postsynaptic kinases may act by modifying glutamate receptor function (Raymond et al., 1993), whereas a role for presynaptic kinases in modulating transmitter release is well established (Greengard et al., 1993; Malenka et al., 1986).

If LTP is maintained in part by presynaptic mechanisms, then an intercellular signal must be released from the postsynaptic cell (the site for LTP induction) and travel "retrogradely" to influence presynaptic function. The two most recently proposed candidates for this retrograde messenger are arachidonic acid and nitric oxide, although there are problems with the evidence supporting either candidate (Bliss and Collingridge, 1993).

LTP IN NEOCORTEX

It has long been assumed that synaptic plasticity in the cerebral cortex must contribute to the storage of information (Hebb, 1949) and to the development of neural circuits (Stent, 1973). However, examination of activity-dependent synaptic plasticity in the neocortex poses challenging experimental obstacles primarily because of the complexity inherent in its neural circuitry. A common approach using *in vitro* slice preparations has been to record extracellular field potential responses from layer II or III in reponse to stimulation at the layer VI-white matter border and to look for changes in the amplitude of the evoked field potential following some repetitive, often prolonged, afferent stimulation protocol. As more studies were performed, it became clear that the LTP of this field potential could be elicited, although the synaptic site at which change occurred was impossible to determine since many polysynaptic pathways were recruited by the afferent stimulation and contributed to current sinks and sources composing the field potential. Of greater concern and frustration, there was significant variability among the studies in terms of the probability of eliciting LTP, the pattern of stimulation necessary for LTP induction, the requirement for pharmacological blockade of GABA-mediated inhibition to elicit LTP, and the age-dependence of the phenomenon under study. This often confusing literature has been the subject of several excellent reviews (Bear and Kirkwood, 1993; Bindman et al., 1991; Teyler et al., 1990; Tsumoto, 1992). The following discussion focuses on the similarities and potential differences between hippocampal LTP and that found in neocortical preparations. It is important to remember, however, that LTP is a generic term used to describe any long-lasting increase in synaptic efficacy and that phenomena that appear similar in form may actually use distinct mechanisms.

The Role of NMDA Receptors and Ca^{2+}

Many studies have addressed the question whether NMDA receptor activation is required to trigger LTP in various regions of neocortex. Much of this work has been performed in rat or kitten visual cortex, but synaptic plasticity has also been examined in somatosensory cortex, frontal cortex, and motor cortex. At this early stage of analysis, no obvious differences between these areas have been observed, and thus the focus of the following review is on underlying mechanisms, ignoring the exact cortical area in which studies were performed.

As in the hippocampus, NMDA receptors are activated by synaptically released glutamate since evoked EPSPs consist of two components, one mediated by non-NMDA receptors and a second voltage-dependent component mediated by NMDA receptors (Artola and Singer, 1987; Artola et al., 1990; Hirsch and Crepel, 1990; Jones and Baughman, 1988; Luhmann and Prince, 1991; Sutor and Hablitz, 1989). Although some of the AP5-sensitive portion of the evoked EPSP is polysynaptic and is caused by activation of NMDA receptors on intervening neurons, it is clear that both non-NMDA and NMDA receptors are found on pyramidal and stellate neurons and are likely co-localized at the same synaptic site (Agmon and O'Dowd, 1992; Carmignoto and Vicini, 1992; Stern et al., 1992). In most studies on cortical LTP, activation of NMDA receptors was required for LTP induction as AP5 blocked LTP or strongly reduced its probability of occurring (Artola and Singer, 1987; Artola et al., 1990; Bear et al., 1992; Hirsch and Crepel, 1991; Kimura et al., 1989; Lee et al., 1991). Consistent with these results, pairing low-frequency afferent stimulation with post-synaptic depolarization can elicit LTP both *in vitro* (Bindman et al., 1988; Hirsch and Gilbert, 1993) and *in vivo* (Baranyi and Szente, 1987; Baranyi et al., 1991), although the percentage of cells in which LTP was elicited by a pairing protocol was lower than that normally observed in the hippocampus.

A puzzling difference between cortical and hippocampal LTP is the optimal frequency and duration of tetanic stimulation for LTP induction. In neocortex, many investigators use prolonged (2 to 60 minutes), relatively low-frequency (2 to 10 Hz) tetanic stimulation to elicit LTP, whereas in hippocampus short (0.1 to 1.0 second) bursts of high frequency (50 to 100 Hz) tetanic stimulation are quite effective. The level of inhibition also seems to be a much more critical variable in neocortex than in hippocampus. Many studies report an inability to generate cortical LTP unless the preparation has been disinhibited by application of small doses of $GABA_A$ receptor antagonists. A final possible difference is that in cortex LTP often develops slowly and (STP) has not been observed, whereas in hippocampus, LTP normally exhibits an initial decaying phase that may correspond to STP.

A recent paper suggests that these apparent differences may have much to do with anatomical considerations and little with real mechanistic differences (Kirkwood et al., 1993). By changing the stimulation site from its normal position at the layer VI-white matter border to layer IV, NMDA receptor-dependent LTP could be generated in adult rat and kitten visual cortex using "theta burst" stimulation in the absence of $GABA_A$ receptor antagonists, conditions that reliably induce LTP in the hippocampus (Larson et al., 1986). However even in these experiments, the potentiation grew slowly and did not exhibit an early decaying phase.

Although NMDA receptor activation often contributes to the triggering of cortical LTP, several studies indicate that it is not necessarily a mandatory step (Aroniadou and Teyler, 1992; Bear et al., 1992; Komatsu et al., 1991). Because loading cells with Ca^{2+} chelator blocks cortical LTP (Brocher et al., 1992; Hirsch and Crepel, 1992; Kimura et al., 1990), it seems likely that these cases of AP5-resistant LTP are due to activation of VDCCs. One potentially important distinction is that activation of L-type VDCCs generates LTP in hippocampus, whereas in the only study performed to date, T-type VDCCs seem to play a role in LTP in kitten visual cortex (Komatsu and Iwakiri, 1992).

One theoretical advantage of using NMDA receptors to trigger LTP is that they provide a mechanism for the input specificity observed both in hippocampus and neocortex (Artola et al., 1990; Baranyi et al., 1991; Kirkwood et al., 1993). However a nagging problem with using activation of NMDA receptors as the main trigger for LTP is that they may be involved in the transmission of sensory information during the normal operation of the cortex (Daw et al., 1993). Requiring that LTP induction depend on some critical threshold level of $[Ca^{2+}]_i$ that is not reached during normal synaptic operations is an obvious way around this problem. Nevertheless, it is impossible to state at this time whether the increase in $[Ca^{2+}]_i$ required to generate LTP in neocortex is due to influx through the NMDA receptor channel or through VDCCs, the activation of which depends on the additional local depolarization provided by NMDA receptor activation. It is also conceivable that the NMDA receptor-mediated and VDCC-mediated rise in $[Ca^{2+}]_i$ result in activation of distinct biochemical processes (Bading et al., 1993).

Cortical Development and LTP

From the classic work of Hubel and Wiesel, it is well established that a "critical period" exists at which time the development of visual cortex circuitry is highly susceptible to environmental manipulations (Wiesel and Hubel, 1965). This immediately raises the question whether the ability to generate LTP also exhibits a strong developmental profile. A general consensus is that in the absence of pharmacological disinhibition LTP is elicited with much higher probability in slices prepared from kitten or young (2 to 6 weeks) rat visual cortex (Bear and Kirkwood, 1993; Tsumoto, 1992). However, it is also clear that reduction of inhibition with $GABA_A$ receptor antagonists or direct depolarization of postsynaptic cells during intracellular recording permits the reliable induction of LTP in adult preparations (Artola and Singer, 1987; Artola et al., 1990; Baranzi and Szente, 1987; Barayni et al., 1991).

Recent work on the development of NMDA receptor-mediated and GABA receptor-mediated synaptic transmission in the neocortex may explain the developmental profile of LTP. In rat neocortex, GABAergic inhibition matures slowly over the first postnatal month (Luhmann and Prince, 1991), resulting in a greater expression of the NMDA receptor-mediated component of excitatory synaptic transmission (Agmon and O'Dowd, 1992; Kato et al., 1991; Luhmann and Prince, 1991). In addition, the properties of the NMDA receptor itself change during development in the rat (Carmignoto and Vicini, 1992; Hestrin, 1992a). Specifically, at early developmental stages NMDA receptor-mediated EPSCs are prolonged, and their duration does not decrease to adult levels until 5 to 6 weeks of age. This change may be due to a developmental switch in the subunit composition of the native NMDA receptor (Williams et al., 1993). The combination of delaying the development of inhibition and prolonging the duration of NMDA receptor-mediated synaptic currents provides elegant mechanisms by which synaptically driven depolarization and the influx of Ca^{2+} are promoted early in cortical development, presumably facilitating LTP induction.

LONG-TERM DEPRESSION IN HIPPOCAMPUS AND NEOCORTEX

The utility of LTP as a developmental and information storage mechanism would be severely limited if mechanisms for decreasing synaptic strength did not also exist. In a classic paper Stent (1973) described how combining LTP- and LTD-like mechanisms might account for ocular dominance plasticity (Wiesel and Hubel, 1965). More recently theoreticians working with neural networks have convincingly demonstrated the power of algorithms that permit both increases and decreases in synaptic strength (Bienenstock et al., 1982; Sejnowski, 1977; Willshaw and Dayan, 1990). Thus, an understanding of LTD and its underlying mechanisms has become an important component of all work on synaptic plasticity. Once again it should be emphasized that LTD, like LTP, may not be a single uniform phenomenon, but rather is a generic term used to desribe any long-lasting decrease in synaptic strength. LTD has been the subject of several recent reviews (Artola and Singer, 1993; Bindman et al., 1991; Malenka, 1993; Tsumoto, 1992) and therefore this discussion is brief, touching on a few salient issues.

LTD can be homosynaptic, occurring only at those synapses activated by the afferent stimulation, or heterosynaptic, occurring at adjacent, inactive synapses. In the hippocampus, either homosynaptic or heterosynaptic LTD can be elicited, and both forms seem to require some change in postsynaptic $[Ca^{2+}]_i$ (Artola and Singer, 1993; Malenka, 1993). Surprisingly, homosynaptic LTD requires activation of NMDA receptors (Dudek and Bear, 1992; Mulkey and Malenka, 1992), whereas heterosynaptic LTD requires activation of VDCCs (Christofi et al., 1993; Wickens and Abraham, 1991). Similarly in neocortex, LTD can be elicited by repetitive low-frequency (1 Hz) activation of NMDA receptors (Kirkwood et al., 1993) or by tetanic stimulation in the presence of NMDA receptor antagonists (Aroniadou and Teyler, 1991; Artola et al., 1990; Hirsch and Crepel, 1991). The AP5-insensitive form of cortical LTD is normally blocked by loading cells with Ca^{2+} chelators (Brocher et al., 1992; Hirsch and Crepel, 1992), indicating that activation of VDCCs or release of Ca^{2+} from intracellular stores is required for its generation.

How can a rise in postsynaptic $[Ca^{2+}]_i$ be responsible for LTD, as well as LTP? One possibility is that an increase in postsynaptic $[Ca^{2+}]_i$ is simply permissive and that additional, as yet unidentified, factors are required to generate LTD. For example, LTP in the hippocampus may require activation of metabotropic glutamate receptors in addition to NMDA receptor activation (Bashir et al., 1993). Another hypothesis is that the magnitude of the increase in $[Ca^{2+}]_i$ determines the direction of synaptic change, with relatively small changes within some critical window causing LTD (Artola et al., 1990; Cristofi et al., 1993; Hirsch and Crepel, 1992). Taking advantage of the different affinities of biochemical processes for Ca^{2+}/calmodulin, a model (Lisman, 1989) has in fact demonstrated how the level of postsynaptic $[Ca^{2+}]_i$ may control the direction of synaptic changes. Consistent with this hypothesis is the finding that in rat visual cortex the induction of LTD is highly voltage dependent, only occurring when postsynaptic depolarization exceeds a critical level that is below a threshold related to activation of NMDA receptor-gated conductances (Artola et al., 1990).

If $[Ca^{2+}]_i$ alone controls the induction of LTD, an important related question is whether the source of Ca^{2+} matters. All of the studies discussed thus far suggest that either NMDA receptors or VDCCs are the source for the critical Ca^{2+} signal. However, a recent paper (Kato et al., 1993) presents evidence that activation of a subtype of metabotropic glutamate receptor coupled to phosphatidylinositol (PI) turnover and the consequent release of Ca^{2+} from 1,4,5-inositol trisphosphate-sensitive stores is sufficient to elicit LTD in cortex. This finding is somewhat surprising since hippocampal LTP also seems to require activation of a metabotropic glutamate receptor (Bashir et al., 1993), although an obvious explanation is the known heterogeneity in subtypes of this class of glutamate receptor (Nakanishi, 1992).

CONCLUSION

It is clear that the study of LTP and LTD in neocortex is still in its infancy, making it impossible to draw any definitive conclusions about the similarities or differences between the mechanisms responsible for hippocampal and cortical synaptic plasticity. In both structures, changes in postsynaptic $[Ca^{2+}]_i$ are critical and can derive from activation of NMDA receptors, metabotropic glutamate receptors, or VDCCs. The biochemical processes involved in cortical synaptic plasticity have not been examined to any significant extent, but it would be surprising if they were profoundly different from those used in the hippocampus. Thus at this stage, the observed differences in the properties of cortical and hippocampal LTP/LTD may not be due to differences in underlying cellular mechanisms, but rather to dramatic differences in neural circuitry and in the cellular density and distribution of neurotransmitter receptors and voltage-dependent ion channels. This does not mean that the function of long-lasting changes in synaptic efficacy will be the same in hippocampus and neocortex. Even if the mechanisms responsible for synaptic plasticity in hippocampus and neocortex turn out to be the same, the occurrence of LTP or LTD in the behaving animal is sure to be markedly influenced by the activity of an array of extrinsic modulatory inputs, many of which are known to have different actions in the two structures. Of course, as more is learned about the properties of synaptic plasticity in different regions of cortex and in different cell types, it would not be at all surprising to find important mechanistic distinctions at different synapses, as has been demonstrated in the CA3 region of the hippocampus (Harris and Cotman, 1986; Zalutsky and Nicoll, 1991).

III

The cortical neuron as part of a network

8

Overview: Basic elements of the cortical network

EDWARD G. JONES

In any examination of cortical circuitry, it is necessary to commence with the two basic classes of cortical neurons—one richly endowed with dendritic spines and the other effectively lacking them. From years of correlative morphological and physiological work, it has become a commonplace that the spiny and nonspiny cortical neurons are morphological expressions of two fundamental categories of neuron—excitatory and inhibitory—a division that can now be expressed in terms of their individual neurotransmitters: glutamergic and GABAergic. Among both classes, subtypes are now beginning to emerge based on morphological and chemical differences and on differential spiking behavior. Much of the information pertaining to that behavior has come from David Prince and his collaborators (Chagnac-Amitai et al., 1990; Connors and Gutnick, 1990; Kawaguchi, 1993; McCormick et al., 1985).

There are two major transmitter-defined categories of cortical neuron. Quantitative studies involving immunocytochemistry for GABA and, more recently, in situ hybridization histochemistry to localize mRNAs for glutamic acid decarboxylase (GAD) have revealed that GABA cells form approximately 25% of the cortical neuronal population (Fig. 8-1) (Hendry et al., 1988b; Jones et al., 1993). The remaining 75% is largely formed by neurons that utilize an acid amino acid transmitter or transmitters acting at the glutamate class of receptors (Conti et al., 1987). For these cells, new markers are beginning to emerge, as indicated below.

For many years morphological studies led in the search for further subdivisions of the spiny and nonspiny classes of cortical neurons, showing, for example, seven or eight basic categories of nonspiny GABA cells, each with a stereotyped axonal configuration and commonly a different dendritic field architecture as well (Jones, 1975). These cells have come to be known by popular names, among which are basket cells, chandelier cells, double bouquet cells, and neurogliaform cells. Further indications of the specificity and independence of these subclass variants have come from studies of substances co-localized with GABA in the nonspiny cells. These substances include surface proteoglycans identified by monoclonal antibodies or by lectin binding (Hendry et al., 1988b; Morino-Wannier et al., 1992; Naegle and Barnstable 1989; Mulligan et al., 1989; Nakagawa et al., 1986), co-localized neuropeptides (Hendry et al., 1984b), and the cal-

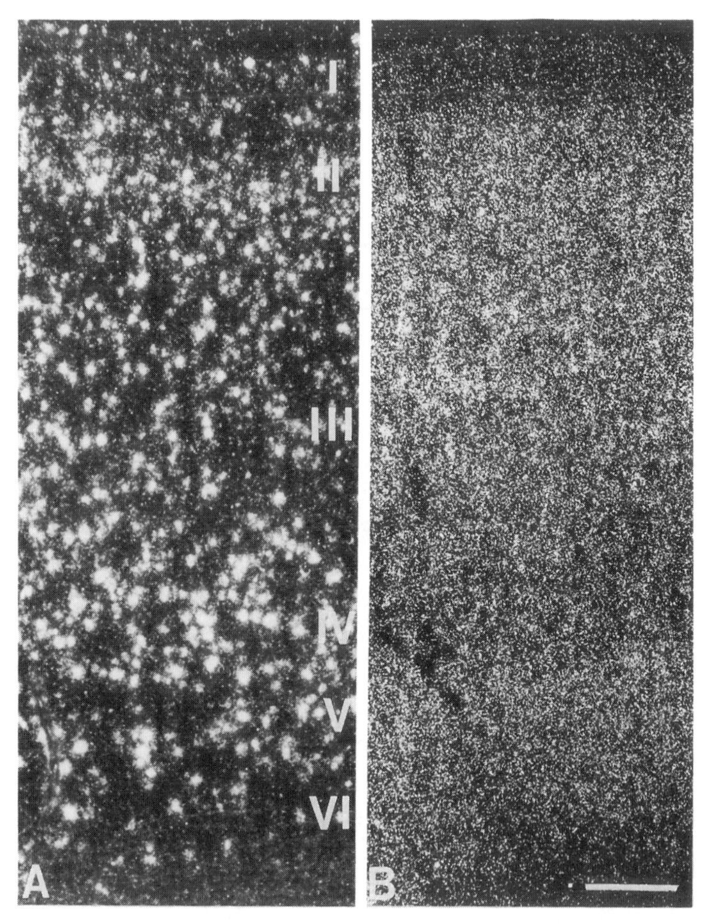

FIGURE 8-1. Darkfield photomicrographs of autoradiographs from sections of the somatic sensory cortex of a monkey, showing localization of mRNAs for GAD_{67} and CAM II kinase α by in situ hybridization histochemistry. Bar 250 μm.

cium-binding proteins—parvalbumin and calbindin-D_{28k} (Hendry et al., 1989).

Immunocytochemical studies have consistently shown the localization of parvalbumin in basket cells and chandelier cells and of calbindin in double bouquet cells, thus providing useful anatomical markers for these categories of nonpyramidal cells. The functional relevance of the differential expression of parvalbumin and calbindin remains uncertain, but recent investigations by Kawaguchi and Kubota (1993) show that their presence can be correlated with different patterns of spike discharge. Recording with whole-cell methods in slices of the motor cortex of rats, they found one population of fast-spiking cells that discharged repetitive trains of spikes when activated by synaptic input from a depolarized state. A second population never showed repetitive firing, but discharged in bursts based on low-threshold spikes induced by synaptic input from a hyperpolarized state. When recovered histologically, after intracellular injection and co-staining for GABA and parvalbumin or calbindin, the two types proved to be aspiny GABA cells. The fast-spiking type was consistently parvalbumin positive, whereas the low-threshold spiking type was invariably calbindin positive. The low-threshold spiking/calbindin cells were characterized by vertical axonal and dendritic distributions spanning several cortical layers.

These cells are probably the rat equivalents of the well-known double bouquet cell of the primate cerebral cortex whose dendrites and axon collaterals form tightly bundled, column-like arrays extending vertically through layers II to V (Fig. 8-2). They are an example of a population of cortical neurons with an axon distribution that implies the capacity to assemble groupings of other cortical neurons into functional ensembles; in this case, in the vertical, interlaminar domain. The presence of large numbers of double bouquet cells is one of the hallmarks of the human cerebral cortex according to Ramón y Cajal (1899). Recent investigations on the double bouquet cells of the monkey cerebral cortex (DeFelipe and Jones 1992; DeFelipe et al., 1990) reveal that the axon bundles of these cells form a regularly spaced series of mini-columns 35 to 50 μm wide and with an approximately 50 μm center-to-center spacing. Since the axons of each bundle give off terminals along their full length, the bundles can be seen as forming a series of mini-columns of inhibition descending vertically through most cortical layers. In a sense, they form the complement of the bundles of apical dendrites of pyramidal cells ascending through the cortex. Contrary to expectations, however, electron microscopy reveals that the terminals of the double bouquet cell axons end in synapses not on the apical dendrites, but on their side branches. This feature would tend to produce divergence of the descending inhibitory effect, rather than restric-

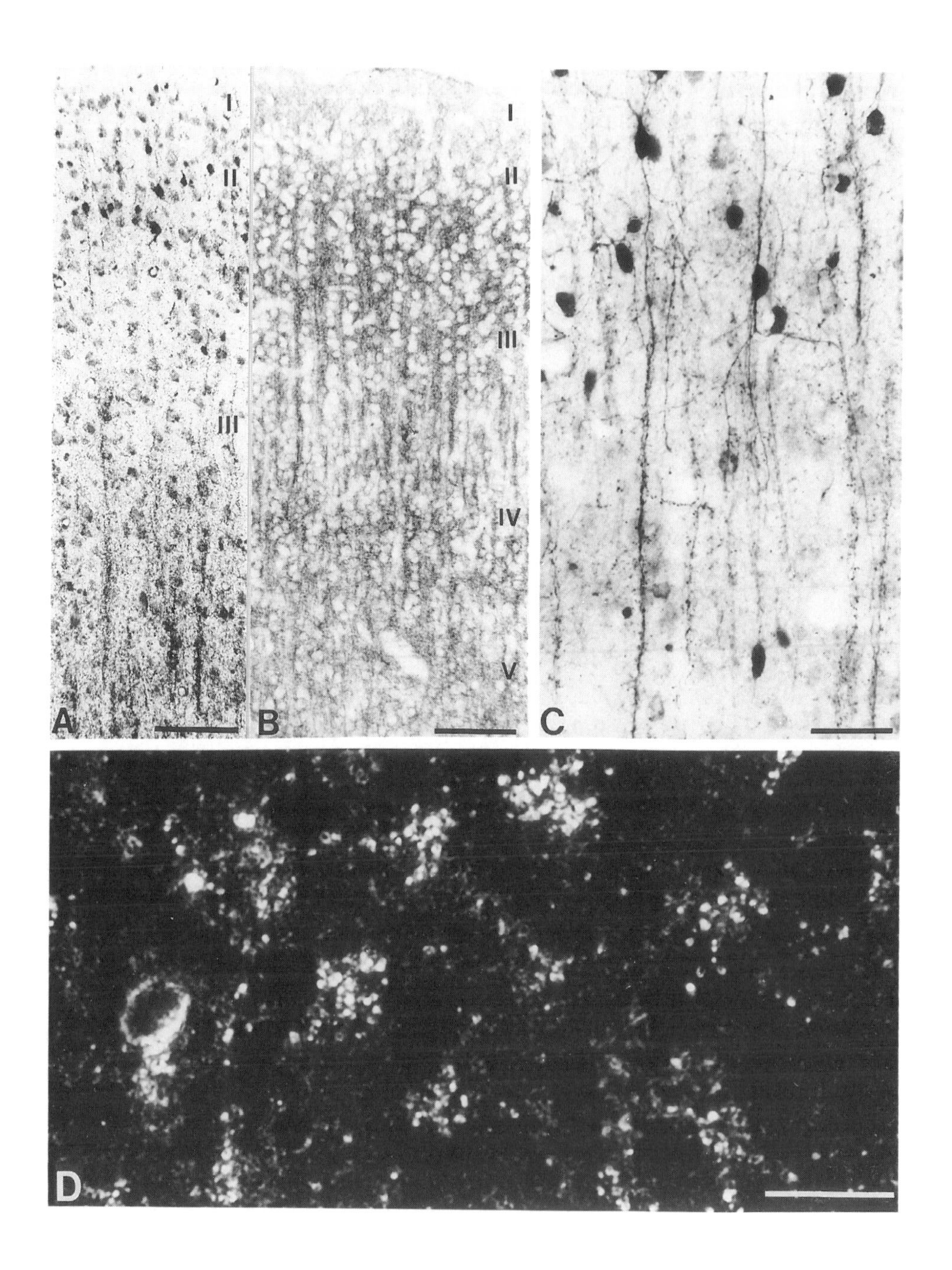
I
II
III
A
I
II
III
IV
V
B
C
D

tion to a tight mini-column of pyramidal cells. Collectively, therefore, the double bouquet cells, most of whose somata are located in layer II and upper layer III, may send a shower of inhibition descending in a blanketing manner through the cortex, rather than being focused on single columns.

Whereas the double bouquet cells represent a cortical cell type involved in vertical, interlaminar inhibitory circuitry, the small spiny nonpyramidal cells with somata typically located in layer IV represent a type involved in the interlaminar flow of excitation. These cells can be specifically labeled by high-affinity uptake and retrograde transport of [^{3}H] D-aspartate injected into the sites of termination of their axons in superficial layers (G. W. Huntley and E. G. Jones, unpublished observations).The spiny nonpyramidal cells ("spiny stellate cells") are far exceeded in number by the major population of cortical excitatory cells, the pyramidal cells. The intracortical collaterals of pyramidal cell axons are probably the major routes for the spread of excitation in the cerebral cortex, especially in the horizontal dimension, and thus form a basis for assembling cortical cells into functional collections across broad expanses of cortex (DeFelipe et al., 1986; Gilbert and Wiesel, 1983). Several varieties of pyramidal cell are now known based on laminar position, extrinsic target of the axon, dendritic structure including spine distribution and branching pattern, extent and direction of axonal collateralization, and spike discharge patterns (reviewed in Connors and Gutnick, 1990; Jones, 1984; Ojima et al., 1991, 1992).

One aspect of the molecular identity of pyramidal and other excitatory cortical neurons may be considered as a marker for these cells much as GAD is the marker for GABAergic nonpyramidal cells. This particular molecule is the multifunctional protein kinase, type II calcium/calmodulin-dependent protein kinase (CAM II kinase). In situ hybridization histochemistry (Fig. 8-1) reveals that the mRNA coding for the α subunit of

←

FIGURE 8-2. **A**. Retrograde labeling of somata and axon bundles of putative double bouquet cells in layer II of the motor cortex of a monkey, after injection of ^{3}H GABA in layer V (from DeFelipe and Jones 1992). Bar 100 μm. **B**. Immunocytochemical localization of β2/β3 subunits of the $GABA_A$ receptor in area 18 of a monkey, showing vertical chains of reaction product that probably reflect receptors associated with GABA terminals arising from axon bundles of double bouquet cells (from Hendry et al., 1990). Bar 100 μm. **C**. Double bouquet cells and their descending axon bundles stained immunocytochemically for calbindin in area 7 of a monkey (from material Hendry et al., 1989). Bar 50 μm. **D**. Darkfield photomicrograph from a horizontal section through layer III of area 17 of a monkey, showing the regular arrangement of calbindin immunoreactive bundles of double bouquet cell axons (from DeFelipe et al., 1990). Bar 50 μm.

CAM II kinase is expressed only in non-GABA cells of the cerebral cortex (Benson et al., 1991; Jones et al., 1993). This pattern of expression can be confirmed by the lack of double labeling for GABA and CAM kinase II α in immunocytochemical preparations (Fig. 8-3), which reveals that the protein is found in both pyramidal cells and in the spiny nonpyramidal cells of layer IV. CAM kinase II α is the major postsynaptic density protein at asymmetric synapses of the forebrain. It also assists in the movement of synaptic vesicles toward the presynaptic membrane by phosphorylating synapsin I and is involved in the induction of long-term potentiation in hippocampal pyramidal cells (e.g. Silva et al., 1992b). In addition to its usefulness as a marker of excitatory neurons in the cerebral cortex and elsewhere, the selective expression of CAM kinase II in non-GABA cells of the cerebral cortex implies that GABA neurons and GABA synapses will not be subject to any long-term changes in synaptic efficiency that CAM kinase II may mediate in the cerebral cortex.

The pyramidal cells, which are the main population expressing CAM kinase II in the cerebral cortex, make major contributions to intrinsic cortical circuitry via their extensive systems of axon collaterals. It is becoming clear that different populations of pyramidal cells may vary in the extent and orientation of the major collateral branches of their axons (e.g. Ojima et al., 1991, 1992), but what most characterizes the collaterals of pyramidal cells is their length. They are clearly the major source of long-range intrinsic connections in the cerebral cortex. In addition to their great horizontal length, which may be 6 to 8 mm, these collaterals are also characterized by the clustered nature of their terminations, usually with intervening terminal-free zones that can be several hundred microns in extent (Fig. 8-4). It is the clustered nature of the terminations of the collaterals that determines the patchy character of retrograde and anterograde labeling of cells and terminals when an extracellular injection of tracer is made in the cerebral cortex (Fig. 8-4).

A question that has always troubled workers in the field is why these seemingly dense projections do not appear to be functionally expressed under experimental conditions; for example, in the case of collateral pro-

→

FIGURE 8-3. Fluorescence photomicrographs from the same microscopic field in layer IV of the somatic sensory cortex of a monkey, showing rhodamine fluorescence indicating immunocytochemical labeling of CAM kinase II-α-containing small excitatory neurons **(A)** and fluorescein fluorescence indicating immunocytochemical labeling of large GABA-containing cells (arrows); **(B)**. Bar 20 μm. **C.** Immunoperoxidase labeling of CAM II kinase-a-containing cells in the somatic sensory cortex of a monkey. Many of the cells are clearly pyramidal (from Jones et al., 1993). Bar 100 μm.

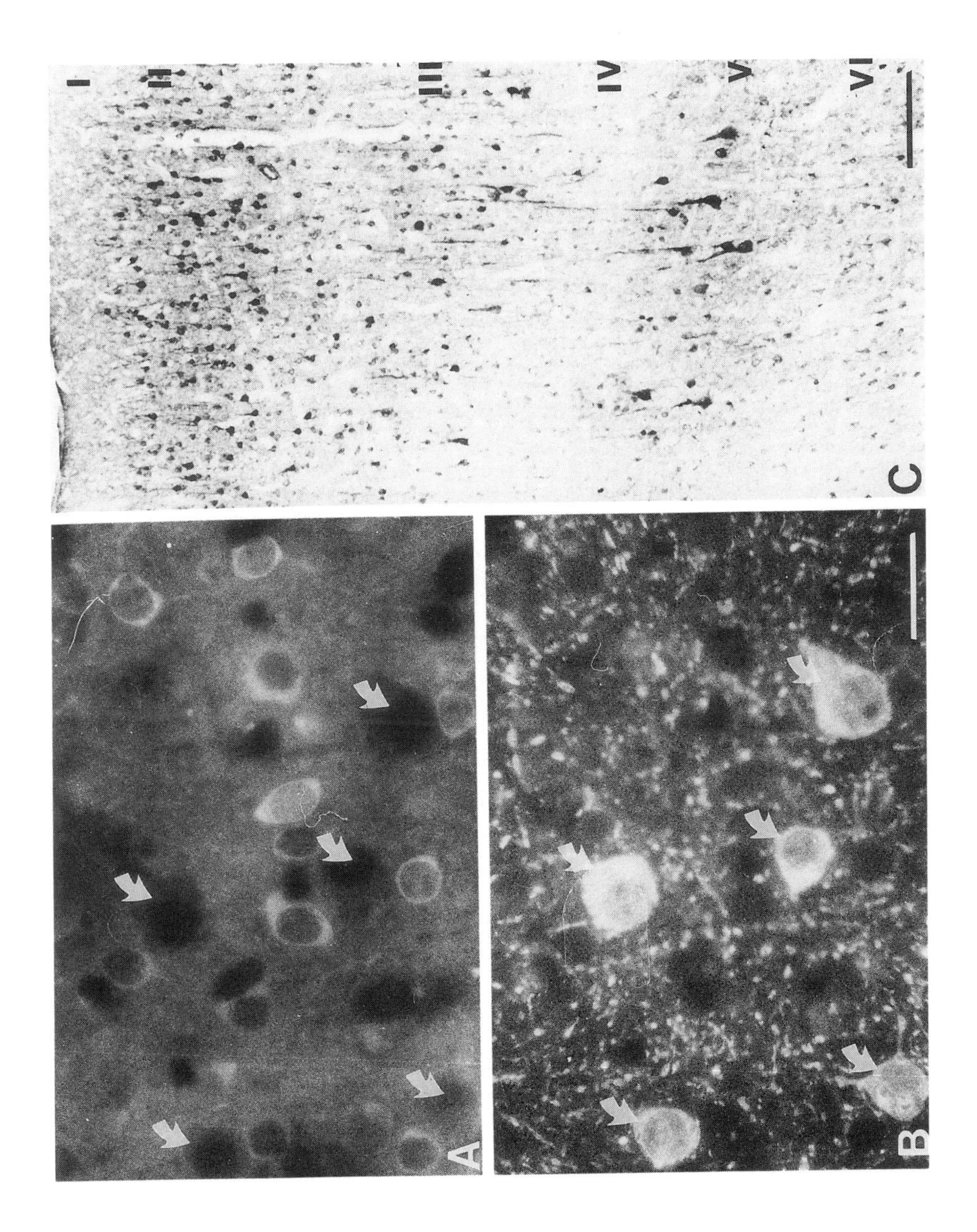
A
B
C
I
II
III
IV
V
VI

jections spreading from one part of a representational map in a sensory or motor area into another. Huntley and Jones (1991) addressed this issue in a series of experiments conducted on the primary motor cortex of monkeys (Fig. 8-5). The upper limb representation was mapped using intracortical microstimulation and recording the movements elicited by threshold stimulation at points approximately 1 mm apart. A small injection of wheat germ agglutinin-conjugated horseradish peroxidase was made at one of the points—usually that from which abduction of the thumb could be elicited. It was found that the labeled efferents emerging from this point were all made up of long-range collaterals of pyramidal cells and formed a series of far-flung terminal patches throughout much of the upper limb representation, including sites from which like or different movements had been elicited, and avoiding others.

At first glance it is difficult to conceive why these apparently heavy connections, especially those to heterotopic parts of the upper limb rep-

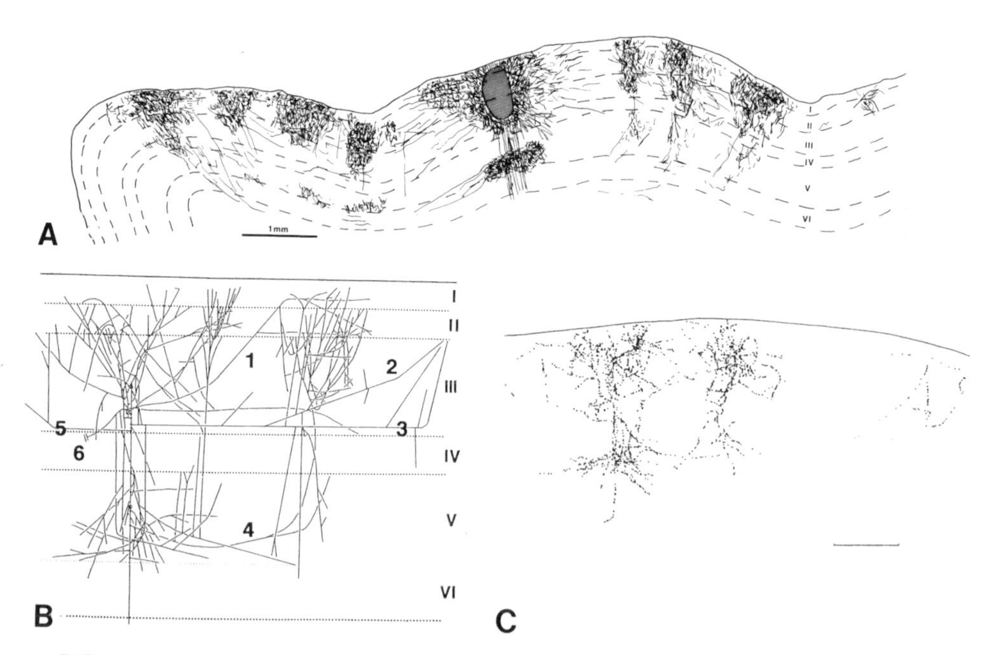

FIGURE 8-4. **A**. Injection of *Phaseolus vulgaris* leukoagglutinin (black) in layer III of the auditory cortex of a cat gives rise to multiple patches of anterogradely transported terminal labeling in other parts of the cochlear representation (from Wallace et al., 1991). **B**. Computer-based reconstruction of the axon collateral system of a cell in layer III of the auditory cortex of a cat, recovered after intracellular injection of biocytin. There are six major horizontal collaterals (1 to 6) that form clusters of terminals that have been plotted independently in **C**. These form the basis of patches of terminations seen in **A** (from Ojima et al., 1991).

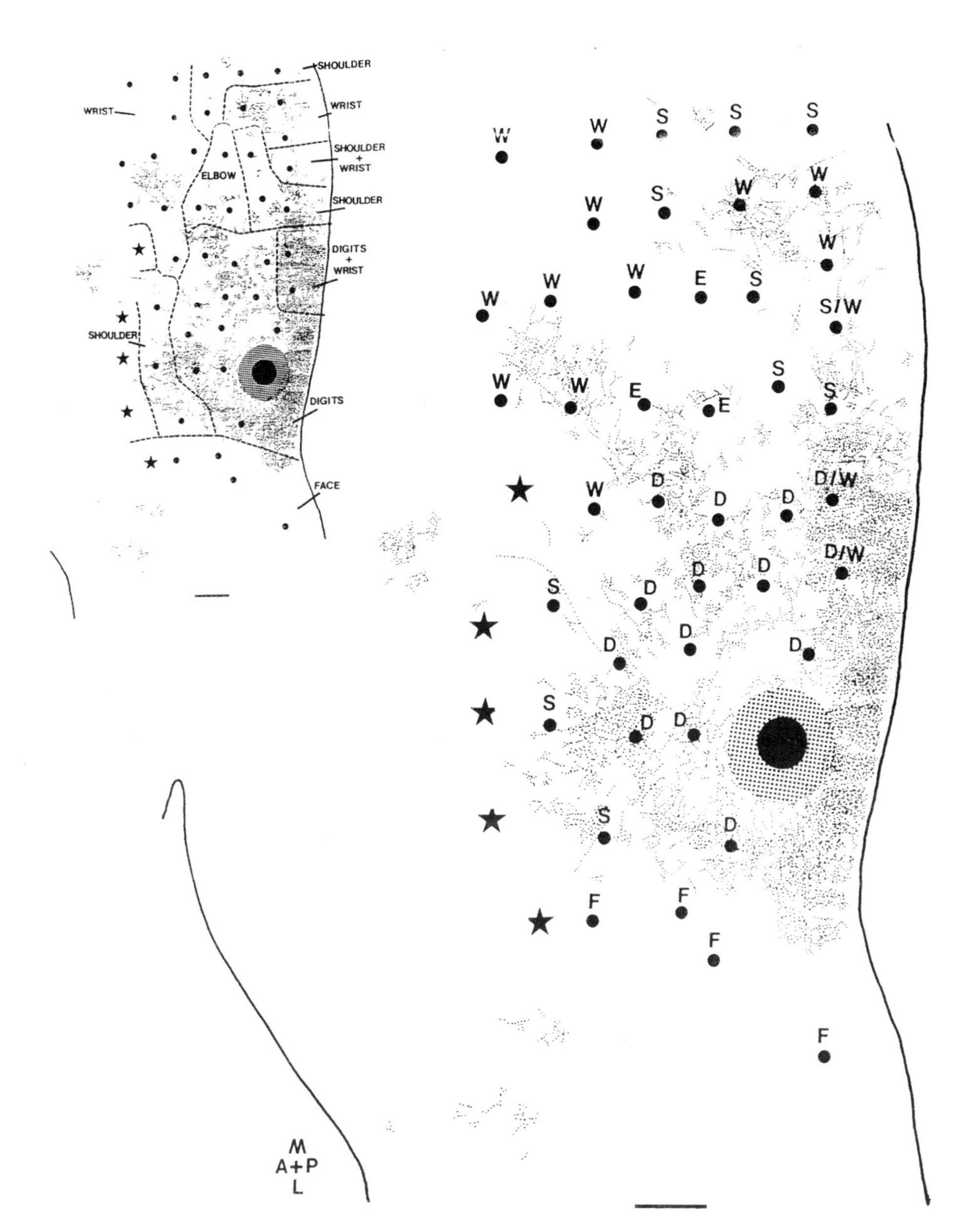

FIGURE 8-5. Surface view reconstruction of primary motor cortex of a monkey showing the positions of electrode tracks (larger dots), the types of movements elicited by threshold intracortical microstimulation (D, digits; E, elbow; F, face; S, shoulder; W, wrist), and the patchy distribution of terminal labeling (small dots) ensuing from an injection of wheat germ agglutinin-conjugated horseradish peroxidase at one of the stimulation sites (black and stipple; (from Huntley and Jones 1991). Bars 1 mm.

resentation, are normally not expressed functionally, leading to blurring of the representational map. One reason may be that the number of synapses made by an individual collateral on a single recipient cell is very small. The small amount of data supporting this idea comes from electron microscopic reconstructions of axons of single-labeled cells in the visual cortex. There, the number of synapses made by a collateral of a single injected cell on another cell at the center of a column of terminations was as little as one (McGuire et al., 1991). Hence, although it is possible to view the extensive network of intrinsic connectivity formed by the horizontal collaterals as providing a basis for the recruitment of an extensive set of motor cortex neurons during the performance of a complex motor task, the low density of the relevant synapses on individual cells would seem to demand that this recruitment occur on a population basis or by enhancing the efficacy of the synapses. One mechanism for increasing synaptic efficiency, leading to the functional expression of seemingly silent synapses, could be based on activity-dependent up- or downregulation of CAM kinase II, which is found on both the pre- and postsynaptic sides of the synapses made by the collaterals on other pyramidal cells.

One situation in which it is thought that previously silent synapses of the horizontal collaterals may be functionally expressed is in the plasticity of cortical representational maps that can be caused by perturbations or repetitive stimulation of the sensory or motor periphery (e.g. Allard et al., 1991; Gilbert and Wiesel, 1992; Jenkins et al., 1990a,b; Pons et al., 1991; Recanzone et al., 1992; Wall et al., 1986). It has been argued that one key mechanism is the activity-dependent, up- or downregulation of transmitters, receptors, and related enzymes that occurs, for example, in the visual cortex of adult monkeys deprived of vision in one eye for 4 or more days by intravitreal injections of tetrodotoxin (Fig. 8-6). These changes, which include downregulation of GABA levels and $GABA_A$ receptors and upregulation of CAM kinase II α in deprived ocular dominance columns (Hendry and Jones, 1986; Hendry and Kennedy, 1986; Hendry et al., 1990), are reversible on restoration of retinal ganglion cell activity (Hendry and Jones, 1988) and depend upon alterations in mRNA levels for the relevant polypeptides or synthetic enzymes (Benson et al., 1991, 1993; Huntsman et al., 1993). It is likely that downregulation of GABA and of $GABA_A$ receptors in a part of a cortical representation, especially when accompanied by upregulation of CAM kinase II and potentially other kinases, will cause the excitatory synapses of the long-range collaterals to be exposed, which may result in apparent spreading of one part of the representation at the expense of that which has been compromised. This may be a key to understanding how changes in representational maps can occur

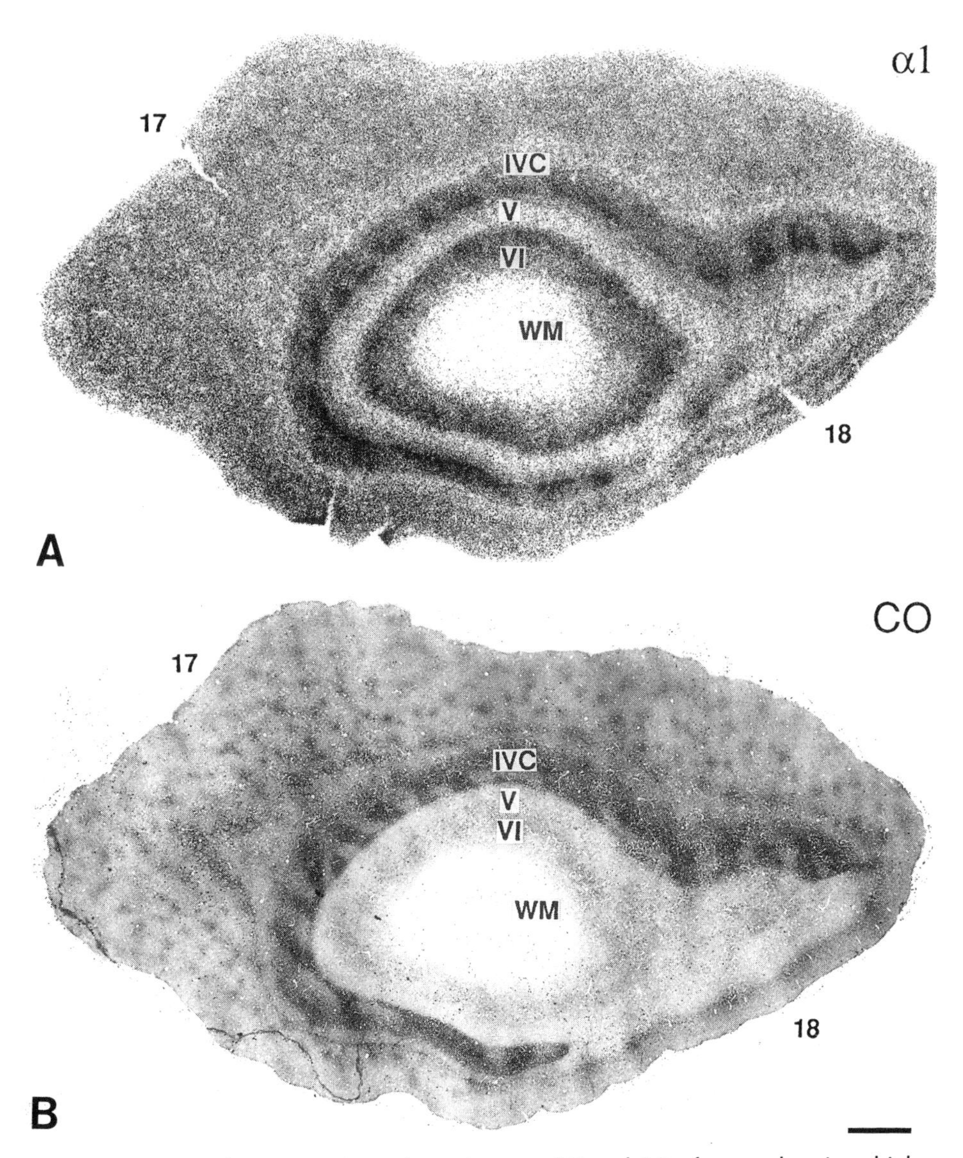

FIGURE 8-6. Adjacent sections through areas 17 and 18 of a monkey in which impulse activity had been blocked in one optic nerve for 14 days by injections of tetrodotoxin in the eye. **A.** An autoradiograph showing downregulation of mRNAs for the α_1 subunit of the $GABA_A$ receptor in deprived ocular dominance columns of layer IVC and in deprived rows of periodicities of layer III. The deprived regions can be identified by reductions in the stainings for cytochrome oxidase (CO) seen in **B** (M. M. Huntsman and E. G. Jones, unpublished observations).

with such rapidity in the cerebral cortex and in the absence of overt axon terminal sprouting and new synapse formation.

Acknowledgments. The work reported in this chapter was supported by grants numbers NS21377 and NS22317 from the National Institutes of Health, United States Public Health Service. I thank Drs. D. L. Benson, G. W. Huntley, M. M. Huntsman, and P. J. Isackson for their many contributions.

9

Functions of local circuits in neocortex: synchrony and laminae

BARRY W. CONNORS and YAEL AMITAI

THE PREMISE

The cortical neuron never works alone. Although no vertebrate neuron may ever perform solo, cooperative interactions are especially important in neocortex. The neocortex is an interconnected web of neurons, as are other brain structures. Yet, it is striking in the degree to which its interconnections are divergent and sparse, its single synapses have only a meager and fleeting influence, and its neurons require the convergence of many inputs to convince them to fire an action potential. A single pyramidal cell in human neocortex sends and receives about 40,000 synapses (Abeles, 1991), and at least 70% of the excitatory synapses received by pyramidal cells originate within the cortex itself (Gruner et al., 1974; White and Keller, 1989). It takes cooperative interactions of a large aggregate of local cortical neurons to yield significant output.

Local cortical circuits are complex, and the search for principles in their organization has inspired numerous conclusions since the time of Ramón y Cajal. Models of neocortex have taken many forms, some of them explicit enough to be implemented computationally (Abeles, 1991; Douglas and Martin, 1992; König and Schillen, 1991; Lehky and Sejnowski, 1990; Szentagothai and Arbib, 1974; Wehmeier et al., 1989). Most models do not attempt to be general in scope, but instead focus on one set of sensory or motor functions. However, a common view is that neocortex, regardless of area or species, does follow certain general principles of architecture and function, ambiguous though those may be. This is embodied explicitly in the "canonical microcircuits" of neocortex proposed by Douglas and Martin (1991, 1992).

Our intention here is not to review the current models of neocortex nor to propose another comprehensive one. Rather, this chapter focuses on one feature of the local cortical circuit that has been relatively neglected but that is particularly important and probably widespread. The neurons that provide output from the neocortex are largely found in layer V. This location alone makes it indispensable, but there are several other characteristics of structure and function that distinguish layer V. Our basic hypothesis is that the neural network within layer V is uniquely capable of ordering, controling, coordinating, and synchronizing the activity of the neocortex. In this chapter we outline some reasons for proposing this hypothesis and speculate about its significance.

THE UNIQUE ANATOMY OF LAYER V

Layer V is the origin of cortical output. With the exception of layer VI cells that project to the thalamus (which, in turn, simply provides input back to the cortex), neurons that project to subcortical areas of the brain reside in layer V (Jones, 1984). Layer V neurons are the primary means by which the neocortex communicates with the tectum, striatum, pons, midbrain, spinal cord, and thus with the world. In a broad sense, then, all other cortical cells are interneurons and exist only to modulate the output cells of layer V.

Ideas about the laminar parcellation of cortical output cells (in the deep layers) and associational cells (in the superficial layers) have a long history (Bolton, 1910; Campbell, 1905; see Jones, 1984). The ubiquitous and central role of layer V was succinctly summarized by Diamond (1979): "every area of the cortex can be viewed as a motor area, or layer V itself could be termed the 'motor cortex'." A corollary is that the associational areas of cortex are most logically placed in the supragranular layers (Neafsy, 1990) and in layer VI, the primary origins of the cortico-cortical projections.

Layer V has some of largest neurons of the neocortex (Fig. 9-1; Feldman, 1984; Sholl, 1956), as well as great diversity of cell size and shape. Large

FIGURE 9-1. Morphology of intrinsically rhythmic pyramidal cells of layer V in the rat neocortex. Neurons were recorded intracellularly *in vitro*, stimulated with depolarizing current pulses, and stained with biocytin. **A.** A rhythmically single-spiking cell. **B**. A rhythmic bursting neuron. Calibration bars = 100 μm (from Silva and Connors, 1992).

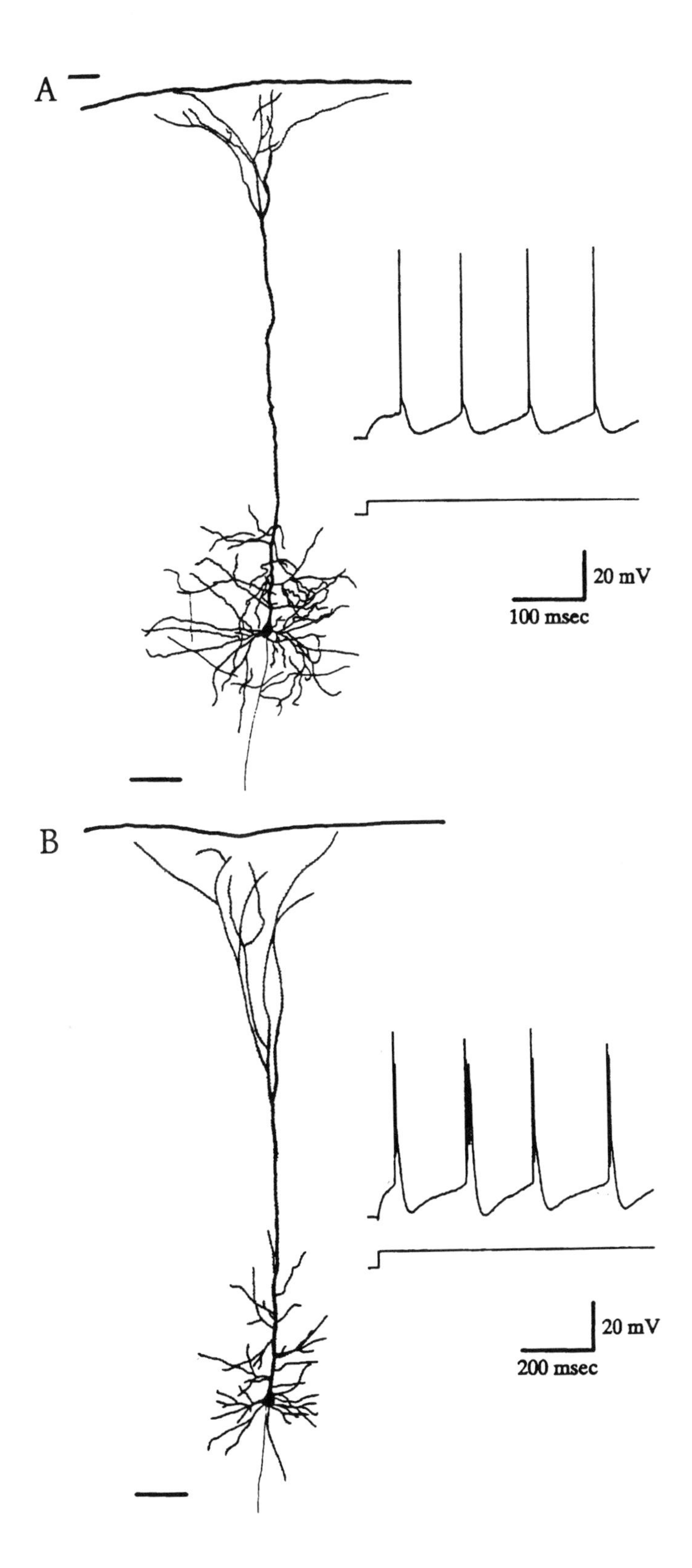
A
20 mV
100 msec
B
20 mV
200 msec

soma size probably follows from a cell's need to support an exceptionally large mass of axons and dendrites. The most striking feature of a large layer V pyramidal cell is its apical dendrite, which spans layers V to I. Apical dendrites may reach lengths of 2,500 μm in human cortex (Marin-Padilla, 1967). In many types of cortex the vertical shafts of these apical dendrites bundle together, forming the core of a dendritic cluster that accumulates the apical dendrites of cells in layer II/III as they all rise toward layer I (Peters and Kara, 1987; Peters and Sethares, 1991). Clusters are spaced about 30 to 60 μm apart and include about 140 neurons in the monkey visual cortex. The neurons of a cluster have strongly overlapping dendritic fields, and it is likely that they receive similar synaptic inputs. Peters and Sethares (1991) have proposed that each such cluster is a fundamental and irreducible unit of neocortex, with neurons that tend to have synchronized activity.

Anatomical studies show that many neurons of layer V receive direct synaptic contacts from thalamic axons (White and Keller, 1989). In the rat somotosensory system, layer IV and Vb neurons are the first to be excited by thalamic input (Armstrong-James et al., 1992). Thus, only layer V provides a rapid, monosynaptic pathway from thalamic input to cortical output. Of course, the much more extensive input to layer V from other cortical layers also provides a compelling influence. The long apical dendrite, combined with an extensive skirt of basal dendrites, allows some cells of layer V to traverse all layers of cortex (Larkman, 1991). No other neurons do this. Electrophysiological studies show that even the most distal inputs to layer I dendrites can strongly excite the soma of a layer V cell 1 mm away (Cauller and Connors, 1992). The afferent systems of the cortex tend to have laminar specificity, and so a single layer V pyramidal cell can, in principle at least, uniquely collect monosynaptic inputs from most of the afferent systems to its cortical area. It can also receive direct input from the local axons of neurons in all other cortical layers. This is not to say that layer V cells sample afferents and local connections uniformly or randomly; the distribution of their dendritic branches, spines, and synapses gives the impression that they are much more interested in some input systems than others (Larkman, 1991; White and Keller, 1989).

The anatomy of layer V clearly distinguishes it from other layers. Its unique structural features include the size, shape, and inputs of its neurons. Most importantly, it provides the output from neocortex to the rest of the brain.

THE UNIQUE PHYSIOLOGY OF LAYER V NEURONS

The intrinsic excitability of a neuron is determined largely by the characteristics of its membrane, and the ion channels and receptor systems therein. Neurons of the neocortex have a melange of physiological properties. The interesting point in this context is that neurons of layer V are more physiologically diverse than those of other layers, and they include a particularly excitable class of neurons that is rarely observed in other layers. Detailed reviews are available (Amitai and Connors, 1994; Connors and Gutnick, 1990; Schwindt, 1992; see Chapter 3).

Intracellular recordings from layer V neurons allow intrinsic membrane properties to be tested in isolation from the surrounding neural network. Current pulses can be delivered to a single cell, and in response it generates a particular pattern of action potentials. Whether tested in vivo or in vitro, the intrinsic patterns of layer V pyramidal cells can take several distinctive forms (e.g. Baranyi et al., 1993; Calvin and Sypert, 1976; Connors et al., 1982; Koike et al., 1968b; Stafstrom et al., 1984a,b). Most commonly, cells fire a train of spikes that begin at high frequency and decline within 100 to 200 msec to much lower frequencies, i.e. they adapt. These are sometimes called *regular-spiking (RS)* neurons, and their rate and degree of adaptation vary. Some RS cells maintain moderately high firing rates indefinitely, as long as they are stimulated, whereas others adapt completely and stop firing. The temporal patterns of spike onset are also variable, and depend further upon the membrane voltage that precedes a stimulus (Spain et al., 1991a). After hyperpolarization, some RS cells respond with faster than normal repetitive firing, whereas the firing of others is transiently inhibited. None of the various RS patterns seems to be exclusive to neurons of layer V, although all are represented there (Amitai and Connors, 1994).

Layer V does have a set of signature pyramidal cells, which we have called *intrinsically bursting (IB)* neurons. Unlike RS cells, these fire tight clusters of spikes (bursts) when stimulated with threshold current. Bursts are followed by relatively long afterhyperpolarizations and spike silence, so that the temporal patterns of IB neurons are nonmonotonic. Many IB cells in layer V can also produce repetitive bursting, in rhythmic patterns with burst frequencies of about 5 to 15 Hz (Fig. 9-1B; Agmon and Connors, 1989; Silva et al., 1991; Wang and McCormick, 1993). Rhythmic firing with single spikes is also common in layer V (Fig. 9-1A). Some neurons of layer IV can generate bursts, but their prevalence seems lower than in layer V (Amitai, unpublished observations; Connors et al., 1982).

The inhibitory nonpyramidal cells of the neocortex also have a distinctive electrophysiology. They apparently fire either nonadapting fast-spiking (FS) patterns (McCormick et al., 1985) or adapting patterns that begin with a low-threshold spike (LTS; Kawaguchi, 1993). There is no evidence that FS or LTS cells have any laminar specificity. However, there is a general dearth of data on interneuron function in neocortex.

Why are repetitively bursting pyramidal neurons concentrated in layer V? Studies in rodents that combine measurements of cell morphology and physiology give some clues. Compared to RS cells, IB cells are larger, have longer and more profusely branching apical and basal dendrites, and have wider horizontal axonal arbors (Chagnac-Amitai et al., 1990; Larkman and Mason, 1990). The differences in shape imply a difference in axonal target, and indeed this is the case. By prelabeling neurons with fluorescent dyes to reveal their projection site, Kasper et al., (in press) showed that cells in rat visual cortex that send an axon to the superior colliculus were invariably IB cells. Neurons projecting through the corpus callosum, in contrast, were RS cells. Wang and McCormick (1993) found that all tested cells of guinea pig layer V that projected subcortically to the colliculus or pons could generate intrinsic bursts or quasi-burst spike "doublets." Corticospinal cells of the rat displayed a range of firing patterns from simple RS to rhythmic IB (Tseng and Prince, in press).

The striking apical dendrites of layer V cells are impressive physiologically as well. Intracellular (Amitai et al., 1993) and whole-cell patch recordings (Kim and Connors, 1993, Stuart and Sakmann, 1994) from apical dendrites show that their membranes are quite excitable. The expression of this excitability varies from cell to cell. Some generate relatively small and slow (compared to the soma) Na^+-dependent action potentials, whereas others fire Na^+ spikes in addition to large, broad Ca^{2+} spikes and plateaus. The more complex Na^+- and Ca^{2+}-dependent forms of excitability occur in the layer V cells with the largest, most elaborate apical dendrites (Kim and Connors, 1993); presumably most of these would be classified as IB cells when recorded from the soma. Nonlinear dendritic membranes facilitate synaptic events and allow interesting interactions between synapses and postsynaptic membrane.

The data imply that layer V has a set of uniquely excitable neurons, the repetitive bursting cells. Most or all of the IB cells of layer V are subcortical projection cells. In other words, a significant fraction of the output of the neocortex is filtered through, and transformed by, the bursting neurons. Yet, layer V projection neurons discharge dual control. They also make extensive connections locally and so can be expected to exert strong influence upon the cortex itself.

SYNAPTIC EXCITATION IN LAYER V

Excitation comes either from outside the cortex, primarily from the thalamus, or from within the cortex. Classical anatomy has provided the general plan for the connections, but information about the specificity of synapses between different cell types has been hard won. Nevertheless, some themes have begun to emerge.

Excitation from the thalamus ends most densely within layer IV, but even there it provides only a small proportion (20% to 30%) of the synapses (Peters, 1987). Synapses of local axons predominate everywhere in neocortex. All layer V pyramidal cells have dendrites within layer IV, but the likelihood that they receive direct thalamic input varies with the cell type (White and Keller, 1989). For example corticothalamic cells, whose somata cluster around the layers V/VI border, receive a relatively large (12% to 20%) proportion of their layer IV synapses from the thalamus (White and Hersch, 1981), whereas layer V corticostriatal cells receive only about 0.5%. By contrast, corticotectal cells of layer V rarely receive monosynaptic input from the thalamus (Ferster and Lindstrom, 1983). An in vitro study (Agmon and Connors, 1992) found that monosynaptic thalamic inputs to RS cells of layer V were common, whereas layer V bursting cells were only indirectly excited. Recall that some IB cells are corticotectal neurons (Kasper et al., in press; Wang and McCormick, 1993). These results are consistent with the idea that thalamic synapses contact some classes of neurons much more than others and that large projection neurons of layer V (including most or all IB cells) come up short. Whether projection cells elude inputs or are avoided by them remains to be determined.

All neurons in the neocortex have axon branches that stay within the area of cortex surrounding their soma and make extensive synaptic connections there. This feature is the basis for calling the neocortex a neural network. Quantitative data on the patterns, specificities, and strengths of these network connections have come only slowly. At the electron microscopic level, pyramidal cells make asymmetric (presumed excitatory) synapses onto a variety of cortical targets, but in general most of their terminations end on other pyramidal cells, with a minority ending onto smooth (inhibitory) interneurons (Kisvarday et al., 1986; White and Keller, 1989). In deeper layers of the cortex, there may be striking differences of connectivity for different cell types. Unidentified layer VI cells of the cat visual cortex (McGuire et al., 1984) and layer V and VI corticothalamic cells of the mouse SI cortex (White and Keller, 1987) make an unusually high proportion of synapses onto dendritic shafts, many of which belong to inhibitory interneurons. However, a reconstruction of two large layer Vb neurons (tentatively identified as corticotectal cells) of cat visual cortex showed that 80% of their intracortical synapses terminated onto spine heads, presumably of other pyramidal cells (Gabbott et al., 1987). Connections were densest to layers V and VI, but also rose into layer IV. Excitation via long horizontal connections from layer V to layer VI has been implicated in building the extended receptive fields of layer VI neurons in visual cortex (Bolz and Gilbert, 1989).

Intracellular injections of such dyes as horseradish peroxidase (HRP) and biocytin reveal the propensity of layer V pyramidal cells to project long distances (up to 6 to 8 mm) while staying largely in layers VI, V, and IV (Chagnac-Amitai et al., 1990; Donoghue and Kitai, 1981; Gilbert, 1992; Gilbert and Weisel, 1983; Martin and Whitteridge, 1984; Miller et al., 1990; Tseng and Prince, in press). However, significant numbers of axons from layer V cells do ascend into all upper layers. Despite their wide connections, the number of synapses from a single layer V cell to any other cell is low, usually less than five (Gabbott et al., 1987; Nicoll and Blakemore, 1993). The general picture is that layer V pyramidal cells make some of the longest horizontal connections within the cortex and that many of these are excitatory interconnections with other pyramidal cells of the same layer. Layer V cells also communicate strongly with all other cortical layers, but the horizontal span of these intralaminar connections is more restricted than those of the deeper layers.

The quantitative anatomy of cortical connections is limited, because it is so hard to do the work. Physiological approaches are also difficult, but the available data show an interesting dichotomy that the anatomy did not predict. Dual intracellular recordings have been made from hundreds of pairs of either layer II/III (Mason et al., 1991) or layer V (Nicoll and Blakemore, 1993) pyramidal cells to look for monosynaptic excitatory potentials (EPSPs). Cells in a pair were no farther than 300 μm (upper cells) or 150 μm (deeper cells) apart. Layer II/III cell pairs had a relatively high probability of connection (0.087), but a low mean amplitude of EPSP (0.4 mV), as compared to layer V pairs ($P = 0.015$ and mean $EPSP = 0.8$ mV). This finding implies that local excitatory connections are sparser but stronger in layer V (Thomson et al., 1988, 1992). Because paired recordings have been limited to closely spaced neurons in slices *in vitro*, it is possible that the low connection probability of layer V cells is balanced by a more widely dispersed set of connections in the horizontal dimension (Ts'o et al., 1986).

There is abundant evidence that glutamate is the major, or perhaps the only, excitatory amino acid transmitter in the neocortex (reviewed by Tsumoto, 1990). Radioligand autoradiography reveals that N-methyl-D-aspartate (NMDA) receptors are not homogeneously distributed in the neocortex. They are densest in superficial layers, although significant levels appear in other layers (Monaghan and Cotman, 1985). This distribution agrees with studies using iontophoretic application of NMDA antagonists; neurons in superficial layers are more sensitive to the NMDA receptor antagonist AP5 than are layer IV cells (Fox et al., 1989; Shirokawa et al., 1989). Many intracellular studies have demonstrated that EPSPs of cortical pyramidal cells in layers II/III and V have both NMDA and nonNMDA components (Higashi et al., 1991; Sutor and Hablitz, 1989a,b; Thomson, 1986, 1990).

Few studies of cortical synapses have specified the physiological or morphological properties of the postsynaptic cells or examined specific input pathways onto them. Jones and Baughman (1988) did find that a layer III to layer V pathway utilized both NMDA and nonNMDA receptors. By stimulating and recording field potentials *in vitro*, Jacobs et al. (unpublished observations) showed that a horizontal excitatory pathway originating in layer V of the rat motor cortex can spread activity at least 1 mm within both layer V and layer III. The rate of spread is about 0.15 m/sec and is dependent mostly upon non-NMDA receptors, with some NMDA receptor involvement. Intracellular recordings, a more sensitive indicator, imply that horizontal pathways within both layer V and layer III can excite cells up to 2 mm away and can travel as fast as 0.4 m/sec (Telfeian, 1993).

To summarize, layer V receives variable amounts of direct thalamic input, dependent upon cell type. There is an extensive set of horizontal interconnections within layer V, much of it excitatory, and it provides excitatory output to all other layers of the cortex.

SYNAPTIC INHIBITION IN LAYER V

Inhibition throughout the neocortex is provided by interneurons that release gamma-aminobutyric acid (GABA). GABA can inhibit by activating $GABA_A$ or $GABA_B$ receptors, which regulate postsynaptic chloride and potassium channels, respectively (reviewed by Connors, 1992). $GABA_B$ receptors may also mediate a form of presynaptic inhibition in cortex (Deisz and Prince, 1989). GABAergic neurons take many forms, and different types target different sets of cells and even different parts of those cells (DeFelipe and Farinas, 1992; Somogyi, 1990).

The cellular mechanisms of inhibition in layer V seem to be similar to those of other layers. However, there are cell-specific variations in the intensity of inhibition that may have repercussions for the network properties of layer V. Local electrical stimuli are less likely to generate IPSPs in IB pyramidal cells than in RS pyramidal cells (Fig. 9-2A; Chagnac-Amitai et al., 1990). In addition, when interneurons of layer V are excited with small applications of glutamate (Silva et al., 1988) or acetylcholine (Nicoll et al., 1993), GABAergic IPSPs are more likely to appear in RS cells than in IB cells, and IPSPs can be evoked from longer horizontal distances for RS cells. Another indication of relatively weak inhibition in IB cells is their behavior in the presence of low concentrations of the $GABA_A$ antagonist bicuculline (Chagnac-Amitai and Connors, 1989a,b). Single shocks to layer VI evoke synchronized, propagating events that include large and long IPSPs onto RS cells of all layers; simultaneously, IB cells of layer V rarely show signs of inhibition, and instead generate transient waves of synaptic excitation (Figs. 9-2B, 9-3A).

The electrophysiological data imply that the inhibitory circuitry contacting RS cells of layer V is somehow more effective than that of IB cells. One possibility is simply that there are fewer GABAergic terminals on IB cells. To test this possibility, White et al. (1994) recently measured the numbers and densities of symmetric (i.e. inhibitory) terminals on somata of RS and IB cells recorded *in vitro* and filled with the dye biocytin. They found that, whereas some IB cells indeed received relatively few inhibitory synapses, others had patterns indistinguishable from RS cells (Fig. 9-3B).

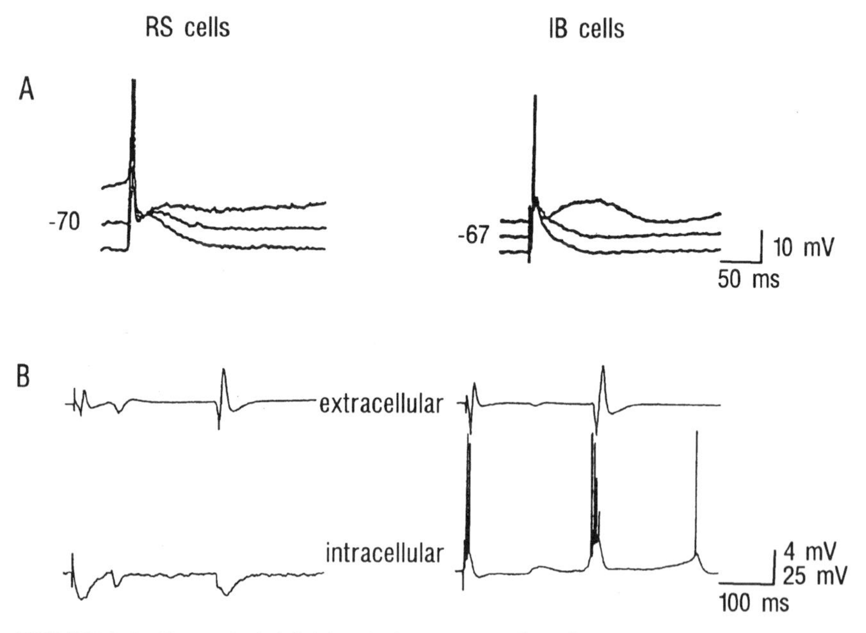

FIGURE 9-2. Synaptic inhibition in layer V tends to be stronger onto regular-spiking (RS; left side panels) cells than onto intrinsically bursting (IB; right side panels) neurons. **A.** Intracellular recordings from RS and IB cells showing shock-evoked synaptic responses. Each cell is shown at three different membrane potentials, with the resting potential indicated in mV. Note the stronger and longer period of inhibition in the RS cell compared to the IB cell (from Chagnac-Amitai et al., 1990). **B.** Synchronized events evoked in the presence of the $GABA_A$ antagonist, bicuculline methiodide (0.9 μm). Top traces show the extracellular field potential and indicate synchronized epileptiform events, and bottom traces show intracellular recordings. Note that during synchronized events, the RS cell was sharply inhibited, whereas the IB cell was excited. Recordings were obtained sequentially from the same region of the same slice. (from Chagnac-Amitai and Connors, 1989b).

Overall there was no significant difference between groups. Since most inhibitory terminals on a pyramidal cell are not on the soma itself but on dendrites, the somatic measurement may have missed a cell-wide difference. It is also possible that the effectiveness of inhibitory synapses on IB cells is less than that of synapses on RS cells. Although at the ultrastructural level the two synapses appear similar (White et al., 1994), GABAergic inhibition is subject to extensive modulation by neurotransmitters (McCormick, 1992). Alternatively, IB cells may be less sensitive to activation of $GABA_A$ receptors than are RS cells (Oka et al., 1993).

RS cells seem to generate stronger inhibition than IB cells within layer V, but the reason is not yet apparent. Nevertheless, the consequence of having a set of IB cells poorly restrained by local inhibition may be dramatic, as discussed in the next section.

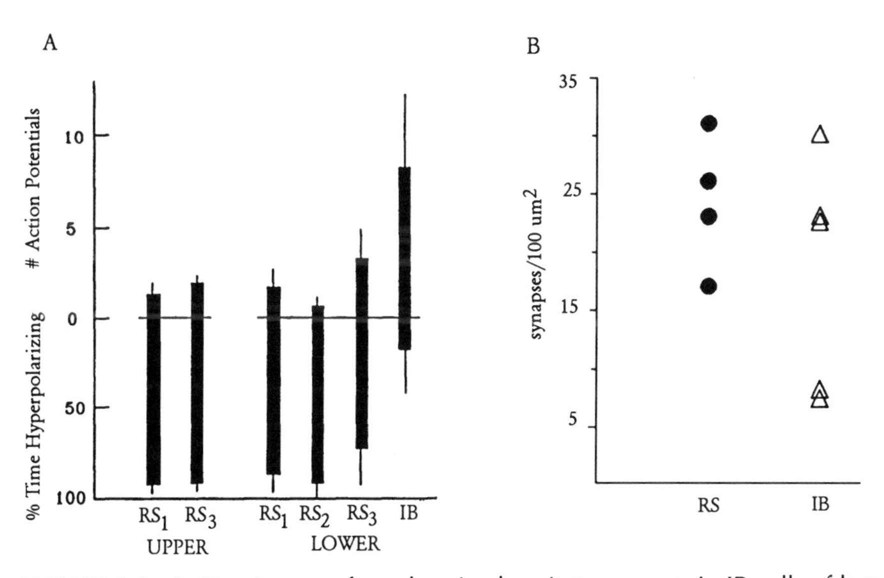

FIGURE 9-3. **A.** Dominance of synchronized excitatory events in IB cells of layer V. Data are derived from experiments similar to that shown in Fig. 9-2B (from Chagnac-Amitai and Connors, 1989b). Slices were exposed to low concentrations (0.8 to 1.0 μM) of bicuculline methiodide to reduce GABAergic inhibition slightly. Downward bars show the mean (±SD) of time spent hyperpolarized (as opposed to depolarized) during the event itself; this was a rough measure of inhibition. Upward bars show the mean number of action potentials generated per event by each cell type. Data from several types of RS cells are compared to IB cells from layer V and show how IB cells are much less inhibited and generate many more spikes during synchronous events. **B.** Densities of somatic (presumed inhibitory) synapses on RS and IB neurons. Cells were recorded intracellularly, filled with either HRP or biocytin, serially sectioned, and then examined in an electron microscope (from White et al., unpublished observations).

ORIGINS OF SYNCHRONY IN LAYER V

Neurons within a region of neocortex fire more or less synchronously, depending upon such factors as the brain's state of arousal (or position in the sleep-wake cycle), the type and specificity of incoming sensory stimuli, and pharmacological and pathological status (Abeles, 1982; Gray et al., 1989; Schwartzkroin, 1993; Steriade et al., 1990a,b; Ts'o et al., 1986). The degree of synchrony has not been completely measured for any of these states; to do so would require simultaneous recordings from large numbers of adjacent neurons, a technology still in its infancy. However, it is clear that the amount, spatial extent, and associated frequencies of coherent activity can vary across a wide range. The characterization of cortical synchrony is a large topic that is not reviewed here. Rather, we summarize the evidence that layer V has an inordinate fondness for synchronous activity under several experimental conditions.

Experiments on slices of neocortex *in vitro* have provided the most direct evidence for the synchronizing abilities of layer V neurons (Connors and Amitai, 1993). Under control conditions, spontaneous activity in slices tends to be very low. However, when inhibition is decreased (by applying antagonists of $GABA_A$ receptors), spontaneous or evoked activity increases tremendously, often in the form of synchronized events. For example, a slight reduction in inhibition results in sharply synchronized, sometimes rhythmic activity that can propagate many millimeters across a cortical slice (Fig. 9-2B; Chagnac-Amitai and Connors, 1989a). Under these conditions synchronized activity is labile, changing its spatial and temporal properties from trial to trial. When $GABA_A$ receptors are more strongly blocked (with high antagonist concentrations) the synchronized events are more stereotyped, are often spontaneous, and propagate unimpeded across the cortex (Fig. 9-4B, control; Chervin et al., 1988; Connors, 1984; Gutnick et al., 1982; Hablitz, 1988). In both cases, layer V seems to be the

→

FIGURE 9-4. Neurons of layer V are important for the horizontal propagation of synchronized activity. **A.** Experimental setup. Neocortical slices were bathed in 50-μM bicuculline methiodide to allow propagating, synchronized events to be evoked by the stimulating electrode (*stim.*) and recorded with electrodes 1 and 2. GABA (2 mM in saline) was pressure ejected in constant quantities along a line of vertical sites, and the effect on propagation was tested after each application. **B.** GABA applied to layer sites within III slowed propagation only slightly, whereas GABA in layer V transiently blocked propagation. **C.** Summary of data from one slice shows that the sensitivity of propagation to GABA peaks sharply in layer V, where synchronized activity could be blocked. Outside layer V the same dose of GABA only slowed propagation (from Connors BW, Chervin RD, Telfeian AE, unpublished observations).

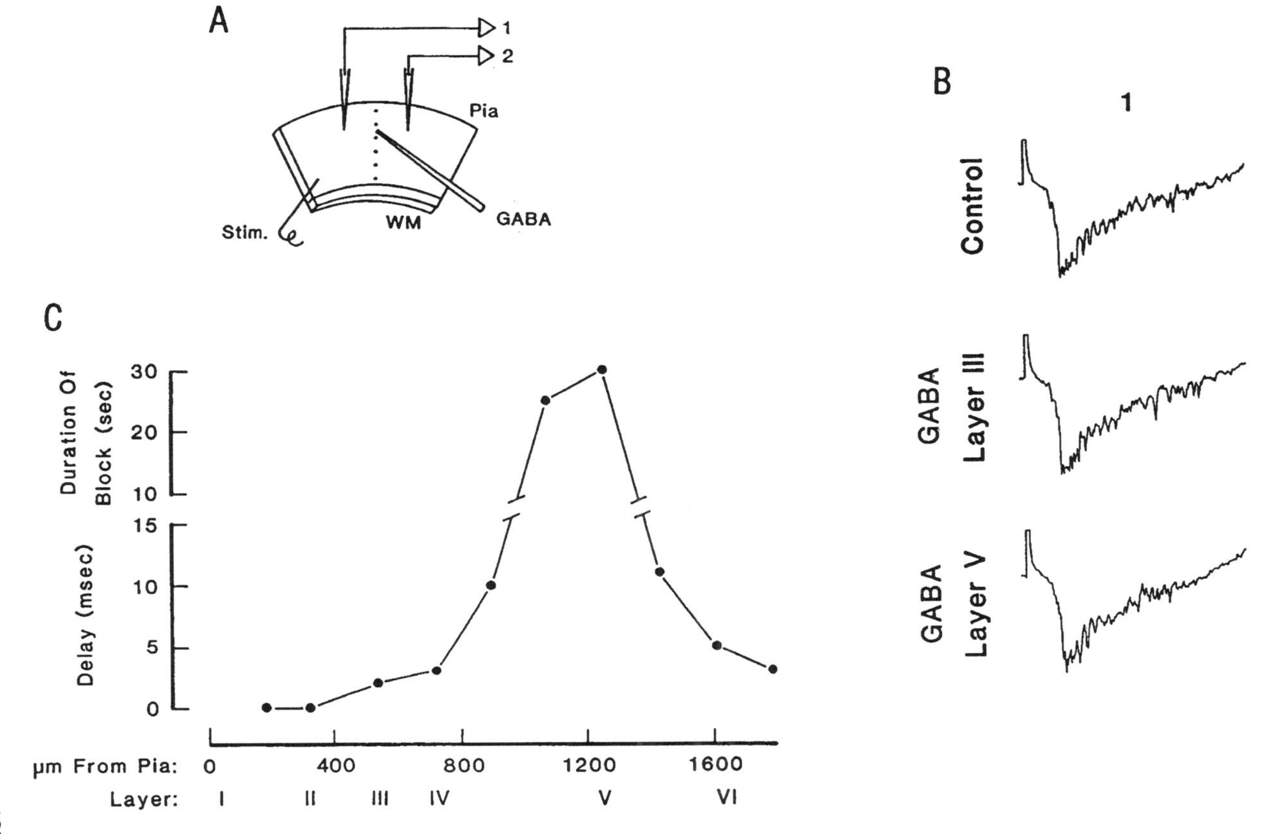
A
1
2
Pia
Stim.
WM
GABA
B
1
2
Control
GABA
Layer III
GABA
Layer V
0.5 mV
30 msec
C
Duration Of
Block (sec)
30
20
10
15
10
5
0
Delay (msec)
µm From Pia:
0
400
800
1200
1600
Layer:
I
II
III
IV
V
VI

site where synchronized activity is initiated. When GABAergic inhibition is only slightly reduced, the IB neurons of layer V are the only class of pyramidal cells that generate synchronized events dominated by synaptic excitation and spike firing; other pyramidal cells tend to be synchronously inhibited and only weakly excited (Chagnac-Amitai and Connors, 1989b). When inhibition is nearly eliminated, layers IV and V are where inward currents associated with synchronous events begin. It is the region most sensitive to $GABA_A$ antagonists, and the area of lowest threshold for event initiation; and synchronous events can be more easily blocked there by GABA (Fig. 9-4A-C; Connors, 1984). In addition, when slices are further dissected with horizontal cuts and bathed in low doses of $GABA_A$ antagonists, those microslices containing layer V (but not other layers) are able to support both the initiation and propagation of synchrony (Telfeian, 1993; Telfeian et al., 1990). Small injections of convulsant drugs into neocortex *in vivo* are most effective in producing epileptiform activity when they are placed in the vicinity of layer IV (Ebersole and Chatt, 1986; Lockton and Holmes, 1983).

Synchronous activity, in a different form, can be generated by boosting synaptic excitation. For example, NMDA receptor-mediated excitation is increased by lowering the [Mg^{2+}] in the bathing solution. Doing so reduces the voltage dependence of the NMDA receptor-operated channel (Nowak et al., 1984) and in neocortex leads to spontaneous, highly synchronous events that occur in rhythmic epochs with 4 to 7 Hz discharges (Fig. 9-5; Silva et al., 1991a; Sutor and Hablitz, 1989a). In horizontally dissected slices in low [Mg^{2+}], a small fragment of layer V is all that is necessary to generate the same type of rhythmic synchrony as is generated by the intact slice (Flint and Connors, 1993; Silva et al., 1991a). Microslices without layer V resemble intact control slices and are incapable of supporting rhythmic, synchronous activity.

Once synchronous events are initiated locally by neurons of layer V, they may propagate both vertically into other layers and horizontally into adjacent cortex (Chagnac-Amitai and Connors, 1989a; Chervin et al., 1988). During moderate disinhibition, the pathway of preference for horizontal movement is within layer V (Telfeian et al., 1990). When inhibition is strongly suppressed, alternative pathways can be found either above or below layer V when layer V is removed. However, it is likely that the pathway through layer V remains the primary mediator of horizontal movement when cortex is intact (Telfeian, 1993).

Layer V is also implicated in other forms of cortical synchrony. Chronic undercuts of the neocortex result in spontaneous epileptiform activity (Burns, 1958). Prince and Tseng (1993) showed with current-source density analysis that the synchronized epileptiform events originate with the neurons of layer V. Their intracellular recordings implied that undercutting disturbs the cellular physiology. The intrinsic excitability of layer V pyramidal cells is increased, synaptic inhibition seems to be depressed, and there may be sprouting of the local axon collaterals of pyramidal cells. Each of these may faciliate the excitability of layer V.

There is at least one flaw in the hypothesis that layer V is *the* major intracortical site of synchronous coordination. A recent study of Flint and Connors (1993) showed that bath application of kainic acid or domoic acid, both agonists of the one class of non-NMDA glutamate receptor, produces spontaneous, low-frequency synchrony that depends absolutely upon neurons of layer II/III and not at all upon layers IV through VI. Kainate-induced synchrony differs from NMDA receptor-induced (i.e. in low [Mg^{2+}]), synchrony in its site of initiation, its timing, and the pharmacology of its modulation by other transmitter systems. This suggests that there are multiple sets of potential synchronizing systems in a single area of cortex, dependent upon different sets of neurons, emphasizing dif-

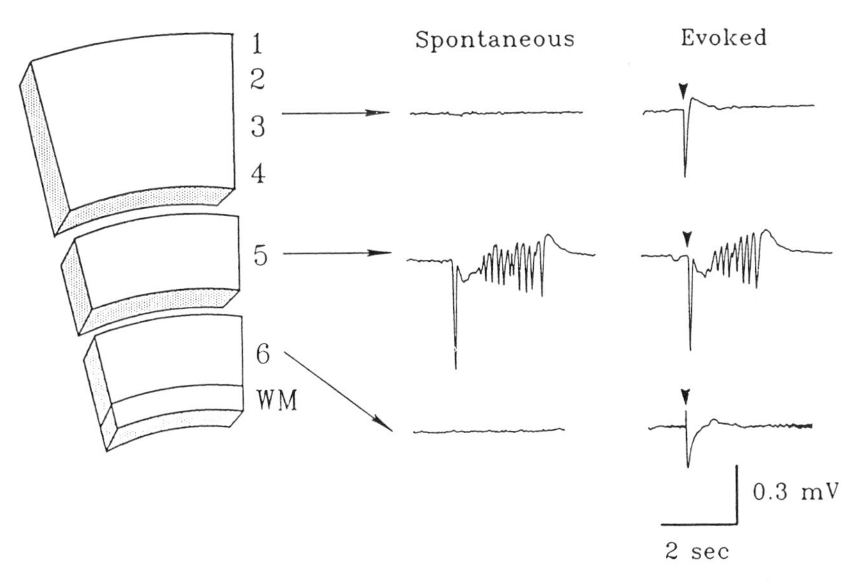

FIGURE 9-5. Layer V is necessary and sufficient for the generation of NMDA receptor-dependent rhythms. Slice of neocortex was trimmed to about 1 mm wide and preincubated in Mg^{2+}-free medium. Recordings showed that it was generating spontaneous rhythmic events. Horizontal cuts then trisected the slice at the layers IV/V border and the layers V/VI border. Each slice fragment was tested for the presence of both spontaneous (left traces) and evoked (right traces) rhythmic synchrony. Although the upper and lower slice fragments were electrophysiologically healthy, only the layer V fragment generated rhythmic events (from Silva et al., 1991).

ferent transmitter receptor systems, and generating different forms of activity.

In summary, layer V has neurons with the appropriate intrinsic membrane properties, intralaminar connections, neurotransmitter systems, and axonal outputs to generate highly synchronized activity and to impose it on the other neurons of the cortex. It may not be the sole synchronizing network in the cortex, but it is a major one. We do not know which properties of layer V are most responsible for its capabilities. A case can be made that IB neurons are the most likely instigators of synchrony: they are prevalent in layer V, they are exceptionally excitable, and they have the requisite local connections (Chagnac-Amitai and Connors, 1989b; Gutnick et al., 1982; Silva et al., 1991). It is tempting to point to the membrane properties of IB cells as the essential feature of layer V synchrony, because intrinsic bursting is so obvious (indeed it defines the IB cell) and so similar to the bursting behavior that is important for synchronization in other structures (e.g., Traub and Miles, 1992). However, bursting may be only part of the answer, or none of it. Unusually strong or dense interconnections between IB cells, relatively weak inhibition, a unique complement of neurotransmitter receptors, or especially excitable dendrites could be more important. It is likely that several characteristics conspire to make layer V a uniquely excitable network in neocortex.

SIGNIFICANCE OF THE LAYER V NETWORK

Electrophysiology is sometimes compared to fishing. In each one begins with a specific goal: catch a neuron or catch a fish, answer a question or relieve one's hunger. Yet, in both tasks the unexpected may turn up, and this is part of their allure. In picking apart the components of the neocortex, myriad facts about layer V have surfaced. One of the most intriguing is its aptitude for organizing the activity of all cortical neurons according to its own whims. The usefulness of this ability is unknown, and indeed there may be none. However recognizing its existence is important. Phenomenology is critical in neuroscience. Brains are so complex that we need to know a great deal about their "natural history" before we can guess at their mechanisms (Bullock, 1993). The phenomenology of layer V synchrony, embedded in the larger structure and function of the neocortex, suggests interesting possibilities.

One function of layer V may be to generate and disseminate rhythmic activity. The many rhythms of the cortex can be driven from different sources, some thalamic and some cortical. There is compelling evidence that the 8 to 12 Hz spindle rhythms, common during drowsiness or barbiturate anesthesia, are initiated within the thalamus and imposed on the cortex (Steriade et al., 1990b; see Chapter 11). The 3 Hz rhythms of absence seizures likewise originate in the thalamus. The origin of other cortical rhythms has not been determined, and it is possible that some arise within the cortex itself (Silva and Connors, 1992). The rhythmicity of layer V might also be important in generating a resonance with thalamic rhythms. Of course, even if we find that layer V has a role in producing or facilitating rhythms, we are still left with a set of interesting phenomena in search of a function.

A variety of functions for cortical synchrony and oscillations have been proposed. One of the most general is that information in cortex is represented by coherently firing assemblies of neurons, rather than by alterations in the mean firing rates of individual cells (Abeles, 1982; von der Marlsburg and Schneider, 1986). In this scheme, a group of cells responding to similar features of a stimulus object discharge together, while the firing of cells that respond to different objects is uncorrelated. Correlated firing might or might not be rhythmic. Evidence from the visual cortex suggests that rhythmic synchrony between groups of neurons can occur across widely separate areas of cortex, as long as the visual stimulus has coherence and specificity for the involved neurons (Eckhorn et al., 1988; Engel et al., 1992; Gray et al., 1989). The anatomical substrates for maintaining synchrony are likely to be intracortical axons. Even with delays of several milliseconds, reciprocal connections can in principle support correlated activity with zero phase lag (König and Schillen, 1991). The layer V network could provide an important component of such a synchronizing system. The roles for cortical rhythms and synchrony in information processing are controversial (Ghose and Freeman, 1992). The binding of neurons into coherent but ephemeral assemblies does seem to occur, but whether it has relevance for brain function remains to be demonstrated. An ironically different purpose for synchronous oscillation has been proposed for the rhythms of sleep. By engaging the thalamus and cortex in coherent activity, which is generated endogenously, the forebrain might isolate itself from the distracting sensory influences of the environment (McCormick, 1992; Steriade et al., 1990b).

Horizontal connectivity and coordinated activity may also mediate dynamic alterations of receptive field properties. The sensory periphery is systematically mapped onto the neurons of the cortex. Under normal conditions, the receptive field of a neuron is much smaller than the intracortical spread of dendrites and axons might be expected to generate (Gilbert, 1992). Presumably, specificity is determined by synaptic selection and inhibition. However, when the sensory periphery is perturbed with a lesion or by manipulation of the environment, receptive fields can expand, shift, and reorganize. There may be several phases to the reorganization, with time courses ranging from seconds to months. Representations in motor cortex are similarly dynamic (Sanes et al., 1988). The extent of these changes implies that they are mediated by long-range intracortical connections. The structure and function of layer V make it an ideal candidate for horizontal reorganization (Bolz and Gilbert, 1989; Jacobs and Donoghue, 1991; Jacobs et al., 1994).

The ability of a set of presynaptic neurons to drive a set of postsynaptic neurons is strongly determined by the temporal coherence of the presynaptic activity. Since layer V pyramidal cells provide the primary output of the neocortex, their coherence must be optimized for maximal effect downstream, i.e., on their subcortical targets. A network of interconnected layer V cells would provide the most direct route to the large-scale synchrony of cortical output. By keeping the connections direct, the layer V network would maximize efficiency, minimize conduction and synaptic delays, and insulate the system from perturbations that might disturb more polysynaptic networks. At the same time, it would be possible to closely control the spatial and temporal aspects of synchrony by modulating the network's synapses and membranes. Thus, synchrony within layer V might be the solution to a general problem in the neocortex; namely, that output must often be coherent but that the degree of coherence must be under dynamic control.

Acknowledgments. The authors' work was supported by the National Institutes of Health, the Office of Naval Research, and the US-Israel Binational Science Foundation.

10

Inhibition in the cortical network

LARRY S. BENARDO and ROBERT K. S. WONG

Neuronal activity in the brain is regulated by a balance between excitatory and inhibitory influences. Population stability requires that GABAergic inhibition be temporally poised to exercise a restraining influence on reverberating synaptic excitation of local neurons due to recurrent connections. Because of this unique circuit organization of the cortex, epileptiform activities arise after compromises in inhibitory strength.

Much is known about several individual aspects of cortical GABAergic inhibition, and many of these topics have been reviewed previously (Alger, 1991; Nicoll et al., 1990; Traub et al., 1987). In this chapter we will attempt to provide as comprehensive a consideration of GABAergic inhibition as space allows, though it is by no means exhaustive for each topic. We highlight specific issues that are especially important to understanding how inhibition operates and how it is regulated. The intention throughout is to provide not only a synthesis of our present knowledge but also to state some of the principles and implications of GABAergic inhibition that are just emerging. The chapter is organized as follows: (1) the morphology and physiology of inhibitory neurons, (2) the recruitment of inhibition, and (3) the plasticity of the inhibitory circuit.

GENERAL SCHEME OF GABAERGIC CIRCUITS

Available evidence suggests that every cortical neuron is subject to inhibition mediated by GABAergic synapses. Two general modes of inhibition have been recognized in cortical structures: feedback and feedforward inhibition. Together these processes effectively limit excessive firing of pyramidal neurons. Evidence for the two types of inhibition is most complete for the hippocampus, but the same scheme is generally presumed to apply to the neocortex as well. The feedback circuit consists of a pyramidal cell with an axon collateral that contacts an inhibitory interneuron and in turn sends reciprocal connections to the pyramidal cell body (i.e., a recurrent loop), causing it to be inhibited. The best example of this circuit is the recurrent inhibitory postsynaptic potential (IPSP) that results from antidromic stimulation of hippocampal CA1 pyramidal neurons, as first described by Kandel et al. (1961; see also Alger, 1991; Andersen et al., 1964a,b; Lacaille et al., 1987). Feedforward GABAergic circuits consist of inhibitory interneurons that are activated by excitatory afferents. GABAergic cells of this type may contact dendrites exclusively (Lacaille and Schwartzkroin, 1988b) or the cell soma, as well as dendrites (Alger, 1991; Alger and Nicoll, 1979, 1982a,b). It should be noted that individual inhibitory interneurons can participate in one or both forms of inhibition (Miles, 1990).

Fast and slow inhibition have been demonstrated in both the hippocampus and neocortex. The pharmacology of both forms of inhibition has been elucidated and seems to be identical in the two structures. There are two general classes of GABA receptors—$GABA_A$ and $GABA_B$—that mediate bicuculline-sensitive and -insensitive responses, respectively. GABA activates both kinds of receptors. The $GABA_A$ response is chloride dependent (reversing about −70 mV) and mediates the ligand-gated fast IPSP (Alger, 1991; Benardo, 1993b, 1994a; Connors et al., 1988; Deisz and Prince, 1989; Kandel et al., 1961; McCormick, 1989; Nicoll et al., 1975, 1990). $GABA_B$ receptors activate a postsynaptic potassium-dependent conductance (reversing about −90 mV; Avoli, 1986; Benardo, 1993b, 1994; Connors et al., 1988; Deisz and Prince, 1989; Gahwiler and Brown, 1985; McCormick, 1989; Newberry and Nicoll, 1984a, 1985), via a G-protein-linked mechanism, and hyperpolarizes cells (Andrade et al., 1986; Dutar and Nicoll, 1988; Thalmann, 1984; 1988). Activation of the $GABA_B$ receptor by synaptically released GABA causes the late, slow IPSP. Support for this hypothesis comes from the similarities between the ionic mechanisms of $GABA_B$ receptor activation and the slow IPSP, the antagonism of the slow IPSP by $GABA_B$ antagonists (Dutar and Nicoll, 1988; McCormick, 1989), and the blockade of both by phorbol esters (Andrade et al., 1986; Baraban et al., 1985).

MORPHOLOGY AND PHYSIOLOGY OF INHIBITORY INTERNEURONS

GABAergic neurons in the cortex are exclusively of the nonpyramidal cell type, i.e. they lack a single dominant apical dendrite. In general they are local circuit neurons, meaning they form symmetrical synapses with neighboring cells confined to the local area in which their cell body resides. However, several exceptions can be noted—contralateral projections in the cat visual cortex (Innocenti and Fiore, 1976), the cat auditory cortex (Code and Winer, 1985), and the rat hippocampus (Seress and Ribak, 1983) and ipsilateral projections in neocortex (Winguth and Winer, 1986) and hippocampus (Seress and Ribak, 1983). GABAergic cells have been identified by histochemical methods (Araki et al., 1984; Nagai et al., 1983); autoradiography (Hamos et al., 1983; Somogyi et al., 1984a,b; Winer, 1986), and immunocytochemistry. Immunocytochemical techniques have utilized an antibody to GABA (Keller and White, 1987; Ottersen and Storm-Mathisen, 1984a,b; Somogyi et al., 1985), or to glutamic acid decarboxylase (GAD), the rate-limiting enzyme in the synthesis of GABA (Houser et al., 1983; Keller and White, 1986; Mugnaini and Oertel, 1985; Ribak, 1978).

Estimates of the proportion of neurons in cortical structures that are GABAergic vary by location and species, for example, from 10% of neurons in human hippocampus (Olbrich and Braak, 1985) to 25% of cells in most areas of monkey neocortex (Hendry et al., 1987). Although all cortical GABAergic neurons are nonpyramidal, the converse is certainly not true. Various anatomical classification schemes exist for nonpyramidal cells, although some general features of GABAergic neurons become apparent after review of the pertinent literature (see Fairen et al., 1984 for an exhaustive review of all nonpyramidal cell types). The two primary systems for classifying nonpyramidal cells are based on soma/dendritic arbors or axonal ramifications. The former system, advanced by Feldman and Peters (1978), classifies cells as multipolar, bipolar, or bitufted according to the shape of their dendritic tree and spiny or smooth (nonspiny) according to the concentration of spines emerging from their dendrites. Multipolar cells have dendrites that emerge from multiple points on the somal surface and radiate out in all directions. Bipolar and bitufted neurons project, respectively, either single or multiple dendritic processes from the opposite poles of cell bodies that tend to be ovoid or spindle shaped. GABAergic cortical neurons are non- or sparsely spiny cells and may have any of the soma/dendritic patterns previously noted (White, 1989).

A more useful organizational scheme, at least for our purposes, is that based on the form and distribution of nonpyramidal cell axonal ramifications. According to this classification GABAergic neurons include chandelier cells, basket cells, local plexus neurons, and several types of vertically oriented neurons. These are in general purely descriptive terms, but are extremely useful for identifying GABAergic neurons. Chandelier cells are named for the resemblance of their axonal arbors and vertically oriented arrays of axon terminals to the branches and candles of a chandelier. Cell bodies of these cells are ubiquitous within the cortex. They are present in layers II through V in a variety of species in several areas including visual cortex (Fairen and Valverde, 1980; Marin-Padilla, 1987; Muller-Paschinger et al., 1983; Peters et al., 1982; Somogyi et al., 1982; Valverde, 1983), the somatosensory cortex (Jones, 1975; Tombol, 1978; Valverde, 1983), the auditory cortex (DeCarlos et al., 1985; Fairen et al., 1981; Szentágothai, 1975), the temporal cortex (Kisvarday et al., 1986); the cingulate cortex (Vogt and Peters, 1981), subiculum and piriform cortices (Somogyi et al., 1982), and the hippocampus (Somogyi et al., 1983). These cells can be additionally classified as bitufted or multipolar, and their dendrites may span one or several layers, but have few branches. Generally they form a profuse plexus in the vicinity of the cell body, coextensive with or somewhat above or below the distribution the cell's dendritic tree. Most remarkably these cells form synapses only with axon initial segments on pyramidal neurons (DeFelipe et al., 1985; Fairen and Valverde, 1980; Freund et al., 1983; Peters et al., 1982; Somogyi, 1977; Somogyi et al., 1979) and spiny stellate cells (Lund, 1987). This termination site would allow chandelier cells to exert a powerful inhibitory influence upon their postsynaptic cells, conferring the ability to gate all postsynaptic neuronal output.

Basket cells are similarly recognized by the distribution of their axonal branches, which form nests or baskets around cell bodies and proximal dendrites of pyramidal cells (Fairen et al., 1984). Each of these arrangements is composed of branches from more than one axon collateral, and each axon collateral contributes to several pericellular nests (Cajal, 1909). These nests arise from multipolar cells mainly in layers III and V (Marin-Padilla, 1969, 1970, 1972), but are also found in hippocampus (e.g., Schwartzkroin and Mathers, 1978). Some cells have a bitufted appearance. Cells of this type may have axons that initially ascend or descend, then run horizontally for up to a millimeter before giving off a pericellular nest (e.g., Jones and Hendry, 1984), whereas others have more restricted axonal ramifications (DeFelipe and Fairen, 1982; Fairen et al., 1984).

Vertically oriented cells, present in most areas of cortex (White, 1989), have also been called "double bouquet" because their dendritic fields are bitufted or bipolar in shape. Cells of this class that form symmetrical synapses and are GABAergic (Somogyi and Cowey, 1981, 1984; Somogyi et al., 1981), have their cells bodies only within layers II and III (Fairen et al., 1984), but send axonal ramifications descending well beneath their dendritic trees.

Local plexus neurons, which may be smooth or sparsely spiny, comprise a group of cells that have been previously referred to as short axon, local circuit, or Golgi type II neurons. These cells are multipolar, and their axonal arborizations end in a local plexus occurring in the immediate vicinity of their soma and dendrites. Some of these cells may have additional axonal branches that ramify in cortical layers above or below the lamina containing the cell body, dendrites, and local axonal plexus. The terminal arborizations of these cells are less well characterized than those considered above. Moreover, they form GABAergic inhibitory synapses (Hendry and Jones, 1981; Keller and White, 1987; Kisvarday et al., 1986; Somogyi et al., 1981) with a variety of postsynaptic elements, including cell bodies, dendrites, and axon initial segments of pyramidal neurons, the cell bodies and dendrites of nonpyramidal neurons; and spines of cells of unknown origin (LeVay, 1973; Peters and Fairen, 1978; Peters and Proskauer, 1980). These cells are the likely source for most of the axon terminals that form symmetrical synapses in rodent brains (Peters and Proskauer, 1980). The diagram shown in Figure 10-1 is a highly schematic composite representation of inhibitory circuits in the cortex.

Further subdivision of inhibitory neurons may be possible by cell-surface markers (e.g. Naegele et al., 1988) or by their content of neuroactive substances. Many nonpyramidal neurons contain one or more neuroactive peptides co-localized with GABA (Gamrani et al., 1984a; Hendry et al., 1984a; Lin et al., 1986; Papadoulos et al., 1987; Penny et al., 1986; Schmechel et al., 1984; Somogyi et al., 1984b). In general, neuropeptide Y (NPY), somatostatin (SRIF), and cholecystokinin (CCK) have been detected in association with GABAergic neurons (i.e. 90% to 100% are GAD or GABA positive), usually in small (8 to 12 μm diameter) local plexus neurons, also described as bipolar cells (Hendry et al., 1984a; Somogyi et al., 1984b). GABAergic cells containing CCK do not co-localize SRIF or NPY. Yet, NPY-containing cells often co-localize SRIF (Hendry et al., 1984a), though the converse is not true. Specifically, CCK immunoreactivity in GABAergic cells is seen in all layers in neocortex (mainly in layers II, III, and VI), stratum oriens, stratum pyramidale, stratum moleculare, and the polymorph layer of the dentate gyrus in hippocampus (Greenwood et al., 1981; Hendry et al., 1984b; Somogyi et al., 1984b; Vanderhaeghen et al., 1980). GABAergic cells immunoreactive for SRIF appear with the same distribution in neocortex, but in hippocampus they are seen in stratum oriens, stratum pyramidale, and the hilus (Finley et al., 1978, 1981; Hendry et al., 1984b; Somogyi et al., 1984b). Interestingly, these peptide-containing GABAergic cells make up less than half of the total GABAergic neurons present in layers II, III, and VI, where they are most abundant, with a lower proportion in layers IV and V (comprising 4 to 11 of 100 GAD-positive cells in layer IV; Hendry et al., 1984a). Moreover, peptide-containing cells were not seen in association with large pyramidal cells in layers IIIB-VI (Hendry et al., 1984a). Aside from distinguishing a subpopulation of inhibitory neurons, the role of these peptide-containing interneurons remains speculative at best (see Nicoll et al., 1990 for a general discussion of these peptide actions), although some modulatory action for GABA inhibition itself would seem most logical.

Correlation of the morphology and physiology of inhibitory neurons is difficult, even in in vitro slices (to which our discussion is confined). Stud-

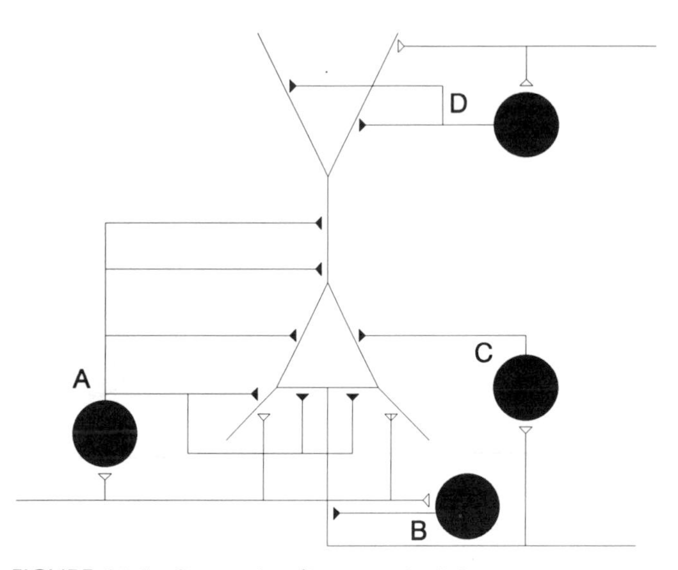

FIGURE 10-1. Composite diagram of inhibitory circuits in the cortex. Pyramidal neurons receive excitatory (open terminals) and inhibitory inputs (closed terminals). Feedforward (**A** and **D**) and feedback (**C**) circuits are shown. **A**. Basket cells form pericellular nests on soma and dendritic regions yielding feedforward inhibition (Cajal, 1909; DeFelipe and Fairen, 1982; Fairen et al., 1984; Jones and Hendry, 1984; Marin-Padilla, 1969, 1970, 1972; Schwartzkroin and Mathers, 1978). **B.** Chandelier cells characteristically form inhibitory synapses on axon initial segments throughout cortex (DeCarlos et al., 1985; DeFelipe et al., 1985; Fairen and Somogyi et al., 1982; Fairen and Valverde, 1980; Fairen et al., 1981; Freund et al., 1983; Jones, 1975; Kisvarday et al., 1986; Marin-Padilla, 1987; Muller-Paschinger et al., 1983; Peters et al., 1982; Somogyi, 1977; Somogyi et al., 1979; Somogyi et al., 1982; Somogyi et al., 1983; Szentagothai, 1975; Tombol, 1978; Valverde, 1983; Vogt and Peters, 1981). Their innervation is not clearly defined, but they probably participate in feedforward and feedback circuits. **C.** Recurrent loop provides somatic inhibition (Alger, 1991; Andersen et al., 1964a, b; Kandel et al., 1961; Lacaille et al., 1987). The recurrent circuit is activated by pyramidal cell axon collaterals (of this and neighboring cells). **D**. Feedforward circuit provides dendritic inhibition. Vertically oriented inhibitory neurons in lacunosum-moleculare of the CA1 hippocampal region (Lacaille and Schwartzkroin, 1988a, b) provide an example of this circuit.

ies of this nature have tended to examine nonpyramidal cells (not necessarily identified as inhibitory neurons, although this is implicit) or groups of inhibitory neurons that are localizable to a general anatomical region or zone but that do not necessarily comprise a homogeneous group by other criteria, e.g. hilar neurons in hippocampus. Thus, a general consideration of inhibitory neuron intrinsic physiology reveals inconsistencies, as is explicated below, that must be based on this heterogeneity. Resting membrane potentials for these cells tend to be lower, which has been attributed to poorer impalements relative to pyramidal neurons (Lacaille et al., 1987). In large part this speculation may be correct, but it should be noted that many of these neurons have active baseline synaptic depolarizing activity that contributes to lower resting potentials. These cells tend to have higher input resistances and shorter membrane time constants relative to pyramidal neurons (Knowles and Schwartzkroin, 1981; Lacaille, 1991; Lacaille and Schwartzkroin, 1988a; Lacaille and Williams, 1990; Lacaille et al., 1987; McCormick et al., 1985; Miles and Wong, 1984; Schwartzkroin and Mathers, 1978; Thomson et al., 1993b).

Identified interneurons in neocortex (McÇormick et al., 1985; Thomson et al., 1993b) and certain classes of identified interneurons in hippocampus (Knowles and Schwartzkroin, 1981; Lacaille and Williams, 1990; Lacaille et al., 1987; Schwartzkroin and Mathers, 1978) have short duration action potentials compared to pyramidal cells in these structures (but see Dykes et al. 1988 for contradictory evidence). However, this property is not shared by all GABAergic interneurons (Lacaille and Schwartzkroin, 1988a,b; Miles and Wong, 1984), even those yielding the same postsynaptic response type (Miles and Wong, 1984). It remains unclear whether cells with this property might define a subgroup of GABAergic interneurons that are related on other bases, such as morphology.

In interneurons, large fast afterhyperpolarizations are seen in tandem with short duration action potentials, both in hippocampus and neocortex (Knowles and Schwartzkroin, 1981; Lacaille and Williams, 1990; Lacaille et al., 1987; McCormick et al., 1985; Schwartzkroin and Mathers, 1978; Thomson et al., 1993b). Yet, some hippocampal cells definitively identified as GABAergic do not display this property (Miles and Wong, 1984). This finding suggests that the presence of some specific cellular properties may not have predictive value regarding a neurons' postsynaptic action, but again it is unclear whether cells having properties in common might be related by other characteristics.

The voltage-current relationship gives information about the presence and voltage-dependence of active membrane currents, as indicated by nonlinearity of the relationship (rectification). Inward rectification in GABAergic neurons has been noted in hippocampus in stratum lacunosum-moleculare interneurons (Lacaille and Schwartzkroin, 1988b), but not in stratum oriens/alveus and stratum pyramidale cells (Lacaille, 1991; Lacaille and Williams, 1990). Interneurons in both hippocampus and neocortex purportedly show linear stimulus current-firing frequency relationships (Knowles and Schwartzkroin, 1981; Lacaille and Williams, 1990; Lacaille et al., 1987; McCormick et al., 1985; Schwartzkroin and Mathers, 1978; Thomson et al., 1993b). Yet, some inhibitory neurons in hippocampus show burst firing (Lacaille and Schwartzkroin, 1988a,b; Lacaille et al., 1987; Miles and Wong, 1984). Thus, GABAergic interneurons may have heterogeneous firing characteristics, indicating that these types of cells must be comprised of more than one class and may be capable of shifting firing patterns similar to other central neurons (Benardo and Foster, 1986; Jahnsen and Llinás, 1984; McCormick and Wang, 1991; Wong and Prince, 1981).

RECRUITMENT PROPERTIES OF INHIBITORY NEURONS

In this section the activation of inhibitory neurons is considered from two vantage points. One assesses the presynaptic innervation of these cells, and the other is directed toward analyzing the recruitment of inhibition as seen postsynaptically in principal (pyramidal) neurons. Our present knowledge base is patchy, but some of the trends that are developing are discussed here.

Presynaptic effects on inhibitory neurons have previously been emphasized with regard to the action of neuromodulatory substances such as acetylcholine. The initial cholinergic inhibition of hippocampal pyramidal neurons that follows acetylcholine application was found to result from presynaptic excitation of inhibitory interneurons (Benardo and Prince, 1982). This result has been confirmed in hippocampus many times (Haas, 1982; Pitler and Alger, 1992), as well as in neocortex (McCormick and Prince, 1986). Interestingly, evoked $GABA_A$ IPSPs appear depressed during this same period (Haas, 1982; Pitler and Alger, 1992; Segal, 1982). Similar actions have been observed in response to norepinephrine application in the CA1 region (Madison and Nicoll, 1988a). The specific mechanisms involved in these effects have not been completely defined, although it has been suggested that these compounds have direct excitatory effects on inhibitory interneurons and may have a presynaptic inhibitory effect on the afferent excitatory fibers to the interneurons (Alger, 1991; Madison and Nicoll, 1988a). Presynaptic inhibition of inhibitory neurons has been implicated for other transmitters, including enkephalin (Madison and Nicoll,

1988b) and serotonin (Sugita et al., 1992). GABA itself is involved in inhibiting its own further release and may regulate its release via activation of presynaptic $GABA_B$ (Connors et al., 1988; Deisz and Prince, 1989; Thompson and Gähwiler, 1989b) and $GABA_A$ receptors (Benardo, 1993a). This regulation presumably leads to some component of use-dependent IPSP depression (Benardo, 1993a; Deisz and Prince 1989; McCarren and Alger, 1985; Thompson and Gähwiler, 1989b; Wong and Watkins, 1982).

Another focus of study has been the presynaptic innervation of inhibitory neurons by glutamatergic afferents. Synaptic excitation of inhibitory neurons giving rise to $GABA_A$ IPSPs may be quantitatively different from that arising in pyramidal cells. For example, excitatory postsynaptic potentials (EPSPs) in inhibitory neurons are larger and have a faster time course (Miles, 1990; Thomson and West, 1993a) than recurrent EPSPs elicited in pyramidal cells by other pyramidal cells (Miles and Wong, 1986; Thomson et al., 1993a). Orthodromic stimulation of interneurons at normal intensities can cause very large, often burst-like postsynaptic responses (Lacaille et al., 1987; McCormick et al., 1985). Moreover, unlike pyramidal cells, single excitatory synapses made onto inhibitory neurons can be strong enough to cause the neuron to fire (Miles and Wong, 1984). These differences may derive from differences in intrinsic properties—for example, the faster time constant and higher input resistance seen in inhibitory neurons (Miles, 1990)—or to differences in the site of termination of excitatory afferents onto inhibitory neurons (i.e. somatic; Frotscher, 1985; Kisvarday et al., 1985; Schwartzkroin and Kunkel, 1985; Seress and Ribak, 1985) as opposed to pyramidal neurons, i.e. dendritic spines (Hamlyn, 1963).

Excitation of inhibitory neurons is beginning to be assessed in terms of the respective proportion of non-NMDA versus NMDA components. It seems that fast IPSPs may be preferentially recruited by non-NMDA transmission. Evidence for this speculation takes a variety of forms, although at present it is somewhat circumstantial. For example, in CA1 stratum pyramidale interneurons only one-third of cells have a "late" NMDA-mediated EPSP (Lacaille, 1991). In addition, EPSPs recorded in CA3 inhibitory interneurons decay with time constants indicative of non-NMDA transmission (Miles, 1990). A recent study of hippocampal hilar neurons provided evidence for two populations of interneurons, differentiable on the basis of their expression of non-NMDA versus NMDA receptors (Brown and Dingledine, 1992). More recently, interneurons in the dentate hilus and CA3 have also been differentiated on the basis of the non-NMDA receptors they express; that is, they have different AMPA subunits, which produce distinct postsynaptic potentials (Livsey et al., 1993; McBain and Dingledine, 1993). Similar findings will be forthcoming in neocortex, where it has already been shown that non-NMDA receptor subtypes differ in layer II/III pyramidal cells versus interneurons in visual cortex (Hestrin, 1993).

Recent experiments that are related to these issues examined the effects of reducing extracellular magnesium on inhibition in layer V pyramidal cells in somatosensory cortex (Benardo, 1993b). Magnesium-free solutions, which enhance excitation, yielded significant increases in the size and duration of EPSPs and slow IPSPs, but fast hyperpolarizations became significantly smaller (Fig. 10-2A). When cells were recorded with microelectrodes containing cesium, which blocks the $GABA_B$ receptor-mediated potassium conductance, slow hyperpolarizing responses were blocked, and depolarization of cells to 0 mV revealed an isolated fast IPSP following synaptic stimulation. The amplitude of this response was unchanged after exposure to magnesium-free solutions (Fig. 10-2B). Thus, increased activation of NMDA receptors did not result in increased feedforward activation of fast inhibition. Given these results, stimulus-response characteristics of neocortical neurons were reassessed under control conditions. With higher-intensity stimuli, larger EPSPs and slow IPSP responses were evoked, but fast hyperpolarizations showed a decremental response (Fig. 10-2C).

Thus, when excitatory activity is enhanced by exposure to magnesium-free solutions or electrical stimulation, the amplitude of EPSPs and slow IPSPs is increased, but the capacity of fast inhibitory responses for incremental recruitment is limited. This finding suggests that fast inhibition is saturated (maximal) at submaximal levels of excitation and can be overcome by increasing levels of excitation.

The limitation on fast inhibition may serve to allow the system to retain a high degree of plasticity. If the capacity for recruitment of inhibition equaled that of excitation, processes involving potentiation of excitatory events would be nullified by an equivalently enhanced IPSP. Limited fast inhibition thus is highly permissive, since it allows the activity-dependent modifications of excitatory synaptic transmission, which we associate with higher cortical functions, to occur. Interestingly, it is this same property that renders the cortex so vulnerable to epilepsy and excitotoxicity (Ascher et al., 1991).

Whether or not a single GABA interneuron activates both postsynaptic $GABA_A$ and $GABA_B$ receptors in cortex is still an intriguing but unresolved issue. In the hippocampus, two functional sets of GABAergic interneurons have been postulated to account for the observations that (1) antidromic activation of pyramidal cells produces only a simple $GABA_A$ IPSP; (2) orthodromic stimulation of afferent fibers produces a complex IPSP, with both $GABA_A$ and $GABA_B$ components (Alger, 1984; Alger and Nicoll,

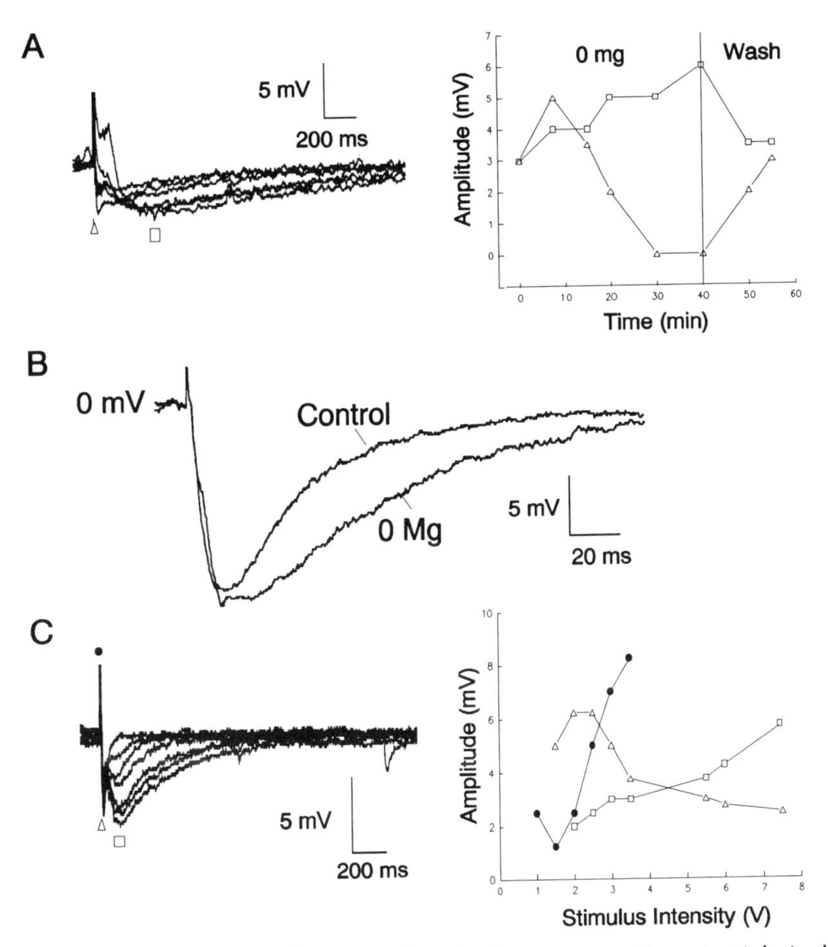

FIGURE 10-2. Effects of increased excitation on synaptic potentials in layer V pyramidal neurons. **A**. Effect of magnesium-free (0 Mg) on fast (triangles) and slow synaptic hyperpolarizations (squares). Right panel plots data obtained from traces in left panel. Note fast hyperpolarizations decrease, whereas slow hyperpolarizations increase in amplitude. **B**. Fast hyperpolarizations triggered in cells recorded using cesium-acetate-filled microelectrodes before and after exposure to magnesium-free solution. The magnitude of the fast hyperpolarization is essentially unchanged. **C**. Stimulus-response characteristics for the depolarizing (circles), fast hyperpolarization (triangles), and slow hyperpolarizing potentials (squares). With increasing stimulus intensity the depolarizing and slow hyperpolarizing synaptic potentials increase, whereas the fast hyperpolarizing response decreases. Data in left panel are plotted in right panel. Note the different thresholds. (Modified from Benardo, 1994a.)

1982; Dutar and Nicoll, 1988; Fugita, 1979; Newberry and Nicoll, 1984); and (3) $GABA_A$ and $GABA_B$ responses seen in the same postsynaptic cell can be separately activated (Otis and Mody, 1992b; Solis and Nicoll, 1992; Williams and Lacaille, 1992). Simultaneous recordings from interneuron-pyramidal cell pairs have revealed either slowly developing IPSPs after activation of some interneurons (Knowles and Schwartzkroin, 1981; Lacaille and Schwartzkroin, 1988b) or rapidly rising (interestingly never mixed) IPSPs after activation of other interneurons (Lacaille et al., 1987; Miles and Wong, 1984). Correspondingly different morphologies and electrical properties of the involved interneurons support the suggestion that different cells mediate $GABA_A$ and $GABA_B$ IPSPs. Drop application of 4-aminopyridine onto the distal dendritic field of CA1 neurons can elicit a $GABA_B$-mediated response without an accompanying $GABA_A$ response (Segal, 1987, 1990; see also Williams and Lacaille, 1992). Similar results are available in the lateral amygdala and ventral tegmental area (Sugita et al., 1992) and thalamus (Crunelli and Leresche, 1991; Steriade et al., 1985). A reasonable interpretation of these data is that there is a distinct class of GABA interneuron that is activated by the restricted 4-aminopyridine application and that produces only dendritic $GABA_B$ IPSPs.

An alternate theory that has been advanced (Isaacson et al., 1993) is that the slow IPSP represents the activation of diffusely distributed receptors by a "spill-over" phenomenon, based on the fact that higher levels of stimulation or other conditions favoring large amounts of transmitter release are often required to evoke a slow IPSP (Dutar and Nicoll, 1988; Otis and Mody, 1992b). This view would seem to ignore much of the evidence provided above. A more parsimonious explanation is that $GABA_B$ interneurons are a distinct class of inhibitory cell with a higher firing threshold. Future experiments are sure to resolve this controversy.

The physiological details of inhibitory circuits in neocortex are an area of even greater deficiency in our knowledge. However, very recent experiments may now shed some light on this issue. Benardo (1994a,b) used focal microapplications of glutamate to stimulate inhibitory neurons while recording from layer V neurons in rat somatosensory cortical slices. Both fast $GABA_A$ and slow $GABA_B$ IPSPs were generated, but they were observed as separate events, and mixed responses were never seen. Glutamate delivery often resulted in bursts of fast IPSPs, as shown in Fig. 10-3A, which could be resolved as single fast IPSPs. These fast IPSPs reversed at about −70 mV (Fig. 10-3B) and were blocked by picrotoxin. Slow IPSPs reversed at more negative potentials (−90 mV) and were blocked by saclofen. These findings suggest that the two distinct types of IPSPs resulting from GABA interaction at $GABA_A$ and $GABA_B$ receptors on neocortical neurons may be mediated by separate classes of inhibitory neurons, just as has been suggested to be the case in hippocampus.

PLASTICITY IN THE INHIBITORY CIRCUIT

Recent studies suggest that GABAergic inhibition in the hippocampus may show use-dependent plasticity. Specifically, experimental results suggest that repetitive afferent stimulation reduces the efficacy of GABAergic inhibition (Merlin and Wong, 1993; Miles and Wong, 1987b). Using the

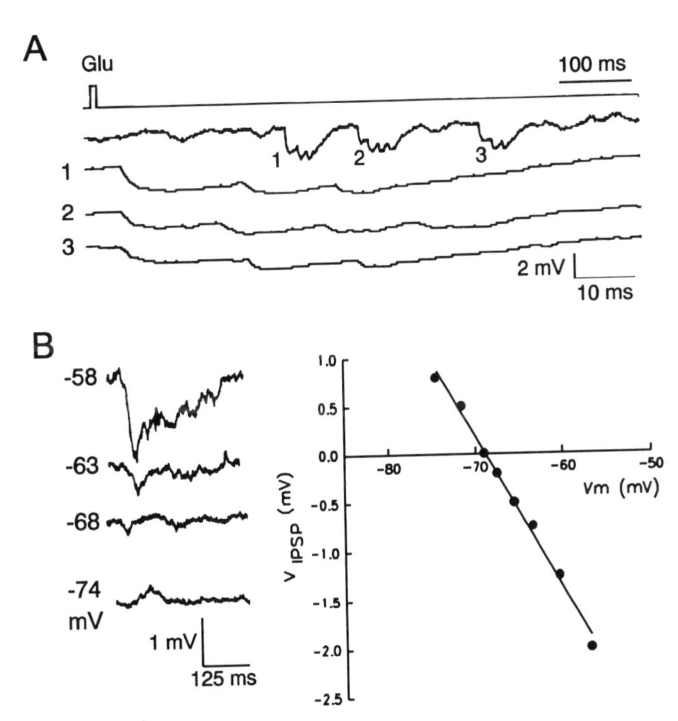

FIGURE 10-3. Fast IPSPs triggered by glutamate microapplication. **A**. Response of a layer V neuron to a 15-msec pulse of glutamate (5 mM). Lower traces are expanded sweeps of events designated 1 to 3 in slower sweep, and show events are composed of bursts of individual fast IPSPs. Membrane potential was maintained at −60 mV. **B**. Determination of reversal potential for glutamate-elicited fast IPSPs. Left trace, IPSPs triggered at different membrane potentials as indicated. Right trace, peak amplitude of individual IPSPs plotted against membrane (Vm). A straight line fitted by linear regression analysis indicates the IPSP reversal potential. These events were blocked by picrotoxin. Glutamate microapplications (5 mM) were made in the presence of the non-NMDA ionotropic glutamate receptor blocker CNQX (5 μM) and the NMDA blocker CPP (10 μM).

slice preparation, intracellular recordings from a pair of cells in the CA3 region of the hippocampus were obtained. These cells showed no synaptic interaction. Tetanic stimulation applied to the mossy fiber afferents to the region allowed the activities of one of the recorded cells to begin to produce polysynaptic EPSPs in the other cell. Additional data revealed that disynaptic IPSPs activated by one pyramidal cell to the other, via intercalating GABAergic neurons, were suppressed following tetani. The results demonstrated that, within a recurrently connected population, disinhibition caused previously non-communicating members to become connected by recurrent synapses via polysynatic pathways. Possible modifiable sites include (1) postsynaptic $GABA_A$ receptors, (2) the presynaptic input to the GABA interneurons, (3) interneurons themselves, and (4) the presynaptic terminals of GABAergic neurons.

The modifiability of postsynaptic $GABA_A$ receptors was examined using acutely dissociated cells obtained from the hippocampi of adult guinea pigs. During whole-cell patch-clamp recording experiments, $GABA_A$ responses progressively decreased in amplitude (run down) after breakthrough of the cell membrane by the recording pipette. The run-down was not caused by desensitization. When ATP and Mg^{2+} together with the "fast" Ca^{2+} buffer, BAPTA, were included in the patch pipette solution, $GABA_A$ responses stabilized. The data suggest that the $GABA_A$ receptor must be phosphorylated to remain functional. The phosphorylation process, which confers stability to the functional state of the receptor, is counteracted by a dephosphorylation process activated by elevation of intracellular Ca^{2+} (Chen et al., 1990; Stelzer et al., 1988). At present, the identity of the kinase remains unclear. Additional experiments are needed to assess the physiological significance of the intracellular site regulating the $GABA_A$ response. Thus, this study emphasized modification at the postsynaptic site. Other possible plasticity mechanisms at the presynaptic level—interneurons (Miles, 1990) and presynaptic release (see the section on Recruitment Properties)—are plausible and are illustrated by other studies.

Other modifiable sites are suggested by recent findings regarding the interconnections between GABAergic neurons and the postsynaptic effects of GABA. As was indicated in Fig. 10-1, we know that GABAergic neurons are activated via feedforward and feedback mechanisms during normal activity. Excitation through these pathways is mediated by glutamatergic synapses. Recent studies in the cortex (Aram et al., 1991; Michelson and Wong, 1991) suggest that another mechanism, independent of glutamate transmission, may be be involved in the synchronous activation of inhibitory interneurons. This conclusion is based on studies performed in neocortical and hippocampal slice preparations maintained in the presence of the convulsant 4-aminopyridine (4-AP) and glutamate receptor blockers CNQX and CPP. Under this conditions, rhythmic, synchronized IPSPs can be recorded in pyramidal and granule cells. Synchonous IPSPs are larger than unitary ones (up to 15 mV in amplitude and 900 msec in duration) and occur at frequencies ranging from 0.1 to 0.3 Hz.

Intracellular recordings from GABAergic interneurons in the hippocampal hilar region revealed that synchronous burst firing occurred simultaneously with the synchronized IPSPs recorded in pyramidal cells. These bursts were sustained by large-amplitude synchronized EPSPs. Since the recordings were made in the presence of glutamate receptor antagonists, synchronized EPSPs in interneurons could not be accounted for by activation of glutamate ionotropic receptors. Instead the data showed that the application of $GABA_A$ receptor blockers (picrotoxin or bicuculline), blocked the synchronized EPSPs (Michelson and Wong, 1991).

These results identified surprising features of the GABAergic inhibitory system. They suggest that GABA can function as an excitatory neurotransmitter onto inhibitory interneurons; that is, recurrent connections between GABAergic neurons can cause excitation of interneurons. Through recurrent GABAergic synapses, synchronized firing of the GABAergic interneurons can be activated much the same way as for pyramidal cells. The consequence of the excitatory coupling between GABAergic cells is an augmented output from interneuron aggregates, giving rise to synchonized IPSPs in pyramidal cells. In this way the excitatory action of GABA onto interneurons remains consistent with the role of GABA as an inhibitory transmitter.

Thus, in an attempt to identify modifiable sites in the GABAergic inhibitory pathway, we realize that there is much more to learn about GABAergic inhibitory neurons and their circuit organization. Continued work at this fundamental level is necessary if we are to understand the role of inhibition in epileptogenesis.

11

Thalamocortical interactions

JOHN R. HUGUENARD, DAVID A. MCCORMICK,
and DOUGLAS COULTER

All sensory information, except for olfaction, passes through the synaptic network of the thalamus just before reaching the cerebral cortex, suggesting that the thalamus may exert a powerful influence on sensory and motor processing. However, the connection is not unidirectional: layer VI of the cerebral cortex contributes up to 50% of the synapses on thalamic relay neurons, indicating that thalamus and cerebral cortex are intimately associated in some form of reciprocal loop. Many previous studies have shown the thalamus to be critically involved in a wide variety of phenomena, including the generation of the electroencephalogram (EEG), the blocking of transmission of sensory information during slow-wave sleep, and the generation of generalized seizures (reviewed in Andersen and Andersson, 1968; Avoli et al., 1990a,b; McCormick, 1992; Steriade and Deschenes, 1984; Steriade et al., 1990a,b, 1993a,b). Since these phenomena occur within the same synaptic network (thalamocortical interactions), they exhibit strong interrelations and interdependency. For example, the transition from waking to slow wave sleep in a child with absence (petit mal) epilepsy is associated with a decrease in responsiveness to sensory stimulation, the appearance of spindle oscillations in the EEG, and a markedly enhanced rate of occurrence of absence seizures (Kellaway, 1985; Steriade and McCarley, 1990; see Avoli et al., 1990b). In this chapter we will examine, at the subcellular, cellular, and network levels, the biophysical mechanisms for all three of these and related phenomena.

UNIQUE PROPERTIES OF THALAMIC NEURONS

Relay Cells

Thalamic relay neurons have an unusual ability to respond to stimuli with two dramatically different types of output, depending on the level of behavioral awareness (see below). Intracellular studies *in vivo* (Deschenes et al., 1982) and *in vitro* (Llinás and Jahnsen, 1982) in cat and guinea pig thalamic neurons have demonstrated how the outputs of these cells can be switched from one mode to another by manipulation of the resting membrane potential. One output type, the so-called *relay mode*, occurs at relatively depolarized resting potentials (positive to ~ −60 mV) and is characterized by a relatively linear input-output relationship. In this mode, relay neurons respond to increasing stimulation intensities with higher frequencies of action potentials in the range of 1 to 200 Hz and higher (Jahnsen and Llinás, 1984a; Fig. 11-1, A1 and B1). Furthermore, there is little or no decrement in the output during sustained inputs, i.e. there is little adaptation of firing frequency (Fig. 11-1B3). As discussed in more detail below, this mode dominates during periods of arousal, in which relay cells act as faithful followers of synaptic input.

In marked contrast to the relay mode is the *burst-firing* mode, which occurs when the membrane potential is negative to −60 mV. In the burst firing mode, the same inputs that produce regular, repetitive firing in relay mode instead result in brief, high frequency (~300 to 500 Hz) bursts of action potentials (Fig. 11-1, A2 and B1). Several notable differences are apparent in the burst-firing mode (in comparison with the relay mode), with the net result being a non-linear, non-coherent input-output relationship. It is non-linear in that the number and frequency of action potentials produced in this mode are only weakly dependent on stimulus strength (Fig. 11-1, B1 and B2) or duration; it is non-coherent because the output is phasic with a brief bursts of spikes followed by a silent period (that can be overcome only with high intensity or prolonged stimulation; Jahnsen and Llinás, 1984a). To summarize, in relay mode output closely follows input in terms of strength and timing, whereas in burst-firing mode output is only weakly dependent on input strength and is also highly dependent on recent history—in the immediate post-burst period relay cells are largely unresponsive to input (see McCormick and Feeser, 1990). This non-linear output mode of relay cells is thought to interfere with the coherent flow of sensory information to the cortical level during periods of drowsiness and delta wave sleep, thereby inhibiting arousal (reviewed by Steriade and Llinás, 1988).

The ionic mechanisms underlying burst firing in thalamic neurons have been extensively studied. Llinás and Jahnsen (1982) demonstrated that

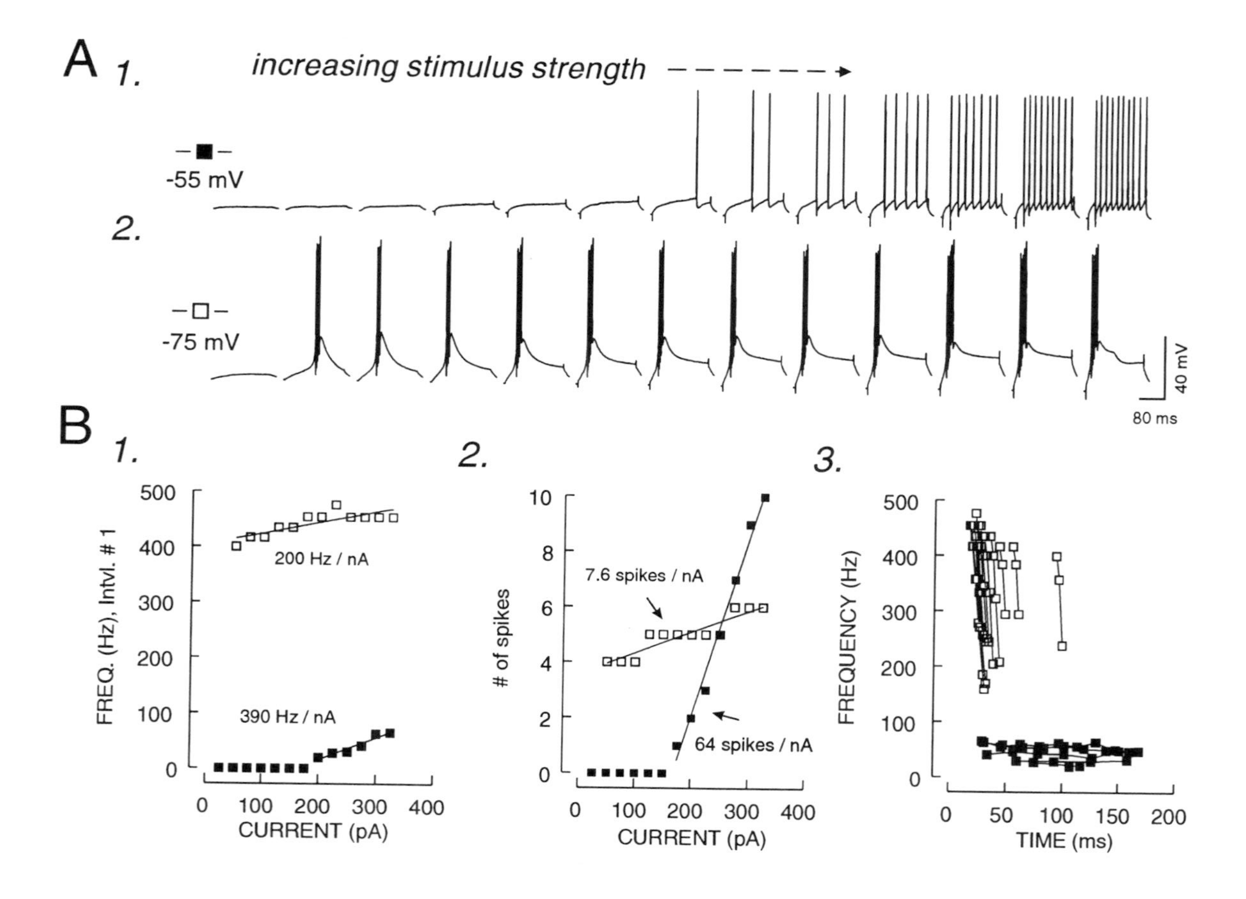
A
1.
increasing stimulus strength
-55 mV
2.
-75 mV
40 mV
80 ms
B
1.
FREQ. (Hz), Intvl. # 1
200 Hz / nA
390 Hz / nA
CURRENT (pA)
2.
# of spikes
7.6 spikes / nA
64 spikes / nA
CURRENT (pA)
3.
FREQUENCY (Hz)
TIME (ms)

block of Na^{+}-dependent spikes with tetrodotoxin revealed a slow (20 to 30 msec), small (5 to 15 mV) depolarization with a similar time course to the burst response. This low-threshold spike (LTS) was Ca^{2+}-dependent; it was reduced in the presence of lowered extracellular Ca^{2+}, and was blocked by Co^{2+}, Ca^{2+}, or Cd^{2+}(Jahnsen and Llinás, 1984b). This LTS was thus similar to that seen in inferior olivary neurons (Llinás and Yarom, 1981). Recent studies (Coulter et al., 1989b; Crunelli et al., 1989; Hernandez-Cruz and Pape, 1989; Suzuki and Rogawski, 1989) have demonstrated that a specific Ca^{2+} current, known as the low-voltage activated (LVA) or transient (T) current, is present in thalamic neurons in relatively high density (Fig. 11-2). Owing to a strong T current, low-threshold Ca^{2+}-dependent bursts are much more common in thalamic cells than in some other types of neurons, for example, most neocortical or hippocampal cells.

In the process of LTS generation, T current channels first activate (open) and then close to an inactivated state. There are important functional consequences related to reversal of inactivation (deinactivation). For example, since deinactivation is time and voltage dependent (Coulter et al., 1989b;

FIGURE 11-1. Firing patterns in a typical thalamocortical relay neuron. **A.** Voltage responses in a rat ventrobasal (VB) neuron held at two resting levels by constant current injection with whole-cell recording methods. **A1.** V_m −55 mV. As the depolarizing current injection magnitude is gradually increased, a threshold is reached at 200 pA and one spike is produced. With further depolarization increasing numbers of spikes are obtained. There is little adaption during the current pulse (see **B3**), and spike output is obtained throughout the period of stimulation. **A2.** With the V_m at −75 mV the current threshold is now lower (50 vs. 200 pA), and a burst discharge with four spikes is obtained. Increasing the stimulus beyond this level alters the latency of the burst, but the basic morphology of the response is relatively insensitive to current intensity. For example, currents of 50, 75, and 100 all elicit a burst with four spikes (see **B2**), and a maximum of six spikes is obtained with 325 pA. After the burst, there is complete cessation of firing, even in the presence of maintained depolarization; this can only be overcome with very high stimulus intensities. **B1.** Firing frequency (measured from the first interspike interval) as a function of current injection. Frequencies are very high (> 400 Hz) during a threshold burst from a Vm of −75 mV (□) and increase toward 500 Hz with stronger depolarizations. With a V_m of −55 mV (■) the threshold frequency was <20 Hz and increased linearly with stimulus strength (390 Hz/nA). **B2.** Input/output functions at the two holding potentials in A. With a Vm of −55 mV the total number of spikes per depolarization was linear and steep (64 spikes/nA); however, when the V_m was hyperpolarized to −75 mV the slope was much shallower (7.6 spikes/nA). **B3.** Spike frequency versus time in the two firing modes. In burst-firing mode (V_m = −75) the frequency is initially high and slows during the burst. In contrast, there is little or no adaptation of firing frequency when V_m = 55 mV.

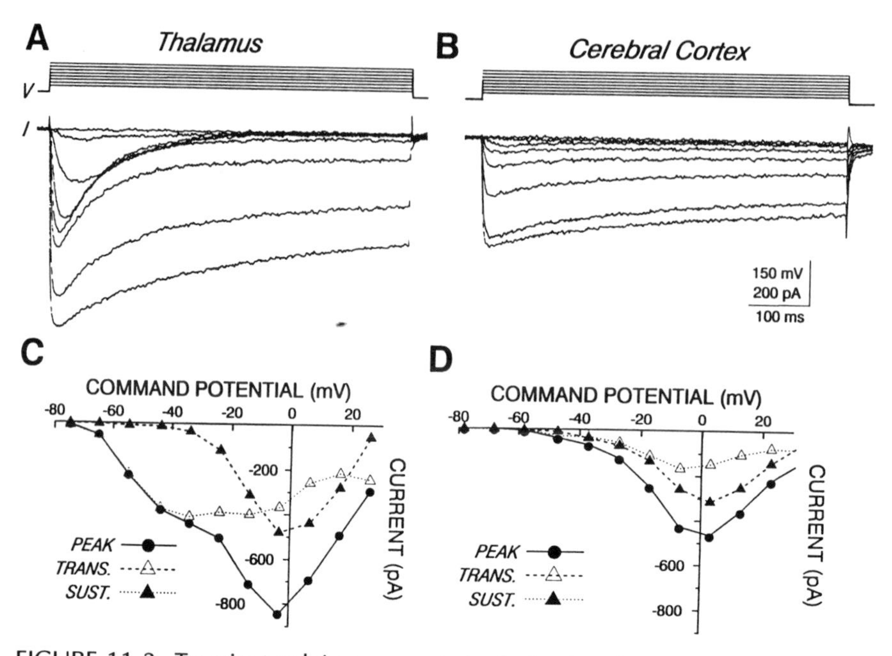

FIGURE 11-2. Transient calcium currents in thalamic relay neurons: comparison to another cell type. Both recordings were obtained from acutely isolated neurons from 8 to 10-day-old rats. **A.** Voltage clamp steps in the range of −80 to + 30 mV evoke a family of currents in a relay neuron of the thalamus. The "threshold" current is transient and inactivates in the presence of maintained depolarization. Larger step depolarizations evoke currents that are largely noninactivating. **B.** Currents in a neocortical pyramidal cell. The threshold current is small, and there is not a prominent transient component. **C.** I-V plots for the various components of Ca current in the thalamic cell. *Trans.* refers to the inactivating component of the current, i.e. the difference between the maximal inward current and the current level at the end of the stimulus. *Sust.* refers to that current still present at the termination of the stimulus. Thalamic cells have a large transient current (Δ) that activates near −60 mV, reaches peak amplitude near −35 mV, and becomes smaller with larger depolarizations. The sustained current activates near −40 mV and peaks near 0 mV. Because thalamic cells express a relatively high level of I_T, the ratio of peak transient to peak sustained current is high (average ~ 1.2; see Coulter et al., 1989b). In this particular cell it is 410:480 = 0.85. **D.** The neocortical transient current is smaller than in thalamic cells and peaks at a higher voltage (−5 mV). The sustained current is similar to that seen in the thalamus. The ratio of transient:sustained currents (measured at the same voltages as in part **C**) in this cell is 50:310 = 0.16.

Jahnsen and Llinás, 1984a), there is a limited frequency range at which repetitive LTSs can be generated (up to about 10 to 14 Hz; see McCormick and Feeser, 1990)—hyperpolarization of adequate duration and magnitude is required to remove inactivation of enough channels to produce a regenerative LTS. Also, at normal resting potentials near −60 mV, almost all T channels are in the inactivated state, and thus LTS generation is blocked. It is only with prolonged (> 80 msec) hyperpolarizing inhibitory potentials or with sustained changes in resting potential (see below) that an LTS can be evoked.

An interesting variant of burst-firing behavior that occurs in certain classes of relay cells (dorsal lateral geniculate, medial geniculate, parafascicular, reticular and others) is repetitive burst firing (Fig. 11-3; McCormick and Pape, 1990; McCormick and Prince, 1988; McCormick and Wang, 1991; Von Krosigk et al., 1993). This behavior occurs in individual cells when conditions are optimum to promote oscillations in the membrane potential through interaction of two or more voltage-gated ionic conductances. A basic requirement for such oscillations is the presence of time-dependent ionic conductances that are turned on and off in a reciprocal fashion by the membrane potential (or with time). In thalamic cells this can occur through an interaction of I_T and a hyperpolarization-activated cation current known as I_H (Pape and McCormick, 1989). At normal resting membrane potentials there is a steady, but only partial, activation of I_H (which exerts a depolarizing influence on the membrane since the reversal potential is positive to rest). Activation of an LTS *deactivates* I_H so that upon termination of the burst there is less depolarizing influence than before, and the membrane hyperpolarizes (Fig. 11-3). This hyperpolarization removes inactivation of I_T and gradually turns on I_H, which then depolarizes the membrane until LTS threshold is reached and the cycle repeats itself (Fig. 11-3).

GABAergic Neurons

Within the thalamus there are two main types of GABAergic neurons: those with processes that remain within the nucleus in which they reside (interneurons) and those of the nucleus reticularis or perigeniculate nucleus (nRt/PGN). The electrophysiological properties of local circuit inhibitory interneurons (Pape and McCormick, 1989) in the thalamus are similar to those in hippocampus (Schwartzkroin and Mathers, 1978) and neocortex (McCormick et al. 1985) and are different from principal cells, in that they have relatively brief spikes (usually less than 1 msec) and high-frequency output capability. There is a notable lack of LTS activity in thalamic interneurons (Pape and McCormick, 1989). On the other hand, relay

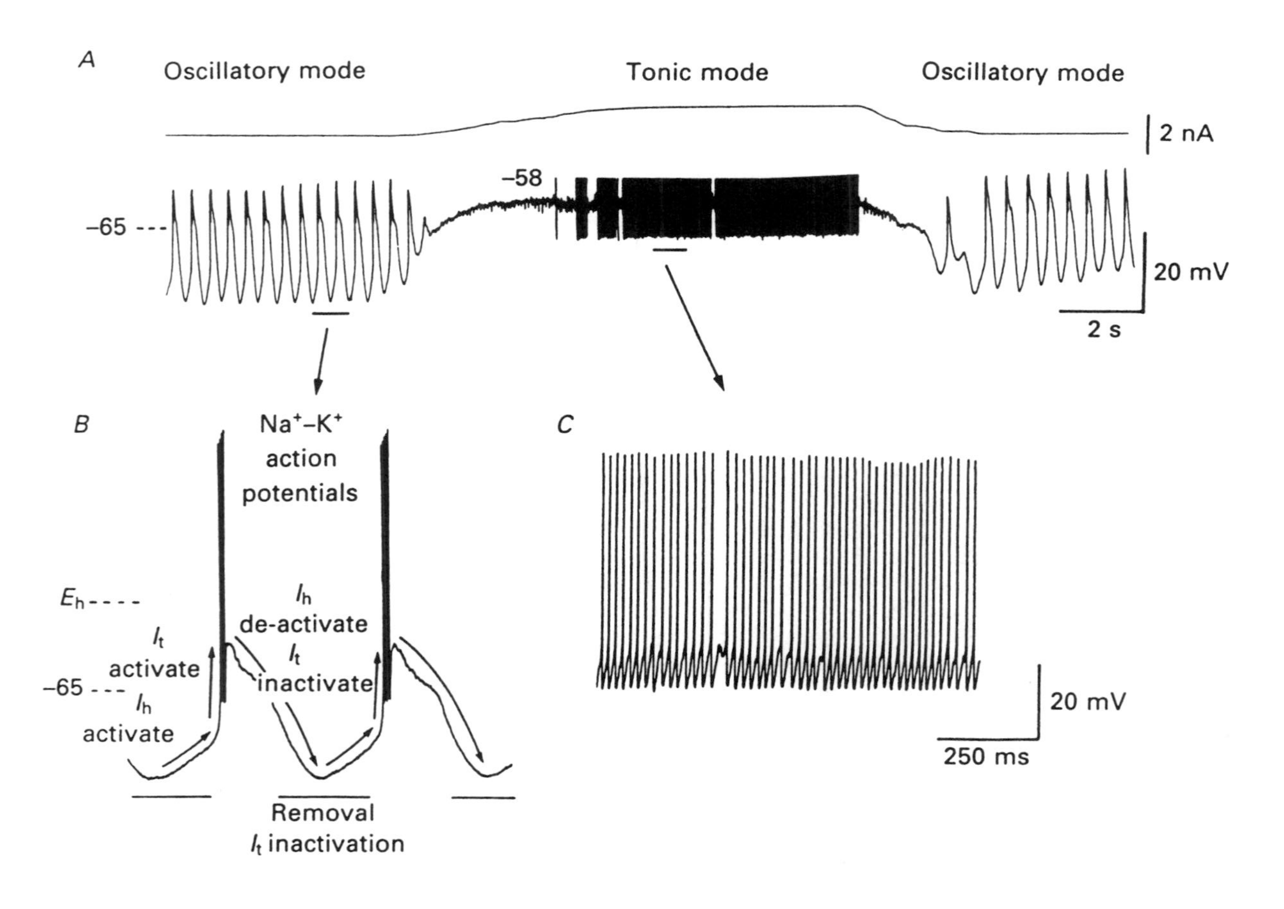

A
Oscillatory mode
Tonic mode
Oscillatory mode
2 nA
−58
−65
20 mV
2 s
B
Na⁺–K⁺
action
potentials
E_h
I_t
activate
−65
I_h
activate
I_h
de-activate
I_t
inactivate
Removal
I_t inactivation
C
20 mV
250 ms

neurons also receive recurrent feedback inhibition from GABAergic nucleus reticularis thalami (nRt) and the perigeniculate nucleus (PGN) (the geniculate structure analogous to nRt), and the neurons in these extrinsic inhibitory nuclei exhibit prominent Ca^{2+}-dependent burst firing similar to that seen in relay nuclei (Avanzini et al., 1989; Huguenard and Prince, 1992a; Mulle et al., 1986). In general, the LTSs in nRt and PGN neurons tend to be longer in duration (Huguenard and Prince, 1992a; Mulle et al., 1986) with higher intraburst spike frequency than their counterpart relay neurons. This difference is caused in part by biophysical differences in the T-type channels in relay neurons versus inhibitory neurons (Huguenard and Prince, 1992a).

NETWORKS UNDERLYING RHYTHMS

Two types of network oscillations involving the thalamus have been described and well studied: spindle waves and absence seizures. These network oscillations involve not only the interplay between ionic currents intrinsic to each cell type but also the interactions of each of the different cell types. Thalamic relay cells and the GABAergic neurons of the nRt/PGN play a key role in the regulation of thalamic activity and the generation of spindle waves (see review by Steriade and Llinás, 1990; von Krosigk et al., 1993), whereas interactions between thalamus and cerebral cortex, as well as the activation of nRt/PGN cells, become particularly important in the generation of absence seizures (see review by Avoli et al., 1990).

FIGURE 11-3. Two different firing modes of thalamic relay neurons and the proposed ionic substrate of rhythmic burst firing. **A.** This cat LGNd neuron generated rhythmic burst firing at a rate of about 2 Hz. Depolarization of the cell to −58 mV with the intracellular injection of current (top trace) halted the rhythmic activity and switched the neuron to the tonic, or single-spike, mode of action potential generation. Removal of the depolarization reinstated the oscillatory activity. **B.** Expanded trace of oscillatory activity and the proposed currents that largely mediate it. Activation of the low-threshold calcium current, I_t, depolarizes the membrane toward threshold for a burst of Na^+- and K^+-dependent fast action potentials. The depolarization deactivates the portion of I_h that was active immediately before the Ca^{2+} spike. Repolarization of the membrane due to I_t inactivation is followed by a hyperpolarizing overshoot due to the reduced depolarizing effect of I_h. The hyperpolarization in turn deinactivates I_t and activates I_h, which depolarize the membrane toward threshold for another Ca^{2+} spike. **C.** Expanded trace of single-spike activity (from McCormick and Pape, 1990a).

The synaptic organization of the thalamus is such that each dorsal thalamic nucleus sends collaterals to a specific area of the nRt. This topographic mapping results in auditory, visual, and somatosensory sections within nRt that have receptive field properties appropriate for their sensory modalities (reviewed in Jones, 1985). The inhibitory output from nRt feeds back largely onto the same nucleus in a reciprocal manner, although there is anatomical evidence for some divergence in nRt axons (Jones, 1985; Mulle et al., 1986). To the extent that there is also divergence in thalamocortical recurrent collaterals within nRt, then there is a physiological basis for substantial long-loop recurrent circuitry that may support synchronized intrathalamic oscillations, such as those occurring during sleep spindles and certain forms of epilepsy (see below). The inhibitory output from nRt can be quite powerful, with both $GABA_A$ and $GABA_B$ components, and can be of relatively long duration (up to 300 to 400 msec), such that rebound LTSs are generated in relay neurons (Fig. 11-4, Hu-

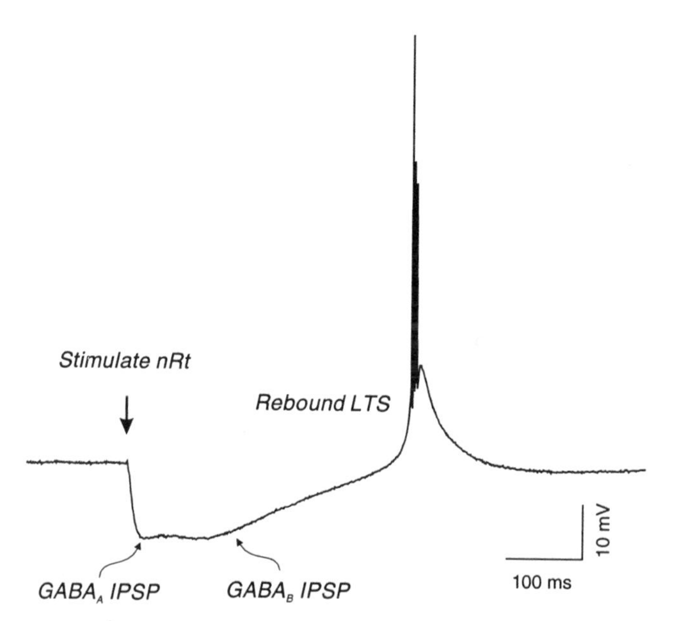

FIGURE 11-4. Activation of nRt results in strong inhibitory synaptic responses in relay neurons. Intracellular whole-cell recording in VB. nRt was activated through extracellular stimulation with a bipolar tungsten-stimulating electrode (250 μA 50 μsec), which resulted in a long-lasting hyperpolarization with an early Cl-dependent, bicuculline-sensitive $GABA_A$ component and a late K^+-dependent, CGP-35348 sensitive $GABA_B$ component. The inhibitory potential was of long enough duration to deinactivate I_T channels such that a rebound low-threshold spike (LTS) was obtained. Since VB output projects back to nRt, this circuit then is the basis for an intrathalamic oscillatory network.

guenard and Prince, 1994a). Of course, since the excitatory output of relay neurons is reflected back, then recurrent excitation of nRt can generate another cycle of the intrathalamic oscillation.

Based upon these reciprocal connections between nRt/PGN and relay cells, the generation of spindle waves has been proposed to follow this scenario: activation of a critical number of relay cells initiates the spindle wave by exciting a critical number of nRt/PGN cells (nRt/PGN cells may also initiate the spindle). The activation of these GABAergic neurons results in the generation of mainly $GABA_A$ receptor-mediated IPSPs in the relay cells. A subset of relay cells then generate a rebound low-threshold Ca^{2+} spike and associated burst discharge following these IPSPs, thereby once again exciting the nRt/PGN neurons and starting another cycle (von Krosigk et al., 1993). The growth of the spindle wave ("waxing") is believed to result from the recruitment of cells into the oscillation. The mechanisms of the ending, or dampening, of the spindle wave ("waning") are less clear, but may be due in part to the hyperpolarization of nRt/PGN cells below the threshold for generation of low-threshold Ca^{2+} spikes (von Krosigk et al., 1993). Through these mechanisms, the resting, or deactivated, thalamus generates periodic synchronized oscillations (spindle waves). Upon activation by ascending transmitter systems, these oscillations disappear and are replaced by a more tonic pattern of action potential generation.

STATE DEPENDENCE OF THALAMIC ACTIVITY

Thalamic activity is state dependent. During slow-wave sleep, thalamic relay and nRt neurons are hyperpolarized and poised to generate rhythmic burst discharges either as spindle waves (Steriade and Deschênes, 1984; Steriade et al., 1990) or perhaps in the frequency range of delta waves (McCormick and Pape, 1990; Steriade et al., 1991). The transition to the awake, attentive state is associated with depolarization of thalamic relay and nRt neurons, a subsequent abolition of rhythmic burst discharges, and the appearance of more tonic, single-spike activity, the pattern of which is related to sensorimotor processing (Hirsch et al., 1983; Livingstone and Hubel, 1984; reviewed in McCormick, 1992; Steriade and McCarley, 1990). This depolarization of thalamic neurons arises from the release of neurotransmitters of extrathalamic origin, particularly from the brainstem and cerebral cortex. The portion of the thalamus that relays visual information from the retina, the dorsal lateral geniculate nucleus (LGNd), receives presumed glutamatergic innervation from the retina and layer VI of the cerebral cortex, cholinergic innervation from the pedunculopontine tegmental nucleus, noradrenergic innervation from the locus ceruleus, serotoninergic innervation from the median raphe, histaminergic innervation from the hypothalamus, and additional brainstem innervation that uses an as-of-yet unidentified neurotransmitter (reviewed in McCormick, 1992; Steriade and McCarley, 1990).

Extracellular recordings from nearly all of these extraretinal inputs to the LGNd reveal a general increase in activity, and therefore presumably an increase in neurotransmitter release, in anticipation of awakening from sleep or during arousal (Aston-Jones and Bloom, 1981; Steriade et al., 1990; Vanni-Mercier et al., 1984). The postsynaptic actions of these various transmitter systems have been detailed on the cellular level using both *in vivo* and *in vitro* techniques. Activation of muscarinic, α_1-adrenergic, H_1-histaminergic, and glutamate metabotropic receptors on thalamic relay cells results in depolarization and a subsequent change from rhythmic burst firing to tonic, single-spike activity through the reduction of a "leak" K^+ current termed I_{KL} (Fig. 11-5; McCormick, 1992; McCormick and Williamson, 1991; McCormick and von Krosigk, 1992). We assume that it is the reduction of I_{KL} that underlies the depolarization seen in thalamic relay cells in the transition from sleep to waking or REM sleep, although an additional cellular mechanism may occur through changes in the voltage dependence of the hyperpolarization-activated cation current I_h, mediated by β-adrenergic, serotoninergic, H_2-histaminergic, and purinergic receptors (McCormick and Pape, 1990b; McCormick and Williamson, 1991; Pape, 1992).

FIGURE 11-5. Thalamic relay cells and nRt cells can generate action potentials either as rhythmic bursts or as tonic, single-spike activity, depending upon the membrane potential of the cell. Activation of muscarinic, α_1-adrenergic, H_1-histaminergic, or glutamate metabotropic receptors results in a depolarization of LGNd relay neurons through reduction of I_{KL} **(D)**. This depolarization subsequently shifts these neurons to the single-spike mode of action potential generation **(D)**. Similarly, activation of α_1-adrenoceptors, 5-HT_2 receptors, H_1-receptors, or glutamate metabotropic receptors has similar effects in the nucleus reticularis thalami **(E)**. In contrast, activation of muscarinic receptors in this nucleus **(C)** or in local GABAergic interneurons **(F)** results in the inhibition of axon output through an increase in potassium conductance (I_{KG}). In the cerebral cortex, activation of muscarinic, α_1-adrenergic, or glutamate metabotropic receptors results in the abolition of rhythmic burst firing in layer V burst-generating neurons and a switch to the tonic, single-spike mode of action potential generation **(A)**. In regular spiking cells, in contrast, activation of muscarinic, β-adrenergic, H_2-histaminergic, serotoninergic, and perhaps glutamate metabotropic receptors results in a decrease in spike frequency adaptation by blocking I_{AHP} (and I_M for ACh and 5-HT) **(B)**. Not illustrated are the inhibitory effects of adenosine and serotonin. These responses allow ascending modulatory transmitter systems to prepare thalamocortical systems for sensory transmission, processing, and cognition.

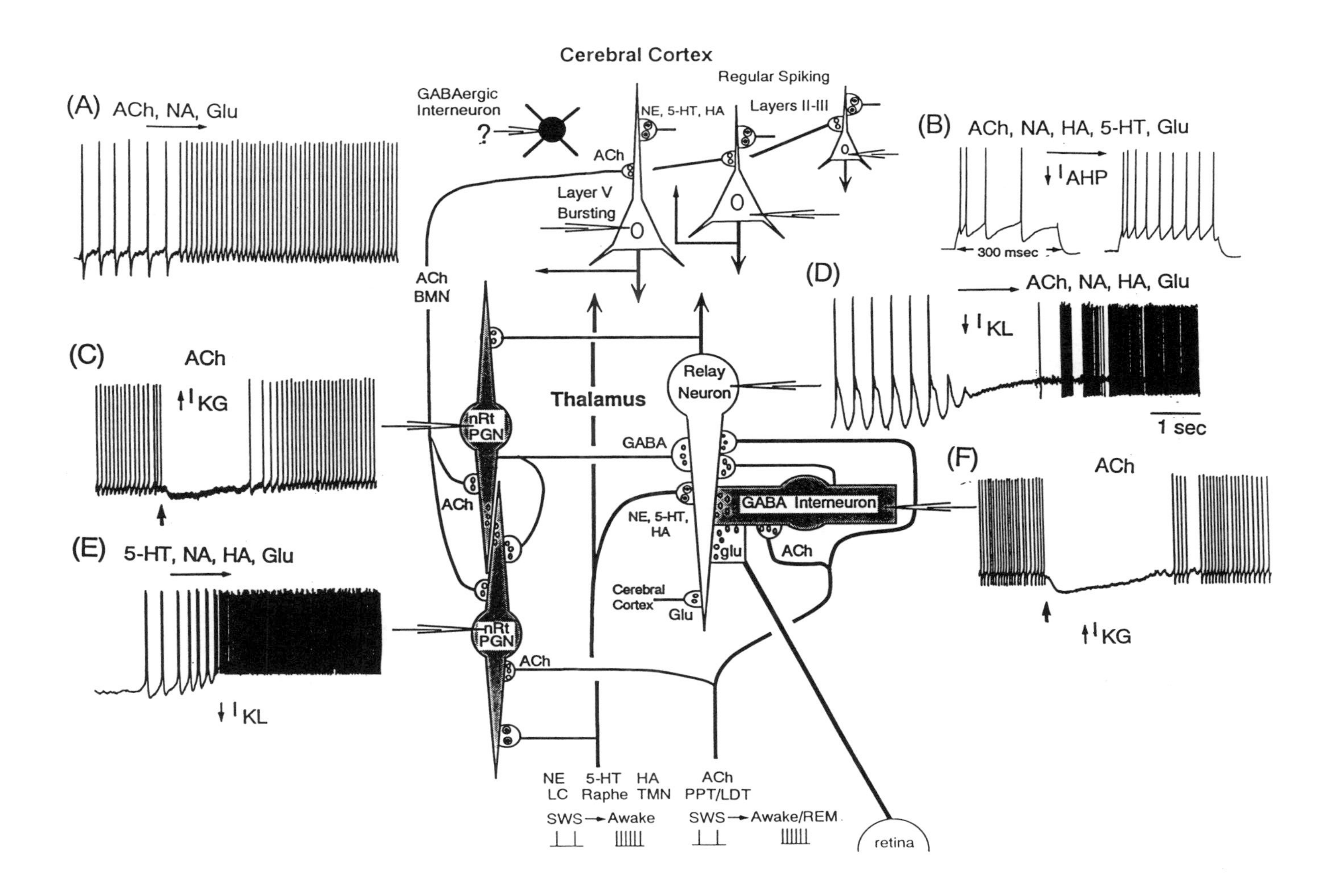
Cerebral Cortex
GABAergic Interneuron
?
NE, 5-HT, HA
Regular Spiking
Layers II-III
ACh
Layer V
Bursting
(A) ACh, NA, Glu
(B) ACh, NA, HA, 5-HT, Glu
$\downarrow I_{AHP}$
300 msec
ACh
BMN
(C) ACh
$\uparrow I_{KG}$
Thalamus
Relay Neuron
(D) ACh, NA, HA, Glu
$\downarrow I_{KL}$
1 sec
nRt
PGN
GABA
ACh
(F) ACh
GABA Interneuron
NE, 5-HT, HA
(E) 5-HT, NA, HA, Glu
glu
ACh
Cerebral Cortex
Glu
$\uparrow I_{KG}$
nRt
PGN
ACh
$\downarrow I_{KL}$
NE LC
5-HT Raphe
HA TMN
ACh PPT/LDT
SWS → Awake
SWS → Awake/REM
retina

The key event in the transition from sleep to waking/REM sleep is the inactivation of the low-threshold Ca^{2+} current (T current) through maintained depolarization of the membrane potential (Fig. 11-5D,E). However, another important phenomenon observed *in vivo* is the reduction of long-duration IPSPs in thalamic relay neurons (Steriade and Deschênes, 1988). Examination of the pharmacological properties of the two major classes of GABAergic neurons in the thalamus—the nRt and local interneurons—reveals that the action potential output of both is inhibited by acetylcholine (Fig. 11-5C,F; McCormick and Pape, 1988; McCormick and Prince, 1986). Another possible mechanism of reduction of long-duration IPSPs in relay cells is the inhibition of burst discharges in nRt neurons through the inactivation of the low-threshold Ca^{2+} current through tonic depolarization (Fig. 11-5E; Steriade et al., 1986; von Krosigk et al., 1993).

IMPLICATIONS FOR THE TRANSMISSION OF SENSORY INFORMATION

Examination of the transmission of synaptic potentials from the retina to the visual cortex through the lateral geniculate nucleus reveals that slow-wave sleep or drowsiness is associated with a markedly reduced faithfulness of transmission (Coenen and Vendrik, 1972; Livingstone and Hubel, 1984). There are at least two cellular mechanisms underlying this degradation of sensory transmission: the lack of responsiveness of relay neurons to EPSPs during periods of rhythmic oscillations and the inhibition of sensory transmission by large and long-duration IPSPs. During periods of rhythmic oscillations, the membrane potential of relay neurons is relatively hyperpolarized, being approximately 10 to 25 mV from the fast Na^{+}/K^{+} spike threshold (McCormick and Pape, 1990a; Steriade et al., 1990b; von Krosigk et al., 1993). Only the largest EPSPs depolarize the membrane potential sufficiently to overcome this hyperpolarized state. In addition, the rhythmic arrival of IPSPs and the presence of intrathalamic or thalamocortical rhythms, (which occur in relation to cellular anatomy and physiology without regard for sensory input), act to "scramble" or degrade the transmission of sensory and motor information through the thalamus. In this way, the thalamus may act as a variable access or gate to the cerebral cortex, as has long been suggested.

THE THALAMOCORTICAL SYSTEM AND EPILEPSY

The Spike-Wave Discharges of Typical Generalized Absence Epilepsy

There is an emerging consensus concerning the intrinsic membrane properties, anatomical circuits, and synaptic mechanisms important in the generation of thalamocortical rhythmicity. As discussed above, these mechanisms involve interactions between neurons within the thalamic relay nuclei and the surrounding GABAergic nucleus reticularis thalami (nRt). Data accumulated from investigations of the best-studied thalamocortical rhythm, spindle discharges, have established that cortical activity is secondary to this interaction, since spindle discharges can be recorded virtually unchanged in thalamus in the absence of cortex or with the cortex functionally inactivated (reviewed in Steriade and Llinás, 1988).

A pathological variant of thalamocortical rhythm is the bilaterally synchronous 3-Hz spike-wave discharges of generalized absence epilepsy. Like spindles, these rhythms involve the synchronous activation of the thalamus and cortex, shown by depth recordings of spike-wave activity in absence patients (Williams, 1953). The T-type calcium current, which generates low-threshold calcium spikes in thalamic neurons (Coulter et al., 1989c), is essentially ubiquitous in thalamic neurons, and virtually defines their intrinsic membrane responsiveness (Jahnsen and Llinás, 1984), has been shown to play a critical role in the generation of spindle discharges (reviewed in Steriade and Llinás, 1988). Recent studies of the anticonvulsant sensitivity of T current in thalamic neurons have shown that the generalized absence anticonvulsants—ethosuximide, dimethadione, and methsuximide—all reduce the T-type calcium current in thalamic relay neurons and that this reduction occurs in concentration ranges overlapping those achieved in free serum levels in patients medicated with these anticonvulsants (Coulter et al., 1989a,b; 1990; Pellock and Coulter, 1994). This type of reduction of T current was not found with the application of anticonvulsants ineffective in the control of generalized absence, nor was it seen for succinimide and tetramethylsuccinimide, structural analogs of ethosuximide that exhibit no activity against absence or are convulsants, respectively. Similar reduction of T current by ethosuximide has been seen in other cell types in culture (Kostyuk et al., 1992, Macdonald and Meldrum, 1989), and there has also been a report that the broad-spectrum anticonvulsant, valproic acid, which is effective in the control of absence and other generalized seizures, blocks T current in cultured neurons (Kelly et al., 1990).

In addition to spindles and spike-wave discharges sharing thalamocortical localization and a dependence on low-threshold calcium conductance within the thalamus, both spindles and spike-wave discharges occur during similar levels of wakefulness. Both of these types of thalamocortical discharges occur primarily during periods of quiet wakefulness, drowsiness, or non-REM sleep, but are reduced or abolished by arousing stimuli or increased levels of alertness or activity in humans (Kellaway, 1985), and in animal models of generalized absence (Lannes et al., 1988). This kind of transition has been hypothesized as being due to activation of ascending activating systems, with associated depolarization of the membrane potential of thalamic relay and nRt neurons and consequent inactivation of the low-threshold calcium current. These actions result in a switch in the thalamocortical system from the phasic, oscillatory mode to the tonic, relay mode, as has been described above (Fig. 11-5).

Our understanding of the mechanisms that give rise to generalized thalamocortical spike-wave discharges of generalized absence epilepsy is based on the assumption that similar mechanisms are involved in both spindle and spike-wave discharge generation. Spindle discharges are hypothesized to be synchronized by large IPSPs directed onto thalamic neurons by the GABAergic nRt. These IPSPs hyperpolarize the thalamic neurons, and deinactivate the low-threshold calcium current, which then can trigger a regenerative low-threshold calcium spike on the depolarizing ramp. This spike occurs as the IPSP decays (Fig. 11-4), reinforcing the next cycle of the rhythm. One consequence of this anomalous "excitatory" action of the inhibitory transmitter GABA in triggering and synchronizing thalamocortical rhythms is that agents that act as GABAergic agonists or that potentiate GABAergic function might paradoxically increase oscillatory synchronization in epileptic discharges that activate and depend on thalamocortical synchronizing mechanisms for expression. This type of anomalous exacerbation of seizure discharges by GABAergic agonists or potentiating agents is evident in generalized absence epilepsy and in animal models of this disorder. Anticonvulsant barbiturates exacerbate generalized absence and also are known to potentiate GABAergic transmission as a primary mechanism of action. Thalamically infused GABAergic agonists—both $GABA_A$ and $GABA_B$ agonists—bring on or potentiate ongoing spike-wave discharges (Hosford et al., 1992; Liu et al., 1992; Vergnes et al., 1984). Conversely, thalamically infused GABAergic antagonists block or reduce ongoing spike-wave discharges. Once again, this is true for either $GABA_A$ or $GABA_B$ antagonists.

Although it seems likely that the intrinsic, synaptic, and anatomical rhythm-generating mechanisms of the nRt-relay nuclei system are intrinsically involved in the generation of spike-wave discharge and spindles, there are important distinctions between the two types of rhythms. These distinctions are evident in studies in the two best-characterized animal models of generalized absence—feline generalized penicillin epilepsy (FGPE), and the genetic absence epilepsy rat from Strasbourg (GAERs). In these models, bilaterally synchronous spike-wave discharges are recorded in the EEG, and these discharges are shown to be generated by rhythmic thalamocortical discharges in depth electrode recordings (Avoli and Gloor, 1982; Avoli et al., 1983; Vergnes et al., 1988). In addition to resembling spike-wave discharges electroencephalographically, these discharges respond to anticonvulsant drugs with a similar sensitivity to typical generalized absence (Guberman et al., 1975; Marescaux et al., 1984; Pellegrini et al., 1978). In addition, during electroencephalographically recorded spike-wave discharges, the animals behave in a manner very similar to that seen in petit mal patients, with loss of consciousness but few other behavioral manifestations of the profound pathological synchronized activity seen in the EEG. As might be expected for a rhythm generated primarily in the thalamic/nRt network, lesioning or inactivation of one thalamic hemisphere blocks spike-wave discharges unilaterally in both FPGE (Avoli and Gloor, 1981) and GAERs (Vergnes et al., 1988). However, unilateral cortical inactivation or lesioning was also found to block thalamically recorded spike-wave discharges in both FPGE (Avoli and Gloor, 1982) and GAERs (Vergnes and Marescaux, 1992; Vergnes et al., 1988), suggesting a cortical dependence for spike-wave discharges generation that is present for spindle-generation mechanisms. Gloor and colleagues have hypothesized that the low levels of penicillin present in the brain during FPGE elicit a diffuse cortical hyperexcitability, which recruits additional intracortical recurrent inhibition, leading to drop-out of every second spindle wave, and causing a transition from spindles to spike-wave discharges (for review see Gloor and Fariello, 1988). In GAERs, preliminary evidence has recently suggested that alterations in cortical excitatory synaptic function may exist, and these differences are not found in nonepileptic controls. This increase in excitatory synaptic drive in epileptic rats was found to be sensitive to N-methyl-D-aspartate antagonists (Pumain et al., 1992). These altered synaptic responses were localized in middle to deep cortical layers, the same areas which receive the main portion of thalamocortical projections.

DEVELOPMENT OF *IN VITRO* THALAMOCORTICAL PREPARATIONS SUPPORTING RHYTHM GENERATION

Recently, studies in our three laboratories have focused on development of *in vitro* preparations that will support the generation of thalamocortical rhythms, facilitating detailed analysis of rhythm-generating mechanisms important in epileptogenesis. Coulter and colleagues (Coulter, 1992a,b, 1994; Coulter and Zhang, 1993, 1994; Coulter and Lee, 1993) have adopted the rodent thalamocortical slice developed by Agmon and Connors (1991) to study mechanisms of epileptogenesis in the thalamocortical system. These slices retain connections be tween somatosensory cortex and thalamus, together with their associated nRt connections. Mouse or rat

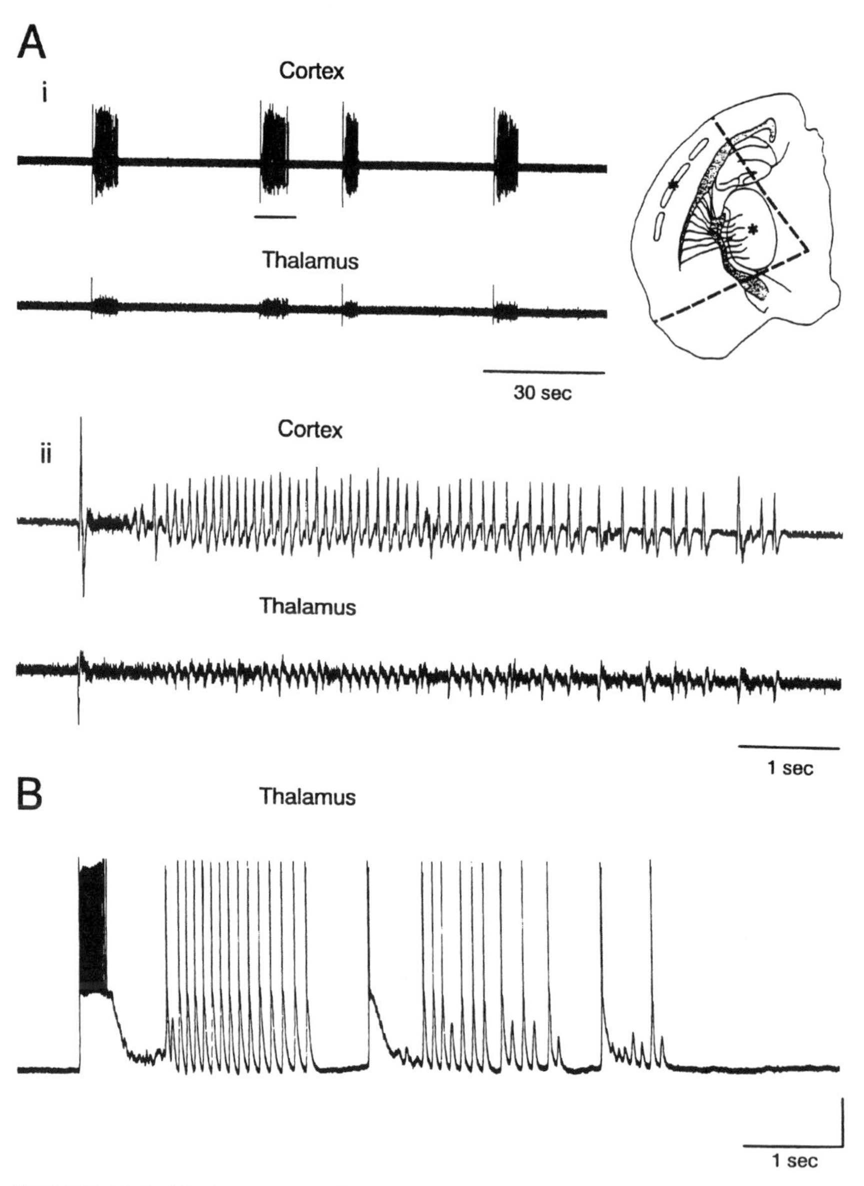

FIGURE 11-6. Thalamocortical activity recorded in vitro in a mouse thalamocortical slice. **A**. Extracellular field potential recordings of spontaneous activity elicited in somatosensory cortex and thalamus in response to perfusion with low-Mg^{++} medium. The spontaneous activity consisted of 2- to 10-second burst complexes, which recurred every 30 to 60 seconds (i). Individual bursts of activity (ii, an expanded view of burst complex indicated by a bar in Ai) were triggered by a single large burst, which was followed by several seconds of rhythmic, 8- to 10-Hz bursting. **B**. Intracellular recording from a thalamic ventrobasal neuron during spontaneous rhythmic activity similar to that recorded in **A**. Note the large triggering burst, which was followed by several seconds of 8- to 10-Hz rhythmic activity. This type of activity was reduced or abolished by the transection of thalamocortical connections. Vertical calibration is 20 mV.

thalamocortical slices, when perfused with a medium containing no added Mg^{2+}, cycle through a series of excitability changes that culminate in the development of spontaneous 8 to 10 Hz rhythmic burst discharges. These discharges physiologically and pharmacologically resemble spindle and spike-wave discharges in some aspects (Fig. 11-6) and are blocked or truncated by lesions severing thalamocortical connections (Coulter and Lee, 1993). Von Krosigk, Bal, and McCormick (1993) studied spontaneous rhythmic discharges in ferret lateral geniculate slices, which retain connec tions between the geniculate and perigeniculate nuclei. These rhythms physiologically resemble spindle discharges and can be transformed into lower-frequency, larger amplitude discharges after application of the $GABA_A$ antagonist bicuculline. These latter rhythms resemble spike-wave dis charges and are blocked by $GABA_B$ receptor antagonists (von Krosigk et al., 1993). Huguenard and Prince (1994a,b) used an *in vitro* preparation to examine reciprocal connectivity between nRt and ventrobasal complex in young rats. They found that rhythmic bursts of IPSPs are observed in VB neurons after localized stimulation in nRt, and that these oscillatory IPSPs were significantly reduced by bath application of the anticonvulsants ethosuximide or clonazepam. Further studies in these *in vitro* systems could prove promising in elucidating mechanisms important in the generation, maintenance, and control of pathological thalamocortical rhythms.

CONCLUSIONS

Thalamic neurons, and circuits involving thalamic neurons, display multiple modes of action potential generation, of which the most prevalent are rhythmic burst discharges and single spike (tonic) activity. The presence of these different firing modes allow for the generation of both normal, such as spindle waves, and abnormal, such as absence seizures, synchronized oscillations in thalamocortical circuits. Through the detailing of the electrophysiological features of the various constituent neurons involved in these circuits, their synaptic interactions, their modulation by ascending and descending transmitter systems, and their response to various antiepileptic agents, we are coming to understand in detail the cellular mechanisms of function of the thalamus and its interactions with the cerebral cortex.

12

Models of cortical networks

PAUL C. BUSH and TERRENCE J. SEJNOWSKI

Our knowledge of the cerebral cortex is increasing at an accelerating pace as new anatomical and physiological techniques become available. It is now possible, with whole-cell patch recording (Hamill et al., 1991), to obtain intimate knowledge of single cortical neurons *in vivo*. Yet, cortical neurons are embedded within circuits that entwine the intrinsic properties of single neurons as they interact through synaptic actions. New properties emerge from these network interactions that are difficult to predict from the properties of the individual neurons. Computer models of neurons and their synaptic interactions may help us dissect these emergent properties in conjunction with experimental studies.

Our models of cortical neurons are based on the pioneering model of the action potential in the squid giant axon by Hodgkin and Huxley (1952) and the dendritic models of Rall (1967). In these models, each element of the neuron is treated as an electrical element having resistive and capacitative properties, and membrane conductances are modeled by voltage-dependent kinetic schemes. In the past it has been necessary to limit the complexity of the model because of the difficulty of solving the large number of coupled differential equations. With the current generation of workstation computers, we can routinely model single neurons with 400 or more compartments. The computer power and the knowledge necessary to model populations of neurons are now becoming available that would allow the same approach to be used to explore cortical networks.

In this chapter we provide an overview of a long-term project that we have undertaken to study a particular population of cells within the cortex—the network of interacting neurons within layer V of the visual cortex. The first step is the development of realistic models of single cortical neurons based on accurate anatomical and physiological data. The responses of these model neurons are matched against single neuron responses to natural sensory and electrical stimuli recorded *in vivo*. The second step is to reduce the complexity of the single-cell model for use in large-scale simulations of networks. The reduction is necessary to make the simulation times tractable. If the essential dynamical elements of the neuron are intact in the reduced models, simulations of large networks of neurons can then serve as a tool for exploring properties that are difficult to approach experimentally. We used such a model to study the synchronization of layer V neurons and its dependence on the pattern of connectivity between columns of excitatory and inhibitory neurons.

DENDRITIC SATURATION

Dendrites have a much higher input impedance than the soma (Rinzel and Rall 1974); even single excitatory postsynaptic potentials (EPSPs) can produce significant local depolarization and reduce the driving force on simultaneous and subsequent EPSPs on the same dendritic branch. In addition, synaptic conductances add to the resting membrane conductance of the neuron (Barrett and Crill, 1974; Bernander et al., 1991; Holmes and Woody, 1989; Rapp et al., 1992). This effectively decreases the membrane resistance (R_m) and increases the space constant (λ), resulting in reduced depolarizations for simultaneous and subsequent EPSPs. We refer to these two phenomena collectively as the *dendritic saturation effect*, although the term "saturation" does not necessarily imply that the membrane is at the synaptic reversal potential or that R_m has been effectively reduced to zero.

There is some direct experimental evidence for saturation during synaptic activation under physiological conditions. Ferster and Jagadeesh (1992) have demonstrated that the size of an EPSP evoked in visual cortical cells by electrical stimulation of the LGN was reduced during depolarizations caused by visual stimulation compared to its size at the resting potential. The reduction in EPSP size was proportional to the somatic depolarization caused by the visual stimulation. Thus, an EPSP that peaked at about 6 mV at rest was reduced to less than 1 mV when the cell was depolarized from a resting potential of −60 mV to a potential of −40 mV by visual stimulation (Fig. 12-1A; Ferster and Jagadeesh 1992).

Ferster and Jagadeesh interpreted their results as evidence that the synaptic sites (dendrites) were significantly more depolarized than the soma

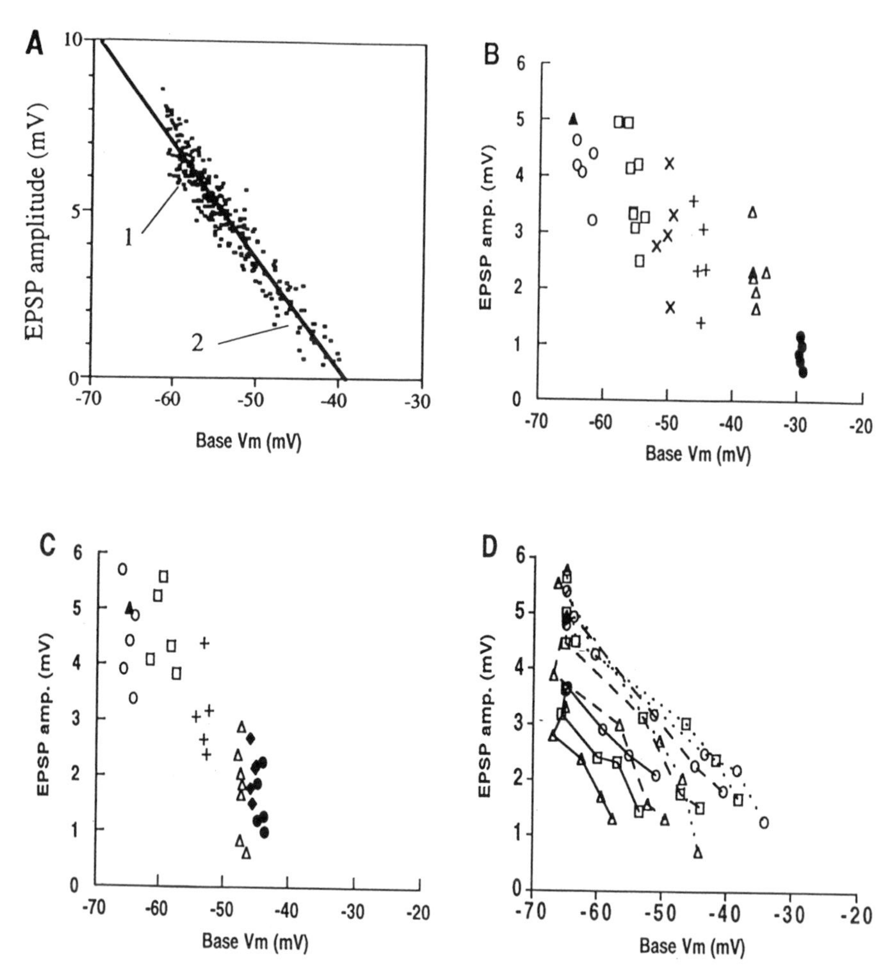

FIGURE 12-1. Dendritic saturation during physiological synaptic activation. **A**. Experimental data: peak amplitude of the test EPSP plotted against V_m just before the EPSP occurred while the visual stimulus was at the optimal orientation for the cell. Arrow 1 indicates control EPSP (before stimulation), arrow 2 indicates test EPSP (at peak of visual response; from Ferster and Jagadeesh, 1992). **B**. Simulation of the experiment in *A*, as detailed in the text. Peak amplitude of the test EPSP plotted against V_m just before the EPSP occurred for a variety of firing frequencies of the 70 excitatory synapses: ▲ = 0 Hz, ○ = 25 Hz, □ = 50 Hz, × = 75 Hz, + = 100 Hz, △ = 200 Hz, ◆ = 300 Hz, ● = 400 Hz. The amplitude of the EPSP decreased linearly with V_m. **C**. Concurrent inhibition is included in the simulation (33 inhibitory synapses at twice the frequency of excitatory synapses). **D**. Simulations in *C* repeated for R_m values of 5 (solid), 20 (dashed), and 100 (dotted traces) kΩ cm^2 and R_i values of 70 (○), 200 (□), and 500 (△) Ωcm. Data plotted for excitatory input frequencies of 0, 25, 100, 200, and 400 Hz. Each point is the average of five trials.

during synaptic activation of the cell, thus producing saturation by the process described above. We have tested this hypothesis by simulating their experiment, and the results are shown in Fig. 12-1. We used a 400-compartment model of a layer II pyramidal cell (see Bush and Sejnowski, 1994 for details) because its morphology is reasonably close to that of the presumed layer IV spiny stellate cells that Ferster and Jagadeesh studied. A constant current of −0.1 nA was injected starting at 30 msec to prevent firing during the synaptic activation, as in the experiments of Ferster and Jagadeesh. Visual stimulation was simulated by 70 excitatory synapses placed randomly on the basal/oblique dendrites, activated by a Poisson process starting at 80 msec. The mean frequency of these visual inputs was varied. A single simultaneous stimulation at 5 msec and again at 150 msec was applied to 35 additional excitatory synapses on the same dendritic segments as the initial 70. The firing of these 35 synapses represented the effects of electrical stimulation of the LGN: the first stimulation gave a control EPSP; the second, occurring during "visual stimulation," gave a test EPSP. Different trials using a different seed for the random number generator produced Poisson-distributed trains of EPSPs.

Figure 12-1B shows the results of multiple trials at different frequencies of visual stimulation. Higher frequencies of excitatory input produced greater somatic depolarization, as would occur in a cell as the visual stimulus was presented at increasingly optimal values of orientation, velocity, and direction. The peak amplitude of the EPSP is plotted against the somatic membrane potential (V_m) just before the EPSP occurs. As in the experimental data (Fig. 12-1A), the amplitude of the evoked EPSP decreased linearly with V_m. The dendrites were depolarized to between −25 mV and −20 mV during maximal synaptic activation, which means that the EPSP driving force was reduced by about 60%. Since the test EPSP was reduced by about 80% during maximal synaptic activation (Fig. 12-1B), a reduction in EPSP size of approximately 20% was due to a reduction in the input resistance (R_{in}) caused by the excitatory synaptic input.

The results shown in Figure 12-1B provide a qualitative match to those of Ferster and Jagadeesh, but the difference between the level of somatic and dendritic depolarization in the model was not great. In addition, the slope of the relationship between EPSP height and somatic V_m was not as steep as in the experimental data: the abscissa intercept (EPSP reversal potential) is about −40 mV in the experimental results (Fig. 12-1A) and −25 mV in Fig. 12-1B. Ferster and Jagadeesh report that visual cortical cells cannot be depolarized by more than 20 mV from rest by visual stimulation (Ferster and Jagadeesh 1992). For our model cell (resting potential −65 mV), this corresponds to complete saturation at a somatic V_m of −45 mV.

Effect of Inhibition on Saturation

The difference between our simulations of dendritic saturation and the experimental results could be due to the omission of inhibitory input in the simulation. Recent studies (Berman et al., 1991; Douglas et al., 1988; Ferster and Jagadeesh, 1992) have suggested that inhibitory input to visual cortical cells is weak during nonpreferred responses and is in fact strongly correlated with the degree of activation of the excitatory cells (Ferster, 1986). This suggestion fits with anatomical evidence that spiny excitatory cells make direct contacts with inhibitory cells, which then make direct contacts back onto the same excitatory population (Douglas and Martin, 1991). We repeated the simulation described above, this time including 33 inhibitory synapses—12 on the soma and 21 on the preterminal basal/oblique dendrites. This is the pattern of innervation characteristic of basket cells, the most common inhibitory cell type in cortex (Martin, 1988). Inhibitory (smooth) cells often fire at much higher rates than typically found in pyramidal cells (McCormick et al., 1985); therefore, inhibitory synapses were activated at a mean frequency twice that of the excitatory synapses. Fig. 12-1C shows the results of including inhibition in the simulation. The EPSP/somatic V_m slope is steeper, with an abscissa intercept of about −35 mV. This is much closer to the experimental data of Ferster and Jagadeesh (Fig. 12-1A). Dendrites were depolarized to about −30 mV during maximal synaptic activation, less than in the excitation-alone case, yet the EPSP/somatic V_m slope is steeper with inhibition. A dendritic depolarization to around −30 mV reduced the EPSP driving force by about 50%. Test EPSPs were reduced in amplitude by up to 80%, so the remaining 30% reduction must be due to decreases in R_{in} caused by the excitatory and inhibitory synaptic conductance changes. Intrinsic subthreshold voltage-dependent conductances during the experiments of Ferster and Jagadeesh may also have contributed to the reduction in R_{in} and hence EPSP amplitude. The simulations of Fig. 12-1 did not include voltage-dependent conductances, so we cannot evaluate the extent of this contribution, but the simulations of Bernander et al. (1991) indicate that it is likely to be relatively small.

Ferster and Jagadeesh (1992) focused on reduction in driving force as the explanation for the decreases in evoked EPSP amplitude that they observed, but we found that decreases in R_{in} due to excitatory inputs made a significant contribution. Furthermore, additional current shunts caused by inhibitory synaptic activity must be included to produce an accurate fit to the experimental data. The data in Fig. 12-1C fall along a straight line because the major component of the saturation effect is due to the linear reduction in driving force. The contribution from increased membrane conductance, which would produce a concave (hyperbolic) curve, is masked by the driving force effect and the variance in the data. When inhibition was included in the simulation, the soma could not be depolarized past −45 mV by the firing of the 70 excitatory synapses, which is the limit of depolarization obtained with optimal visual stimulation as reported by Ferster and Jagadeesh (1992).

Figure 12-1D shows that the decrease in test EPSP amplitude due to dendritic saturation is stable across a wide range of values for R_m and R_i. The rate of decrease of EPSP amplitude with somatic V_m was constant over the full parameter range.

The soma could still be depolarized by 20 mV from its resting level even when the 33 inhibitory synapses are firing at their maximum rate (Fig. 12-1). This finding indicates that firing of the postsynaptic cell could persist despite significant inhibition. It supports the conclusion of Douglas and Martin (1990a), who simulated the effect of a maintained inhibitory conductance on the firing rate of a simplified model neuron driven by intrasomatic current injection. They suggested that inhibition in cortex cannot prevent the firing of a neuron receiving strong excitatory input. In contrast, studies *in vitro* indicate that synaptically evoked $GABA_A$ inhibition is strong enough to briefly suppress the firing of cortical neurons driven by large depolarizing current injections (Connors et al., 1988; McCormick, 1989b). However, these results must be interpreted with caution given the large difference in $GABA_A$ conductance elicited *in vivo* and *in vitro* (Berman et al., 1989).

Effectiveness of Inhibition

Under what conditions could inhibition be strong enough to suppress firing? What is the effect of strong synaptic activation on R_{in} of the target neuron? These questions have been the subject of previous experimental and theoretical studies (Berman et al., 1991; Douglas et al., 1988; Koch et al., 1990). To address these issues the model layer II pyramid was driven by the same concurrently active 70 excitatory and 33 inhibitory synapses as in Fig. 12-1. Active conductances were added to the model to produce adapting trains of action potentials (Fig. 12-2A). Fig. 12-2B shows the firing rate of the model cell as a function of the firing rate of the 70 excitatory synapses. In agreement with Douglas and Martin, the effect of the inhibition was to increase the threshold of the neuron and only slightly reduce the firing rate above threshold.

These simulations ignored the observation that inhibition is correlated with excitation (Ferster, 1986), which would occur if the inhibitory cells were being driven by the excitatory cells that they were inhibiting (Douglas

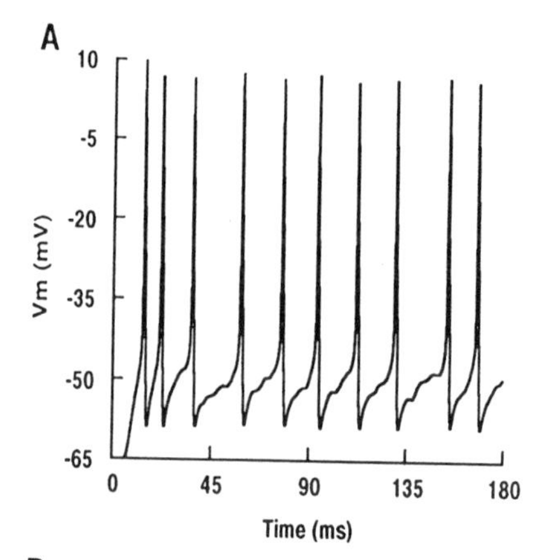

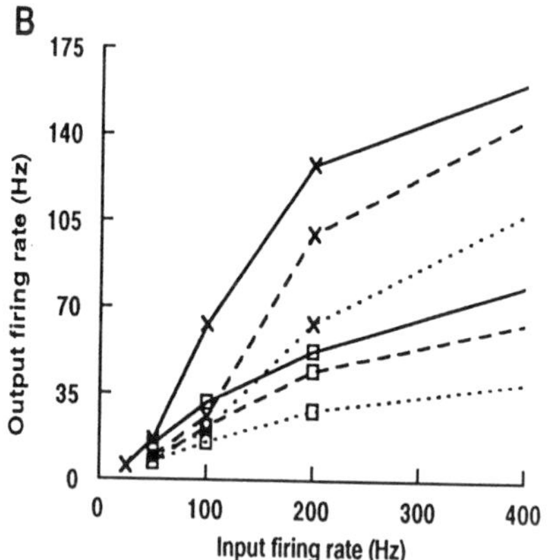

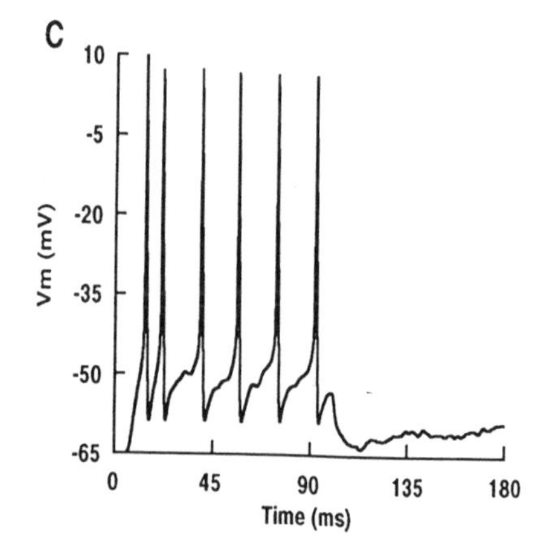

FIGURE 12-2. Effect of inhibition on the firing rate of synaptically activated model layer II pyramid. **A**. Adapting train of action potentials produced by 70 excitatory inputs active at a mean frequency of 200 Hz. **B**. Firing rate of model as a function of the firing rate of its 70 excitatory inputs. × = Initial, peak firing rate. □ = Steady, adapted firing rate. Solid traces are results without inhibition. Dashed traces are results with 33 additional inhibitory synapses firing at 100 Hz. This level of inhibition has little effect on the strongly activated pyramid. Dotted traces are results with the firing rate of the inhibitory inputs equal to twice that of the excitatory inputs in each case. This more realistic level of inhibition causes a significant decrease in the firing rate of the pyramid, although firing is not suppressed completely. **C**. Firing is completely suppressed when an additional 150 somatic inhibitory inputs, which start firing at a mean frequency of 400 Hz at t = 100 msec, are added to the simulation shown in *A*.

and Martin, 1991). It is reasonable to assume that the firing rate of the inhibitory inputs would increase with the firing rate of the pyramidal cell, and hence with the firing rate of the pyramidal cell's excitatory inputs (the 70 excitatory synapses). Therefore, we repeated the above simulations, this time setting the firing rate of the inhibitory synapses to twice that of the excitatory synapses. The results, shown in Fig. 12-2B (dotted traces), demonstrate that inhibitory input does have the potential to reduce significantly the firing rate of the target cell. However, the firing rate of the pyramidal cell was still substantial: the inhibition produced by 33 synapses is not enough to shut off the target cell.

The simulation was repeated with replacement of 33 somatodendritic inhibitory synapses by 25 inhibitory synapses on the first 25 μm of an axon initial segment consisting of seven cylinders with diameters tapered from 2.5 μm to 0.6 μm. This is the innervation pattern characteristic of chandelier cells, a type of cortical inhibitory interneuron (Farinas and DeFelipe, 1991b). Sodium and potassium spike conductances were included on the first 25 μm at the same density as on the soma. The results (not shown) were nearly identical to those shown in Fig. 12-2. Thus, we found no difference between the effect of basket cell inhibition and that of chandelier cells (see also Lytton and Sejnowski, 1991). Similar results were also obtained when using the layer V cell instead of the layer II cell (not shown).

Cortical pyramidal cells receive hundreds of inhibitory synaptic contacts on their somata and proximal dendrites (Douglas and Martin, 1990b; Farinas and DeFelipe, 1991a). Therefore, we increased the number of active inhibitory synapses in our simulation to determine whether the concerted activity of a larger fraction of the cell's inhibitory input was sufficient to suppress firing. We found that the activity of about 200 somatic inhibitory synapses was sufficient to prevent a cell receiving strong excitation from firing (Fig. 12-2C). Consequently, we conclude that strong cortical inhibition (about 10% of the total number of inhibitory synapses on one pyramidal cell) is able to prevent the firing of even strongly driven pyramidal cells, contrary to previous conclusions (Douglas and Martin, 1990a). Firing was completely suppressed by 200 inhibitory inputs whether the R_m of the model layer II pyramidal cell was 20 or 100 kΩcm^2. When R_m was 5 kΩcm^2, the resting R_{in} of the model cell was so low (about 35 MΩ) that the 70 excitatory inputs did not drive the cell very strongly and the activity of just 33 inhibitory inputs was sufficient to suppress firing.

REDUCED COMPARTMENTAL MODELS

The models used in the above studies contain hundreds of compartments and thousands of coupled differential equations that must be solved each time step. To speed up simulations, model networks must use simplified representations of the single neurons that comprise the network. These simplified neurons must retain the electrotonic and synaptic integrative properties of the real cell. We have developed a method for reducing the number of compartments of neocortical pyramidal neuron models from 400 to 8 or 9. It is a simple collapsing method based on conserving the axial resistance, rather than the surface area of the dendritic tree. The cross-sectional area of the equivalent cylinder is made equal to the sum of the cross-sectional areas of all the dendrites represented by that equivalent cylinder:

$$R = \sqrt{\sum_i r_i^2}$$

where R is the radius of the equivalent cylinder and r_i is the radius of dendrite i. The length of the equivalent cylinder is just the average length of all the dendrites represented by the equivalent cylinder.

The surface areas of our reduced models are less than those of the full models. Thus, R_m must be decreased to match the input resistance of the full model and C_m increased by the same factor to match the membrane time constant. The reduced models produced by this method retain the general morphology of the pyramidal cells on which they are based, allowing the accurate positioning of synaptic inputs and ionic conductances on individual model cells, as well as the construction of spatially accurate network models.

We constructed reduced models of the layer II and layer V pyramidal cells used in the above studies (see Bush and Sejnowski, 1993, for details of the reduction process), and the geometries are shown in Fig. 12-3B. Drawings of the HRP-filled pyramidal cells are included for comparison (Fig. 12-3A). To assess the accuracy of the method that produced the reduced models, we compared the responses of the reduced and full models to different types of stimulation. The response of the reduced and the full model layer V pyramidal cell to a continuous somatic current injection of -0.7 nA are compared in Figure 12-4A. The superposition of the two traces shows that both models have the same R_{in} and τ_m. It is relatively easy to match these parameters by tuning R_m and C_m.

However, such a match says little about how faithfully the reduced model captures the synaptic integration properties of the full model (Fleshman et al., 1988). The responses of the reduced and the full model layer II pyramid to a brief somatic current injection are compared in Figure 12-4B; Shelton 1985; Stratford et al., 1989). The response to a transient somatic input is dependent on R_i, as well as R_m and C_m, because it is dependent on how fast current moves from the soma into the dendrites. The response of the reduced model is a good fit to that of the full model. The responses of both models to an EPSP on the soma, with 0.5 nS peak conductance, are compared in Figure 12-4C. This tests essentially the same properties as the brief current pulse; the performance of the reduced model is very close to that of the full model.

Figure 12-5A shows the firing of the full 400-compartment layer V cell in response to a maintained 1-nA somatic current injection. The model cell produced an adapting spike train typical of the regular-firing class of cortical pyramidal cells (McCormick et al., 1985). The conductances underlying this firing behavior, located in the soma only, were put into the nine-compartment model without changing a single parameter. Because both models have the same somatic dimensions, the same conductance densities were used in both. Figure 12-5B shows the firing of the nine-compartment model in response to a 1-nA somatic current injection. The response of the reduced model has the same form as that of the full

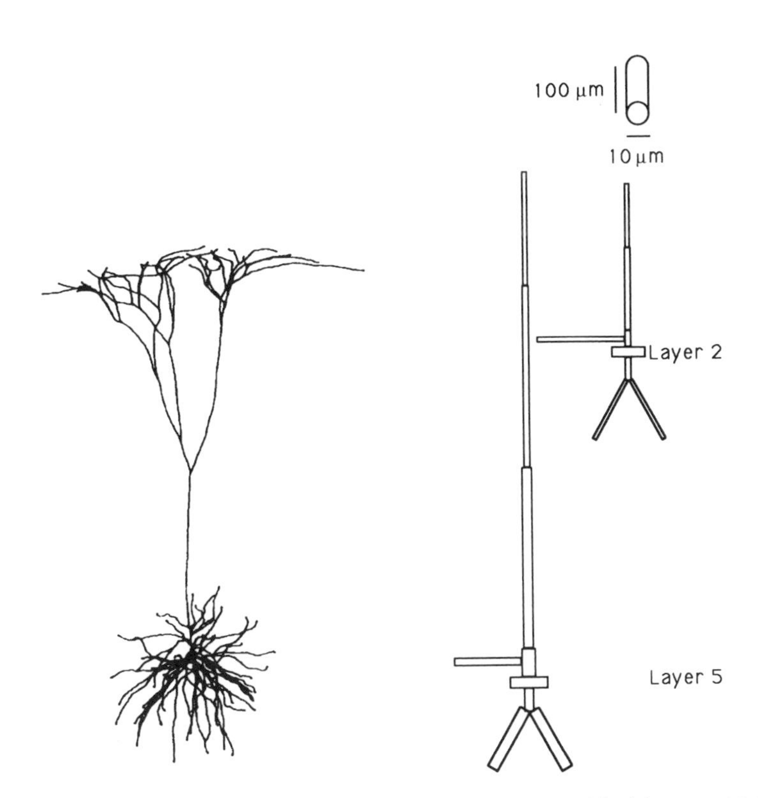

FIGURE 12-3. **A**. Drawings of reconstructed HRP-filled layer II (right) and layer V (left) pyramidal cells. **B**. Geometries of reduced pyramidal cell models.

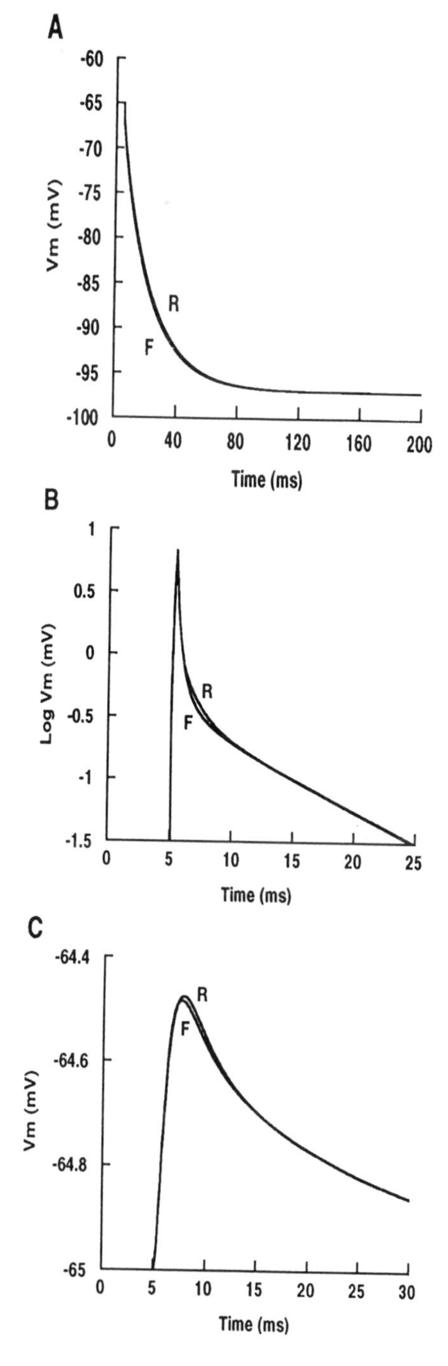

FIGURE 12-4. Comparison of the responses of the reduced (R) and full (F) models to somatic input. **A**. Voltage response at the soma of reduced and full layer V pyramid models to the constant current injection of −0.7 nA. The superposition of the two traces indicates that both models have the same R_{in} and τ_m. **B**. Semi-log plot of voltage response of reduced and full layer II models to a 0.44-msec 0.3 nA somatic current injection at t = 5 msec. **C**. Voltage response of both layer II models to a 0.5-nS somatic EPSP. The close fit of the reduced model with the full model indicates that the dendrites conduct charge away from the soma at the same rate in both models.

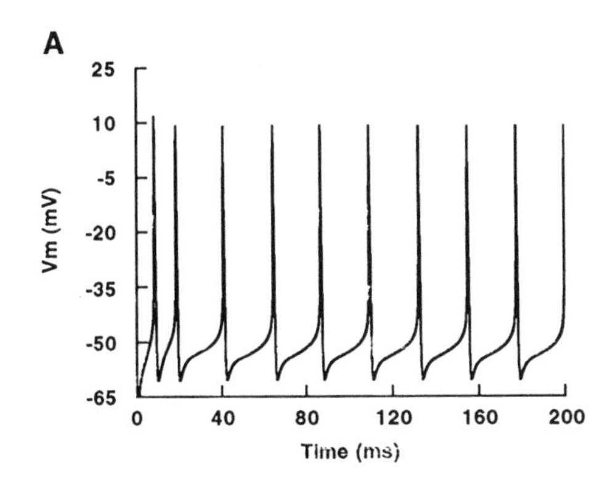

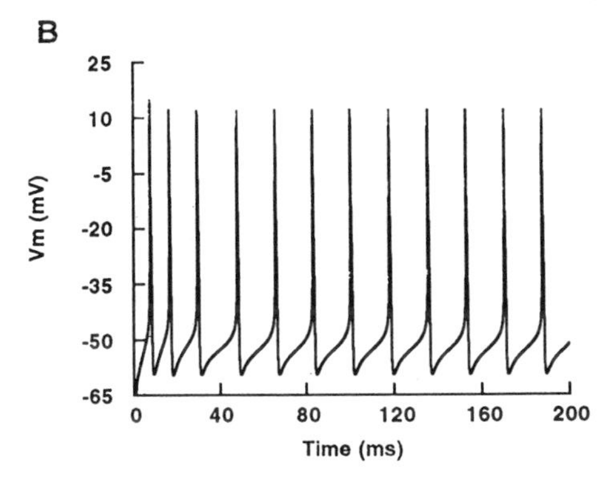

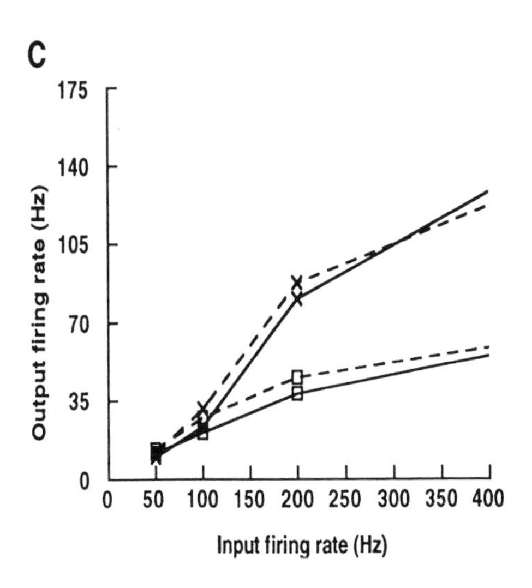

FIGURE 12-5. Comparison of firing responses of reduced and full model layer V pyramidal cell. The somata of both models contain active conductances with exactly the same kinetics and densities. **A**. Adapting spike train of full model in response to a constant 1-nA somatic current pulse. **B**. Spike train of reduced model in response to the same stimulus. The response of the reduced model is of the same form as that of the full cell, but the firing frequency is slightly higher. **C**. Firing rate of full (solid traces) and reduced (dashed traces) models as a function of the firing rate of their 140 excitatory inputs. Each model also receives 45 inhibitory synapses, active at twice the rate of the excitatory ones. × = Initial, peak firing rate. □ = Steady, adapted firing rate. The close fit of the two models demonstrates that the reduced model integrates excitatory and inhibitory synaptic input in the same manner as the full model.

model—an adapting train of action potentials. The firing frequency of the reduced model was slightly higher, but the small difference was within the limits of uncertainty of the model parameters, as well as the variance in response recorded across different pyramidal cells (Douglas et al., 1991; McCormick et al., 1985).

The final test was to compare the response of both models to synaptic input. To produce the same output as the full model, the reduced model must perform the same non-linear integration of dendritic EPSPs and IPSPs as the full model and must display the same dendrites-to-soma transfer characteristics. In other words, the reduced model must have the same input/output function as the full model.

A majority (70% to 90%) of excitatory inputs to cortical pyramidal cells are made on the basal/oblique dendrites (Larkman, 1991a,b). We distributed synapses on the dendrites and soma to reflect these measurements. Thus, 140 excitatory synapses were placed randomly on the basal and oblique dendrites of the full model and 140 on the one oblique and two basal equivalent dendrites of the reduced model. In addition, 33 inhibitory synapses were placed on the proximal dendrites and 12 on the soma of each model, a pattern of innervation characteristic of basket cells, the most common inhibitory cell type in cortex (Martin, 1988). Inhibitory (smooth) cells fire at much higher rates than pyramidal cells (McCormick et al., 1985); therefore, inhibitory synapses were activated at a mean frequency twice that of the excitatory synapses. Figure 12-5C shows the responses of the full and reduced models for the peak (initial) and steady-state (adapted) firing rates as a function of the frequency of the excitatory inputs. The fit is close for all input frequencies, demonstrating that the reduced model shows the same dendritic integration characteristics and response to inhibition as the full model, despite having an extremely simplified structure.

NETWORK SIMULATIONS

We have recently used these reduced compartmental models to construct networks of excitatory and inhibitory neurons and to simulate the synchronization that has been observed to occur within and between columns in visual cortex of cats (Gray et al., 1992).

We simulate one column of layer V of visual cortex using 80 pyramidal cells and 20 basket cells. The pyramidal cells, composed of nine compartments, have passive dendrites and sodium, potassium, high-threshold calcium, and calcium-dependent potassium conductance in the soma. They fire repetitive bursts in response to constant current or synaptic input with a frequency dependent on the intracellular calcium elimination rate. The basket cells consist of seven compartments and have only sodium and potassium conductances in the soma to produce high-frequency nonadapting spike trains. A small amount of current noise is injected into every neuron during each simulation.

Within a column of 100 neurons, connectivity is random all-to-all with a connection probability of 10%. The synaptic delay has a mean of 1.2 msec, and all delays and synaptic conductances have large standard deviations. Due to randomly assigned calcium elimination rates, the pyramidal cells fire at different frequencies and with a random phase with respect to other cells when stimulated in isolation. However, when the network is connected and the pyramidal cells are stimulated with uncorrelated Poisson spike train input, all cells fire synchronously in an oscillatory manner (Figure 12-6). As in the experimental data, the mean phase difference between individual neurons is zero msec, and there is a spike-to-spike jitter of a few milliseconds. The network can be driven to oscillate at about 35 to 60 Hz, depending on the strength of the tonic input.

As in an earlier, simpler study (Bush and Douglas, 1991), the mechanism of synchronization relies on rapid excitatory feedback to cause all the pyramidal cells to fire together. The burst is then terminated by strong inhibitory feedback from the basket cells. Also as in the previous study, not every cell fires on every cycle, but when a cell does fire it is in phase with the population. We found that the cells with a lower intrinsic bursting rate were more likely to "miss" bursts and fire fewer spikes per burst when they did discharge (Fig. 12-6).

Synchronization has been observed not only between neurons in a single column but also between cells in different columns, some of which are several millimeters apart (Gray et al., 1992). We have extended our single-column simulation to attempt to explain these results in terms of known physiological and anatomical data. Neocortical pyramidal cells in one column send axon collaterals to other columns where they contact mostly other pyramidal cells but also some smooth (inhibitory) interneurons. The density of this connectivity is much less than that within a column (Kisvarday and Eysel, 1992). Therefore, we modeled two columns of 100 layer V neurons as described above and connected them together with a 3% connection probability. When not connected, both columns synchronized internally, but were at random phase with respect to each other. When connections from pyramidal cells in one column just to pyramidal cells in the other column were added, the frequency of oscillation increased, and the two columns oscillated 180° out-of-phase, a phenomenon characteristic of coupled excitatory oscillators. However, when connections were made from pyramidal cells in one column to both pyramidal and basket

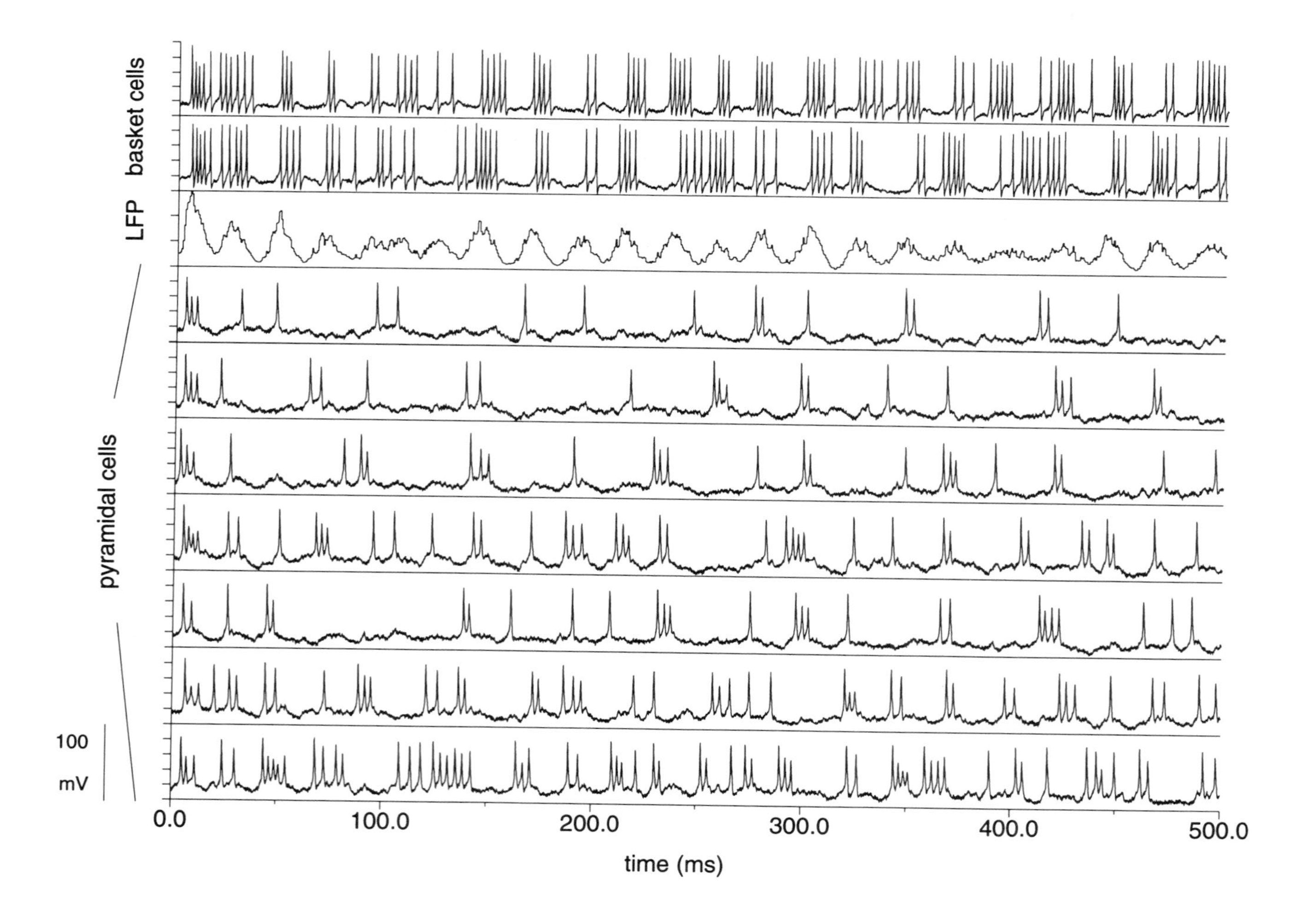

basket cells
LFP
pyramidal cells
100
mV
0.0
100.0
200.0
300.0
400.0
500.0
time (ms)

FIGURE 12-6. Intracolumnar synchronization in a model network of 100 layer V cortical neurons. The top two traces show the somatic membrane potential of 2 basket cells from a population of 20 making up one column. The next trace is an analog of the local field potential (LFP) to show the global synchronization of the whole column. It is the mean membrane potential of the somata of all the pyramidal cells in the column. A flat trace would be expected for no synchronization, whereas the regular oscillation shown here indicates that the neurons in the column were firing synchronously. The bottom seven traces show the somatic membrane potential of 7 sample pyramidal cells from the total population of 80. In this example, the cells' intrinsic bursting frequency increases toward the bottom. There is considerable "jitter" in an individual pyramidal cell's output, but the population synchrony is shown by the LFP and rhythmic firing of the inhibitory neurons, which have no intrinsic oscillatory properties. The synchrony spontaneously dissapears and reappears at 100 and 350 msec. This is seen in real experimental data.

cells in the other column, the two columns synchronized their oscillation within one or two cycles. Thus, the intercolumnar connection from pyramidal cells to inhibitory neurons, although quite sparse, is very important for intercolumnar synchronization.

CONCLUSIONS

Our goal is to understand how networks of cortical neurons are organized to allow long-range contextual inputs to influence local processing. The achievement of synchrony between distant columns of neurons may be one such mechanism. We expect to uncover others as we continue to enlarge and improve on our models.

IV

The developing cortical neuron

13

Determination of cellular phenotype and the fundamental organization of cortical layering: an overview

ARNOLD R. KRIEGSTEIN

The neocortex is composed of a staggeringly large number of neurons differing in phenotype and location in a precise and apparently invariant pattern. The path by which each individual neuron reaches its assigned station and acquires its appropriate phenotype is one of the central mysteries of neuroscience. Recent technological advances have allowed new experiments to address this issue, and the answers are challenging older notions of corticogenesis.

A central question concerning early cortical development involves the steps by which cell fate is determined during corticogenesis. It is generally accepted that cellular development proceeds by way of progressive restrictions imposed upon a pluripotential precursor cell. The key issue concerns the degree to which cell autonomy or environmental influence determines the commitment of each cell to its ultimate fate. Although both factors play a role, it seems that important decisions are environmentally determined at proliferative stages, possibly even before terminal mitosis and "birth" of the neuron.

The cells of the cerebral cortex, neurons and glia, arise during development from ventricular and subventricular zones of dividing precursor cells adjacent to the ventricle of the telencephalon. Precursor cells span the ventricular zone radially, and their nuclei synthesize DNA while in the upper layer and then undergo interkinetic nuclear migration to the ventricular border where mitosis occurs. Then the nucleus of the new daughter cell migrates up once again to re-enter S-phase and continue the proliferative cycle. Eventually each cell undergoes a terminal mitosis at the ventricular border, then migrates out of the ventricular zone, and travels radially to arrive at the embryonic cortical plate; later-generated neurons migrate past already populated deeper layers to take up position in a more superficial layer in accordance with the inside-out pattern of corticogenesis. Glial cells also derive from dividing ventricular zone cells. Initially the non-neuronal cells consist of radial glial cells that will ultimately transform into astrocytes (Rakic, 1975). Other glial cells and neurons derive later in postnatal stages from cells in the subventricular proliferative zone (Hockfield and McKay, 1985; Levit, et al., 1981; Misson et al., 1988).

The homogeneous appearance of ventricular zone cells led to the notion that multipotential cells gave rise to both neuronal and glial cell lineages (for review see Levitt et al., 1981; McConnell, 1988). The radial organization of the cortex into functional columns of neurons led to the theory of ontogenetic columns of clonally related neurons that migrate along radial glial guides to populate successive cortical layers within a restricted radial unit (Peters and Feldman, 1973; Rakic, 1972). Experiments using retroviral lineage markers and brain transplantation are refining these theories.

CELL LINEAGE AND FATE RESTRICTION

Recent advances in genetic engineering have allowed retroviruses to be designed to act as vectors to insert marker genes into precursor cells in the cerebral cortex of experimental embryos (Cepko, 1988). A replication defective retroviral vector is used to introduce a marker gene in the genome of a proliferating precursor cell. The marker gene is inherited by all the progeny of the initially labeled cell, allowing clonal descendants of infected cells to be identified histochemically. In this way precursor cells can be marked, and the fate of their descendants can be followed. This new technology has been applied to test the basic theories of lineage and cell fate determination during corticogenesis.

Data using thymidine labeling to trace cell migration coupled with histochemical labeling techniques had led to the theory that neurons migrated radially to their final destination and that neurons and glia shared common stem cells. It now seems from retroviral lineage analysis that cortical neuron precursors in the proliferative zones give rise to cells that distribute widely across the cortex (Austin and Cepko, 1990; Walsh and Cepko, 1993) and that spatially clustered progeny that may represent subclones of these larger clones are restricted to a particular cell type, such as pyramidal or non-pyramidal neuron or astrocyte (Luskin et al., 1988; Price and Thurlow, 1988). The restricted potential of precursor cells is cell autonomous and does not require interactions in the neuroepithelium, as it remains restricted even in dissociated cell culture (Willams et al., 1991). Curiously, a minority of these spatially restricted subclones contain both neurons and glia (Price and Thurlow, 1988), and approximately 18% of precursor cells in dissociated cell culture from embryonic rat cortex are bipotential at E16 and generate clones of both neurons and oligodendrocytes (Williams et al., 1991).

Since a single progenitor cell contributes neurons to multiple cortical lamina, and cortical layers differ in the numbers of neuron subtypes with cells of different morphologies, physiological, and neurochemical properties, it is possible that a progenitor cell gives rise to several neuronal types. Recent data, however, suggest this may not be the case. Luskin and colleagues (Luskin et al., 1993) used retroviral-mediated gene transfer to label clonally related cells in rat cortex and assigned labeled cells to specific cell types by ultrastructural examination. They found neuronal subclones to be restricted only to pyramidal cells or to non-pyramidal neurons, but not to both. This finding suggests that a degree of lineage restriction has already occurred in the precursor cell population by the onset of neurogenesis.

Although lineage analysis can reveal the fates of clonally related cells, the question of when and how cells become committed to a particular fate has been addressed in heterochronic transplantation experiments. The results suggest that commitment to cell fate takes place before cells migrate out of the proliferative zone. Caviness (1982) proposed that neuronal phenotype is determined before cell migration based, in part, on analysis of the reeler mutant mouse. In this mutant, neurons are generated normally, but migrate abnormally and form an inverted lamination pattern. Nonetheless, the malpositioned neurons develop physiological and anatomical properties appropriate for their cell type. Direct experimental evidence that cell fate is determined before migration comes from heterochronic transplantation studies conducted by McConnell and colleagues (see McConnell, 1991 for review). They removed young neurons destined for deep cortical layers from the proliferative zone of a donor animal and transplanted them into the proliferative zone of a host animal at a stage when neurons destined for superficial layers were being generated. Neurons that had recently undergone their last S-phase in the donor brain migrated to the same layers as they would have in the donor, suggesting that laminar fate is determined before migration out of the proliferative zone.

These experiments indicate that important cell fate commitment decisions are made in the ventricular and subventricular zones during proliferative stages and remain relatively independent of subsequent environmental influence. However, the nature of the specific cellular and environmental signals that interact to seal phenotypic expression and laminar fate remains undetermined. The data summarized above direct our attention to the proliferative zone as the critical region and the period of cell division as the key time frame for exploration of the events that determine cell destiny in the cortex. The term "neuronal migration disorder" has been applied to a variety of developmental brain abnormalities with seizures as a clinical feature. Such studies as those described above expand the etiological possibilities to include errors in signaling within the ventricular and subventricular zones before terminal mitosis and in advance of neuronal migration.

14

Neurotransmitter signaling before the birth of neurons

JO J. LOTURCO and ARNOLD R. KRIEGSTEIN

Recent advances in cellular physiological techniques, particularly the development of *in situ* whole-cell patch-clamp recording, have permitted detailed physiological and pharmacological studies of proliferating cells in the ventricular and subventricular zones of embryonic neocortex. The results are beginning to shed light on the kinds of signals and cellular interactions that may underlie the regulation of cell-cycle events and gene expression in cortical progenitor cells.

GAP JUNCTION CHANNELS PROVIDE AN AVENUE FOR INTRACELLULAR COMMUNICATION AMONG CORTICAL PROGENITORS

When recordings were first made from ventricular zone cells in the embryonic rat, they were found to have unexpectedly low membrane resistances (LoTurco and Kriegstein, 1991). The small, immature cells in the ventricular zone had membrane resistances five to ten times lower than the more mature neurons in the cortical plate, the reverse of what one would expect, and they also had faster time constants (LoTurco and Kriegstein, 1991). One possible explanation was that cells in the ventricular zone were electrically coupled to one another. This was confirmed by recording simultaneously from pairs of nearby ventricular zone cells. When a voltage step was applied to one cell, a corresponding voltage change was recorded from cells nearby (Fig. 14-1). Electrical coupling could result from the persistence of cytoplasmic bridges known to form transient connections between sister cells undergoing mitosis or could be the result of intercellular gap junction channels. Since cytoplasmic bridges form large channels, and gap junctions are much smaller, the first test consisted of injecting the same ventricular zone cell with two dye molecules of different molecular weight—one small, Lucifer Yellow (LY), and one large, horseradish peroxidase (HRP)—and seeing which dye spread from cell to cell. The result was that the smaller-sized LY dye spread to large clusters of ventricular zone cells, whereas the larger HRP molecule was confined to one cell only (LoTurco and Kriegstein, 1991). This result suggested that ventricular zone cells were linked to each other by gap junction channels. It was confirmed by reversibly uncoupling the cell clusters through acidification and the application of halothane, two pharmacological measures known to uncouple gap junction channels (Fig. 14-2).

The number of cells coupled in a cluster decreases as development proceeds and more and more neuroblasts undergo terminal mitosis and migrate out of the ventricular zone. When cells above the ventricular/ intermediate zone border were injected with Lucifer Yellow or biocytin

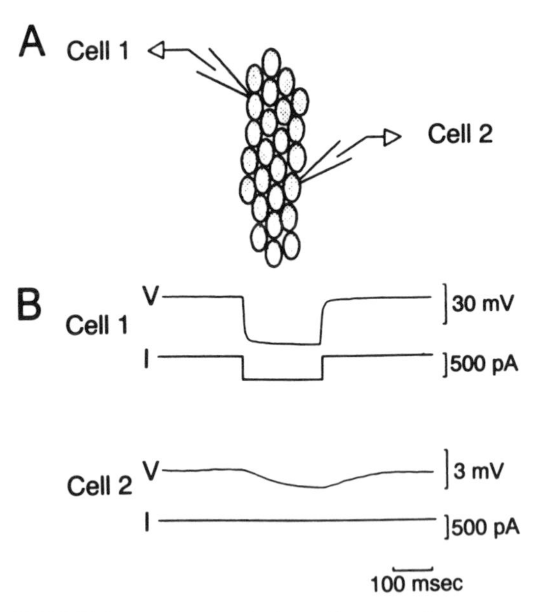

FIGURE 14-1. Dual whole-cell patch-clamp recordings from two cells in the ventricular zone. **A**. A schematic representation of the recording configuration. **B**. Voltage and current records simultaneously recorded from two ventricular zone cells from an E16 slice. Hyperpolarizing current (500 pA) was injected into cell 1 while no current was injected into cell 2. Simultaneous voltage changes were recorded in both cells, indicating that they were electrically coupled.

(dyes that readily pass through gap junction channels) only individual cells were labeled. This finding confirmed that once cells leave the ventricular zone they uncouple from other cells. In the embryonic cortex, therefore, premitotic cells are coupled, and postmitotic cells are not. When biocytin was injected into young neurons in the cortical plate, however, clusters of dye-filled neurons were seen (Yuste et. al., 1992), and they were also shown to be linked by gap junction channels. A likely interpretation is that cells are solitary during migration and that some time after settling in the cortical plate some or all of them recouple into cell clusters. It would be interesting to know whether the antecedents of the cortical plate cell clusters were coupled together in the ventricular zone or if they represent a new cohort of cells linked in a new functional role.

Proliferating ventricular zone cells are coupled while they progress through the cell cycle and uncouple just before or just after their final

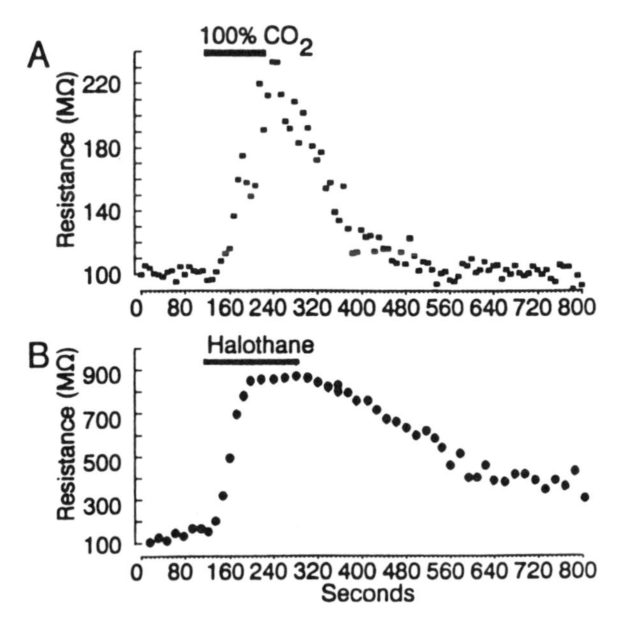

FIGURE 14-2. Coupled cells in the ventricular zone can be uncoupled by acidification and halothane. **A**. Changing the solution bathing the slice from one saturated with 5% CO_2 (pH 7.4) to one saturated with 100% CO_2 (pH 6.6) reversibly increased the membrane resistance of a cell in the ventricular zone. Membrane resistances were determined by measuring the current at the end of a 100-msec voltage step from −70 mV to −90 mV. The data points show the resistance determined every 8 seconds before, while, and after the perfusate was saturated with 100% CO_2. **B**. Halothane (approximately 2 mM), like acidification, reversibly increases the input resistance of ventricular zone cells.

mitosis. The gap junction channels that couple immature neuroblasts provide an avenue for direct intracellular exchange between cells long before the development of synapses. Changes in coupling could influence the time during which cells in the ventricular zone are susceptible to fate-determining signals. For example, well-coupled cells may be less susceptible to fate-determining signals because second-messenger levels would be reduced by diffusion to other cells. Such a mechanism would serve to prevent all cells from becoming committed to a particular fate in the presence of an inducing factor. Selective susceptibility to fate-determing factors is essential in a developing system, such as the cortex, where different fates are generated sequentially.

UNCOUPLING BLOCKS DNA SYNTHESIS

The correlation between mitotic activity and coupling raised the possibility that coupling is essential to DNA synthesis in the ventricular zone or cortex. To test this hypothesis, cortical slabs were exposed to either halothane or acidification and incubated for 5 hours in the presence of [^{3}H]thymidine. Both of these manipulations uncoupled cells in the ventricular zone and decreased the incorporation of [^{3}H]thymidine by over 90% (Fig. 14-3). Since the radioactivity levels remaining after acidification or halothane were similar to those obtained with non-mitotic postnatal or heat-killed embryonic brain tissue, DNA synthesis in cortical explants was terminated by uncoupling. Together with the correlation of mitotic activity and coupling during development, these data support the hypothesis that coupling between cells in the ventricular zone is important to mitotic activity. Since pharmacological uncoupling terminates DNA synthesis, the decision to uncouple from a cluster may be causally related to the decision not to re-enter the cell cycle. Through neurogenesis, cells become progressively less coupled, which may serve to slow and then terminate division in the ventricular zone. It also follows that factors that change coupling —growth factors (Mehta et. al., 1986; Ren et. al., 1990) or retinoids (Zhu et. al., 1991)—would promote either entrance or departure from the cell cycle depending on their corresponding effects on intercellular communication.

THE PRINCIPAL EXCITATORY AND INHIBITORY AMINO ACID RECEPTORS ARE EXPRESSED BEFORE NEURONAL DIFFERENTIATION

In addition to their role in synaptic transmission, amino acid receptors may sense changes in background levels of amino acids (LoTurco et. al.,

1990; Sah et. al., 1989). Such a function could operate during development before the formation of synapses. In culture, it has been shown that activation of amino acid receptors promotes the growth, differentiation, and survival of developing neurons (McDonald and Johnston, 1990; Mount et. al., 1993; Redburn and Schousboe, 1989). Determining when amino acid receptors are first expressed on cortical cells indicates when amino acids can begin to play a role in cortical development. The timetable for insertion of some of the key amino acid neurotransmitter receptors into the membranes of developing neocortical neurons has recently been explored. Even before their final mitotic division, cortical neuroblasts already express functional receptors for the principal excitatory and inhibitory transmitter substances used for synaptic communication by adult cortical neurons.

The $GABA_A$ subtype receptor for the inhibitory substance GABA and the AMPA/kainate subtype receptor for the excitatory transmitter glutamate are already functionally expressed on the membranes of proliferating ventricular zone cells (Figs. 14-4 and 14-5). After mitosis and during migration out of the ventricular zone, immature migrating neurons start to express the NMDA-type glutamate receptor (LoTurco et al., 1991). NMDA receptor activation has trophic effects on neurons, directing neu-

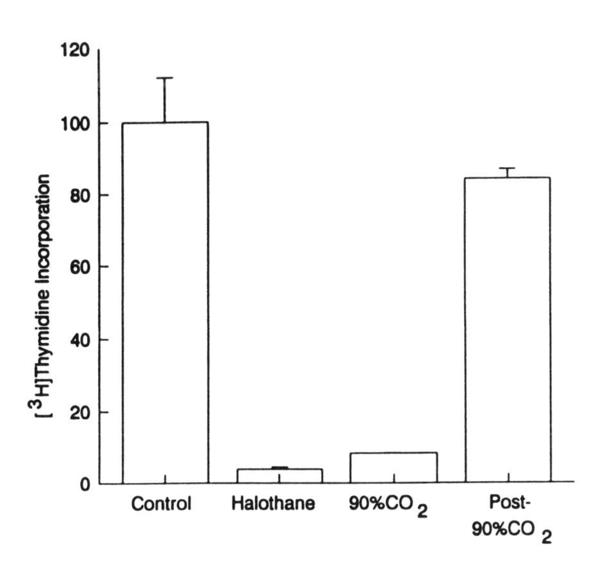

FIGURE 14-3. Halothane and acidification terminate DNA synthesis in embryonic cortex. Normalized [^{3}H]thymidine uptake in four conditions: control, 2 mM halothane, 90% CO_2/10% O_2, and then after 90% CO_2/10% O_2. There is a significant difference between control, halothane, and acidification (p<0.0001) treated explants.

rite outgrowth and preserving neuroarchitecture (Mattson and Kater, 1989). Could neuron differentiation, occuring when neurons enter the cortical plate, be regulated by glutamate activation of these early appearing NMDA receptors? Using *in situ* whole-cell recording from embryonic cortical slices, it was possible to demonstrate spontaneously active NMDA currents in immature neurons in the cortical plate (LoTurco et al., 1991). An endogenous source of NMDA agonist, presumably glutamate, is there-

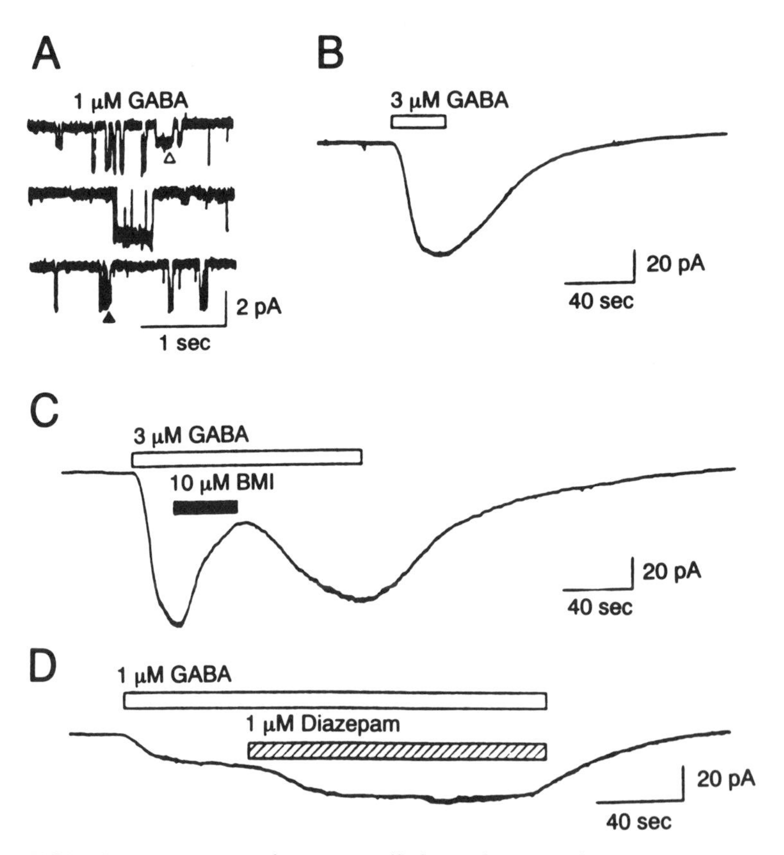

FIGURE 14-4. Ventricular zone cells have functional $GABA_A$ receptors. **A.** GABA-evoked single-channel currents in a patch of membrane excised from a cell in the ventricular zone and exposed to 1 μM GABA. The membrane potential was −70 mV, and two channel conductances were apparent, as are seen in $GABA_A$ receptor channels in mature neurons—an infrequently occurring conductance of 13 pS (open triangle) and a more frequently occurring conductance of 21 pS (closed triangle). **B.** Inward currents in ventricular zone cells are elicited by the application of GABA. **C.** Antagonism of the response to 3 μM GABA by the $GABA_A$ receptor antagonist bicuculline methiodide (BMI; 10 μM). **D.** Potentiation of the response to 1 μM GABA by the benzodiazepine, diazepam (1 μM).

fore present in the developing cortical plate and may be playing a role in guiding differentiation or signaling the end of migration. Recent data in the cerebellum suggest a role in regulating migration along glial fibers (Komuro and Rakic, 1993). The specific role mediated by NMDA receptor activation at this early stage of cortical development remains to be explored.

Electron microscopic analysis of developing cortex indicates that type 1 putative glutamatergic synapses do not form in cortex until the beginning of the first postnatal week, and that type 2 putative GABAergic synapses do not form until the beginning of the second postnatal week (Blue and Parnavelas, 1983), a sequence that is consistent with the electophysiological appearance of spontaneous and evoked synaptic events (LoTurco and Kriegstein, unpublished observation). The specific sequence of receptor expression on immature neurons and precursor cells suggests that GABA and glutamate may have an effect on neuronal development quite apart from their more traditional roles in synaptic transmission.

CELL-CYCLE EVENTS IN THE EMBRYONIC CORTEX ARE INFLUENCED BY GABA AND GLUTAMATE

$GABA_A$ and non-NMDA receptors are already present on progenitor cells in the embryonic ventricular zone, well in advance of their role in synaptic signaling or synaptogenesis. What role, if any, could they be playing at

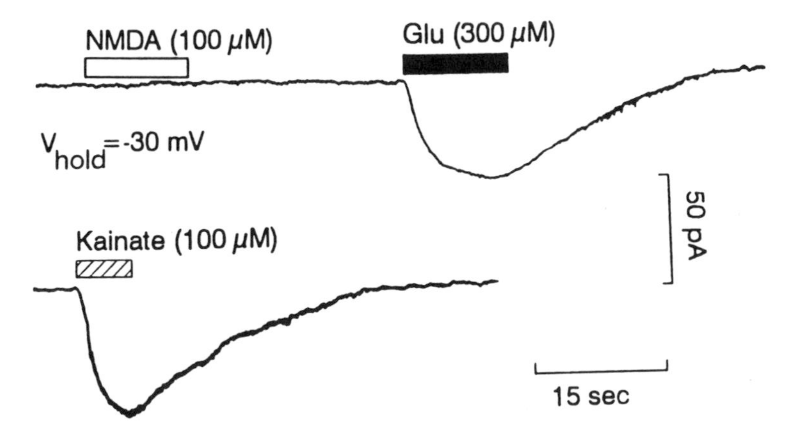

FIGURE 14-5. Ventricular zone cells have AMPA/kainate receptors but not NMDA receptors. Both traces are voltage-clamp recordings from cells in the ventricular zone (E16). NMDA (100 μM) induced no detectable currents, whereas both glutamate (300 μM) and kainate elicited inward currents. These currents were blocked by the AMPA/kainate receptor antagonist CNQX (not shown).

such an early stage? Since neurons differentiate only after leaving the ventricular zone, and the chief activity of ventricular zone cells is cell division, the precocious amino acid receptors might regulate cell-cycle events. Mitotic activity in the ventricular zone of developing neocortex is highly regulated. In the rat, cells in the ventricular zone destined for different layers undergo their last divisions in an overlapping sequence from embryonic day 13 to 20 (Miller, 1988). After E20 there are no more neuronal divisions in the cortex. One hypothesis for the regulation of proliferation in developing cortex is that feedback from differentiating cells terminates division in the ventricular zone (Anderson, 1989; McConnell, 1991). Since transmitter systems differentiate during early cortical development, (Blanton and Kriegstein, 1991; Lauder et. al., 1986; Parnavelas et. al., 1988a,b; Van Eden et. al., 1989) and since developing neurons send growth cones to the top of the ventricular zone (Kim et. al., 1991), neurotransmitters are a candidate for such a feedback signal. The precocious appearance of amino acid receptors on mitotically active cells in the ventricular zone further supports the hypothesis that neurotransmitters regulate mitotic activity.

The rate of DNA synthesis in embryonic cortical cells was indeed found to be regulated by amino acid agonists. An assay of DNA synthesis was developed by bathing embryonic cortical explants in labeled thymidine for fixed intervals and then measuring the amount of thymidine incorporation. Both GABA and kainate caused a significant decrease in the incorporation of [^{3}H]thymidine in embryonic cortical explants. The GABA-induced decrease was blocked by the $GABA_A$ antagonist bicuculline methiodide (BMI), and similarly the kainate-induced decrease was blocked by the AMPA/kainate receptor antagonist CNQX (Fig. 14-6). Therefore, the activation of $GABA_A$ and kainate receptors can decrease DNA synthesis in embryonic neocortex, but is there evidence that such a mechanism is functional during cortical development? This hypothesis was tested by exposing intact embryonic cortical explants to bathing medium containing specific amino acid receptor *antagonists* and measuring changes in DNA synthesis. Consistent with the hypothesis, the antagonists CNQX and BMI *increased* [^{3}H]thymidine incorporation, confirming that embryonic cortex contains endogenous sources of agonist for both GABA and glutamate receptors (Fig. 14-6).

GABA DEPOLARIZES VENTRICULAR ZONE CELLS BECAUSE OF HIGH INTRACELLULAR CHLORIDE CONCENTRATION MAINTAINED BY A CHLORIDE EXCHANGE PUMP

Although GABA and glutamate receptors are coupled to different ionophores and have opposing effects on membrane polarization in adult neu-

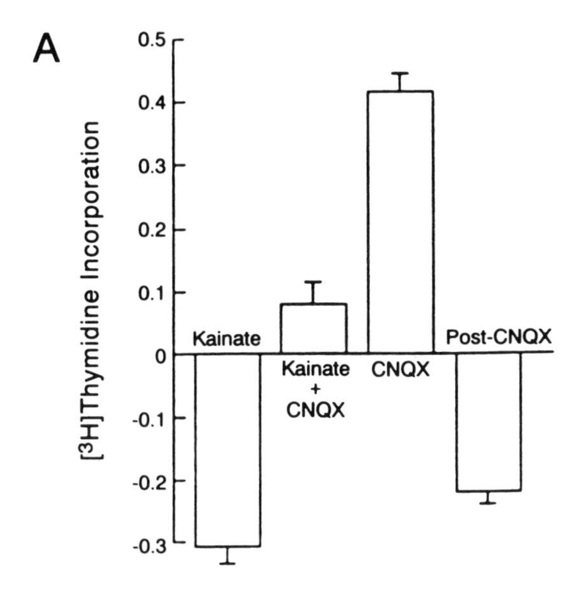

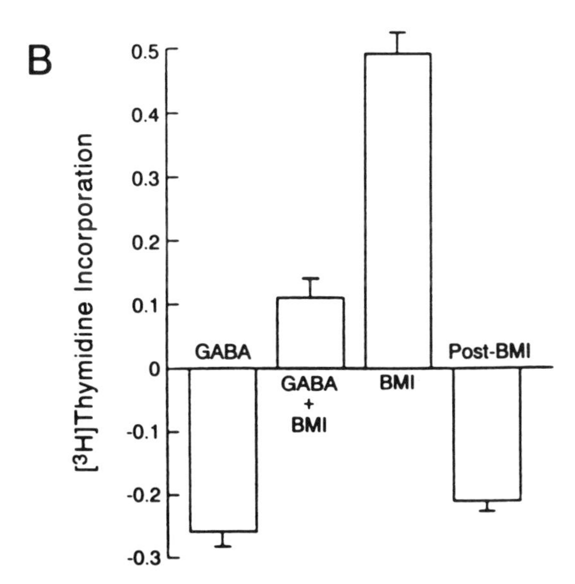

FIGURE 14-6. Thymidine incorporation in embryonic cortex (E19) is decreased by GABA and glutamate and increased by GABA and glutamate antagonists. **A**. Thymidine incorporation is inhibited by exposure to kainate (150 μM) for 5 hours. The decrease in DNA synthesis is blocked by the AMPA/kainate antagonist CNQX (50 μm). Conversely, exposing cortical explants to CNQX alone significantly upregulated thymidine incorporation during a 5-hour exposure. Incorporation during the 5 hours after CNQX exposure was significantly depressed. **B**. Thymidine incorporation is inhibited by exposure to GABA (50 μM) for 5 hours. This effect is blocked by BMI. Exposure to BMI (30 μM) alone for 5 hours increased thymidine incorporation. Incorporation was significantly decreased for 5 hours after the withdrawal of BMI.

rons, they both have the same effect on immature cortical cells—slowing down DNA synthesis. In proliferating cells within the ventricular zone, GABA is depolarizing, the reverse of the hyperpolarizing effect induced by GABA in mature neurons. In fact, the reversal potential of GABA-induced currents was −5 mV (Fig. 14-7A). In mature neurons, GABA usually exerts hyperpolarizing effects because the intracellular chloride concentration is maintained at a low level by a chloride pump (Llinás et. al., 1974; Lux, 1971). This raised the possibilities that either chloride is actively accumulated within ventricular zone cells or that the GABA conductance on ventricular zone cells is not mediated by chloride but perhaps by cations.

The chloride dependence of GABA channels in ventricular zone cells was tested by recording from cells injected with increasing concentrations of chloride. As the intracellular chloride concentration increased, the reversal potential for GABA-induced currents became more depolarized, indicating that chloride was a permeant ion (Fig. 14-7B). In fact, reversal potentials reached as high as 25 mV, with 124 mM chloride in the patch electrodes and 129 mM chloride in the extracellular solution. In order to achieve such depolarized reversal potentials, high intracellular chloride concentrations must be maintained against a concentration gradient. To test for the presence of a chloride exchanger that could maintain a high intracellular chloride concentration, furosemide, an anion exchanger blocker, was applied to cortical explants. In the presence of furosemide, the reversal potentials of GABA currents were shifted from their depolarized levels to approximately −47 mV (Fig. 14-7C). This result is consistent with the hypothesis that an active chloride exchanger keeps the chloride concentration high within ventricular zone cells. Stimulation of $GABA_A$ receptors therefore depolarizes ventricular zone cells as a consequence of the high internal chloride concentration.

DEPOLARIZATION MEDIATES THE DNA SYNTHESIS INHIBITION INDUCED BY GABA AND GLUTAMATE

Activation of either $GABA_A$ or kainate receptors had the same effects on DNA synthesis, even though they conduct different ions. However, as reviewed above, activation of either receptor type depolarizes ventricular zone cells. A process mediated by depolarization, therefore, may cause downregulation of DNA synthesis. To test whether membrane depolarization was responsible for GABA-induced downregulation of DNA synthesis, $[^3H]$thymidine incorporation was determined for GABA alone and for GABA and furosemide combined. Blocking the chloride exchanger with furosemide shifts the reversal potential of GABA-elicited currents to

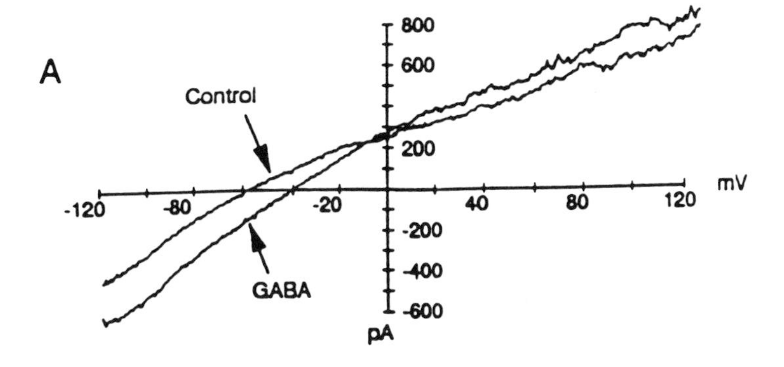

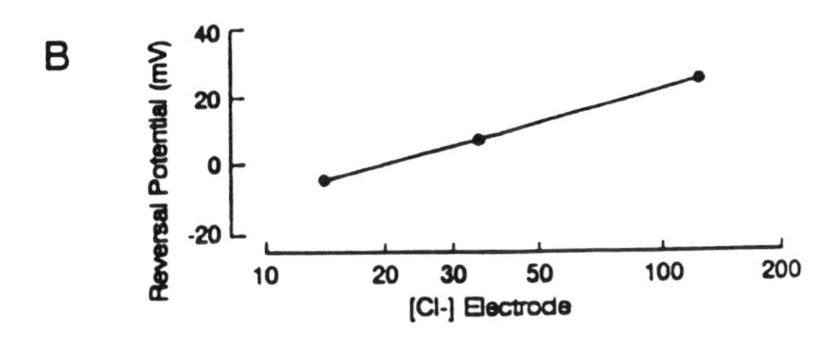

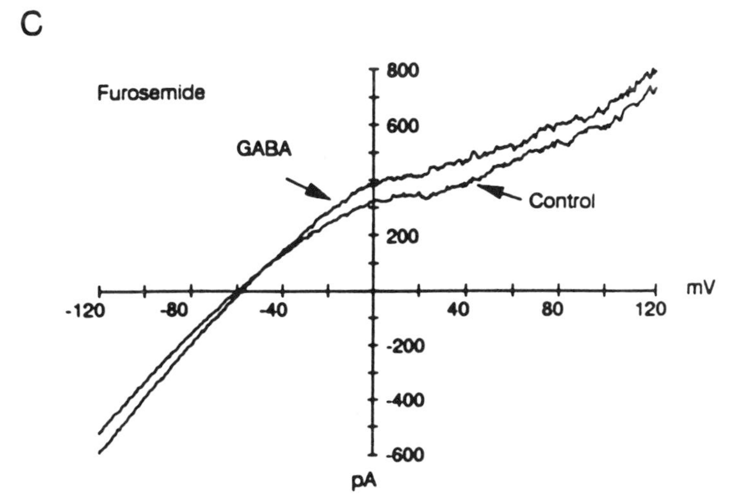

FIGURE 14-7. Chloride dependence and the reversal potential of GABA-mediated currents in ventricular zone cells. **A**. Current versus voltage plot before and after the application of GABA. The I-V relationship was determined by making a ramp voltage change from −120 to 120 mV. **B**. A plot of GABA current reversal potentials versus chloride concentration in the patch electrode. **C**. The chloride exchanger blocker, furosemide (5 mM), shifts the GABA-induced current reversal potential to a more hyperpolarized level.

more hyperpolarized values. This action essentially makes GABA no longer depolarizing on ventricular zone cells. Blocking the depolarizing effects of GABA with furosemide inhibited the GABA-mediated decrease in DNA synthesis (Fig. 14-8). The GABA-induced downregulation in DNA synthesis is therefore mediated by membrane depolarization.

To further test whether depolarization alone could mediate DNA synthesis downregulation, E16 cortical explants were incubated in elevated potassium and [^{3}H]thymidine. Raising the extracellular potassium concentration to 20 mM, a concentration that causes a 15-mV depolarization, reduced [^{3}H]thymidine incorporation by 48% (Fig. 14-8). This result is consistent with the hypothesis that amino acid receptors decrease DNA synthesis by membrane depolarization.

CONCLUSIONS

These experiments raise several intriguing questions. How does depolarization produced by amino acid stimulation modulate cell-cycle control mechanisms? What is the source for GABA or glutamate in the embryonic cortex? Why is extrinsic regulation of the timing of cell-cycle events necessary? At least two phases of the cell cycle—entry of cells in G1 phase into S phase and progression through M—are highly calcium dependent (Hazelton et. al., 1979; Izant, 1983). Amino acid receptor activation leading to depolarization could activate voltage-gated calcium channels, and calcium entry could regulate progression through the cell cycle. Recent

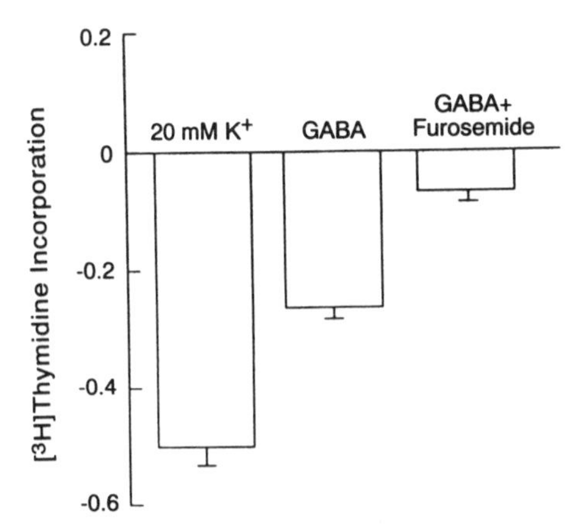

FIGURE 14-8. Depolarization is necessary and sufficient to downregulate DNA synthesis in embryonic cortex. Potassium (20 mM) decreases [^{3}H]thymidine incorporation, as does 30 μM GABA. GABA does not decrease [^{3}H]thymidine incorporation in the presence of 5 mM furosemide.

anatomical studies have shown that small numbers of GABAergic and glutaminergic cells develop in the cortical plate and send axons into the ventricular zone at very early stages (Kim et al., 1991). Transmitters released from either growth cones or migrating cells could activate receptors on dividing cells in the ventricular or subventricular zone and regulate the timing and termination of proliferation. It is intriguing to speculate that there may be "feedback" regulation by already differentiated cortical neurons on the production of new neurons in the ventricular zone. In addition, this regulatory process allows for "fine tuning" corticogenesis and could help explain how prenatal drug exposure can lead to cortical maldevelopment.

Cells in the ventricular zone communicate directly with each other through gap junction channels and respond to changes in their environment through amino acid receptors. Moreover, these interactions regulate cell-cycle events and the proliferation of cortical cells. These interactive influences demonstrate the important role of environment, rather than simply cell-autonomous genetic programs, in early cortical development. A variety of developmental disorders with seizures as a manifestation, ranging from major birth defects to subtle cortical heterotopias, have their origin in early stages of cortical differentiation and migration. As we discover that cortical development involves interactive signaling and we learn more about the nature of the signaling factors, we step closer to the time when therapeutic intervention may become possible to treat or prevent certain disorders of brain development.

15

Gene control of cortical excitability

JEFFREY L. NOEBELS

The properties of all cell membranes change with use, but in neurons the nucleus is critically responsible for the plasticity that distinguishes moment-to-moment changes in synaptic signaling from the lifelong storage of integrated information in the brain. When a single cell is exposed to quantal packets of transmitter, the electrical response depends on the instantaneous spatial arrangement and kinetic behavior of receptor and ion channel proteins residing in the membrane. In the native state, most known membrane currents in cortical neurons recover in less than one second. Within the first few milliseconds of opening, a rapid change in the activation state and recovery period of these primary signaling molecules can be triggered by a reversible network of second-messenger systems beginning at the cytoplasmic face of the membrane. This use-dependent shift in firing properties describes a new signaling mode that persists for the lifetime of the modulated state; typically it is measurable in minutes to days. During this period, further changes in membrane behavior continue according to a second derivative, described by the rate at which both the membrane protein and its cognate ligands in the signaling cascade are further modified or replaced by the products of new patterns of gene expression. In some cases, the pairing of synaptic events may be critical in determining whether or not new gene transcription occurs. Since a typical pyramidal neuron in the neocortex receives afferent input from up to 10,000 synapses, the nucleus is a major integrator of synaptic synchrony and the site where long term decisions regarding the functional significance of complex network activity are made.

The task of predicting which genes contribute importantly to specific cortical excitability phenotypes is difficult for several reasons. First, gene expression, as measured by the diversity of mRNA transcripts, is more heterogeneous in the brain than in any other tissue. In one region of human archicortex alone, the hippocampus, a cDNA library has been constructed containing partial sequences from over 2,375 genes (Adams et al., 1992). Second, the rates of gene transcription are differentially altered by afferent activity, and the stability of their mRNA transcripts and end products varies. Between 500 to 1,000 gene products are induced by the glutamate analogue kainate (Nedivi et al., 1993). Within 3 hours after the hippocampus is stimulated in a pattern producing long-term potentiation, altered levels of 11 gene transcripts are found (Fazeli, 1993), and others undoubtedly emerge and subside at various times following different stimulus paradigms. Finally, the complexity of fetal and adult brain mRNA patterns changes dramatically during development under the control of transactivating factors that link the expression of multiple gene loci into specific constitutive patterns. The promoter regions of these genes also contain elements responsive to transcription factors that can be regulated by non-synaptic signals, such as steroid hormones, cytokines, and other trophic molecules (Catarsi and Drapeau, 1993; Mehler et al., 1993, Yamagata et al., 1993).

Regardless of the factors complicating the search, one group of genes expressed in central neurons can be operationally defined and constitute a category of central concern: *excitability genes*. These genes initiate and maintain the critical capacity for voltage-dependent membrane behavior and synaptic transmission in specific brain networks and participate directly in their ability to display rapid neuromodulation, synchronization, and long-term retrieval of information. It is from within this group that the genetic elements most directly responsible for the heritable component of human cognitive abilities will ultimately be found.

A decade ago, a discussion of excitability genes was confined to genes for ion channels, neurotransmitters, and their receptors (Noebels, 1985); however, steady breakthroughs have created a new list of critical excitability mechanisms that operate in developing cortical circuits. Molecules mediating migration, outgrowth, and connectivity between cortical cell assemblies have been isolated and added to the rapidly emerging tangle of genes for cytoplasmic and nuclear signaling cascades; new effector molecules, such as tyrosine kinases, and the proteins mediating vesicle mobilization, fusion, and docking (Bennett and Scheller, 1993; De Camilli, 1993; Schlesinger and Ulrich, 1992) are now incorporated into model synapses along with non-quantal uptake and release systems, such as neurotransmitter transporters (Amara, 1992). Along with this new biology, nucleic acid probes have contributed the basis for a new molecular anatomy of the brain by assembling synaptically unaffiliated neurons into functional groups based on the shared expression of transcripts for specific excitability properties. The spatial expression patterns of these genes create a new nomenclature for neuronal circuitry, and a key to clarifying the seemingly unrelated components of a neurological phenotype that can result from a mutation in a single gene. The three interlocking attributes—neuronal position, signal transducing properties, and firing patterns—are the source of hereditary variants of cortical excitability. Since the subset of alleles that modulate cortical excitability without producing a lethal phenotype are neither constant nor predictable throughout development, experimental strategies to identify physiologically relevant genes are emerging from several competing technologies, each based on the tactic of examining the effects of one gene at a time in the developing nervous system. As has been found in the dissection of virtually all multistep biological phenomena, mutational analysis has proven to be an exceptionally powerful tool, and this approach is emphasized in this chapter.

GENETICS OF CIRCUIT ASSEMBLY

Neuronal differentiation and brain morphogenesis are mutually governed by the expression of sets of genes in temporally ordered sequences. Cell-cell interactions during brain development therefore provide an integral entry point into the process of identifying excitability genes in cortical circuits. Although the search for rules that govern the interplay between excitability gene expression and cortical cytoarchitecture might be expected to conclude that the two processes are inseparable, there are interesting exceptions to this relationship at different developmental stages.

Division, Migration, Neurite Outgrowth, Synaptogenesis

Mapping of the movements of central neurons from the early embryo to the adult has defined a four-dimensional framework to examine the contribution of specific genes to the processing of information within cortical modules. Various complementary strategies, including the use of mouse mutants, experimental manipulations of cortical neurogenesis, and reconstruction of cell lineages, migration, and axon trajectories, are currently being employed to identify specific molecular determinants.

Cell-Cycle Control of Cortical Connectivity. The neuronal birth date and the maturity of neighboring cells are key predictive factors in determining the fates of neuronal connections. Both the intracortical destination of the migrating neuron and its ultimate identity as a member of a particular projection type are largely defined at the time of the final S phase of cell division, before attainment of the final laminar position and before the outgrowth of specific cortical connections (McConnell, 1991; Takahashi et al., 1993). Since cell identity is heavily influenced by the segregation of undifferentiated neurons into distinctive laminae, gene defects expressed at early rounds of cell division in the ventricular zone could lead to a larger or smaller population of a specific cell type; for example, interneurons. Glial precursors could undergo parallel cell-cycle anomalies. This category of error could lead to numerical imbalances in synaptic excitation and inhibition with deleterious effects on network excitability. For example, a mutation in the tyrosine kinase gene *fyn* results in an increased number of neurons in the hippocampus in the *fyn*⁻ mutant mouse (Grant, et al.,1993). *Fyn, a* member of the non-receptor subtype of tyrosine kinases in the cytoplasmic transduction pathway, seems to play a role in neuronal development and synaptic plasticity, whereas deletions in the related *src, yes*, and *abl* genes have little effect. Members of the second class of tyrosine kinases, the membrane-spanning receptors for neurotrophic molecules, are interesting candidates that might link cellular position with membrane excitability by modulating extrasynaptic transmitter- or contact-activated receptor expression (Catarsi and Drapeau, 1993). The mouse mutations, *white spotting* (W) and *Steel* (Sl), delete the genes for the tyrosine kinase *c-kit* receptor and its ligand, a growth factor expressed widely in neurectodermal tissues including the central nervous system (Reith and Bernstein, 1991). Although the mutations result in hematological and melanocytic abnormalities, no neurological phenotype has yet been reported in either mutant.

Neuronal Migration. Excitability problems may arise from gene-linked defects in cortical migration. *In vitro* studies suggest that a cortical neuron migrating from the proliferative zone may not be confined to a single radial glial guideline, but may select other intracortical paths (O'Rourke, et al., 1992; Walsh and Cepko, 1992). Whether horizontal dispersion occurs by coincidence or design is unknown, but mutations that improperly redistribute cells across cortical modules could lead to defective surround inhibition and abnormal cortical synchronization. The organization of major tel-, rhin- and rhombencephalic regions seems to involve specific genetic signals, such as LH-2 (Xu et al., 1993), Lamp (Barbe and Levin, 1992), and L7 (Oberdick et al., 1993), although the determinants of local boundaries of cortical compartments and their significance are less clear (Purves et al., 1992; Wilson et al., 1993). Transgenic mice with a *lacZ* reporter gene fused to the HPRT promoter reveal the existence of distinct topographic and temporal domains of gene expression in neocortex (Bonnerot et al., 1990), creating the possibility for highly focal inherited migratory lesions in the developing neocortex. Delayed maturation in specific cortical regions could contribute to the transient appearance of focal seizures found commonly in neonates (see Chapter 17).

The molecules for migration guidance in neocortex may coincide with some of those governing the subsequent step of axon outgrowth. One candidate is the gene for Miller-Dieker syndrome, a human lissencephalic malformation with abnormal neuronal migration resulting in a smooth, agyric neocortex (Reiner et al., 1993). The LIS-1 gene shows significant homology in its amino acid sequence to the β-subunits of heterotrimeric G proteins (β-transducins) expressed in cortical pyramidal neurons (Bourne and Nicoll, 1993), implicating the involvement of a critical step in the cytoplasmic signal transduction pathway for neuronal migration during normal cortical development. Blockade of N-type calcium channels (Komuro and Rakic, 1992) and genetic deletion (Li et al., 1994) or blockade (Komuro and Rakic, 1993) of NMDA receptors can also interfere with migration and topography, thus linking cell and axon motility with activity-dependent extracellular glutamate concentrations and voltage-dependent calcium entry.

Neurite Outgrowth. Once neurite outgrowth begins, the chemoaffinity hypothesis remains well supported by the evidence for essential molecules encoding topographic connectivity (Baier and Bonhoeffer, 1992; Goodman and Shatz, 1993; Sanes, 1993). Most data involving axon path finding have been obtained in subcortical projection relay sites, such as the geniculate and collicular nuclei. Candidate genes involved in axon transit and the final positioning and stabilization of synaptic terminals include those for membrane-bound cell adhesion molecules and transmitter receptors. Diffusible signals initiate outgrowth (Klar et al., 1992; Placzek et al., 1990), guide direction by attraction (Zheng et al., 1994) and repulsion (Pini, 1993), and regulate branching and collateral targeting of axons (O'Leary and Koester, 1993). Proteins expressed within the differentiating growth cone are co-participants, since axon growth can be inhibited by antisense oligonucleotides homologous with SNAP-25 transcripts (Osen-Sand et al., 1993).

Analysis of mutations affecting cortical migration and neurite outgrowth in mouse and humans is beginning to provide clues to the dependence of cortical excitability upon neuronal architectonics. In the mouse mutant *reeler*, there is a striking anatomical reversal in cortical lamination patterns, resulting in an inversion of large cortical pyramidal output neurons from deep to superficial layers and a disruption in the radial orientation of apical dendrites (Caviness, 1976). Despite the chaotic cortical organization, spontaneous EEG rhythms in *reeler* neocortex are unremarkable compared with wild-type mice (Fig. 15-1). These mutants show no accompanying seizure disorder. Physiological studies demonstrate functionally intact synaptic connectivity in *reeler* visual cortex (Simmons and Pearlman, 1983), and normal intrinsic membrane properties in sensorimotor cortex (Silva et al., 1991b), and behavioral studies show no major defect in simple learning paradigms (Goldowitz and Koch, 1986). This evidence suggests that the maturation of intrinsic firing patterns characteristic of pyramidal neurons in specific neocortical laminae (Connors and Gutnick, 1990) is not strictly position dependent. In contrast, minor neuronal heterotopias are an increasingly common neuropathological finding in primary generalized epilepsies (Meencke amd Veith, 1992). In a human brain malformation designated as the "double cortex" syndrome, the neocortical mantle is characterized by a bilaminar separation presumed to reflect a genetically defective migration program, and the syndrome is characterized by a clinical phenotype of mental retardation and intractable seizures (Palmini et al., 1992). Although neither *reeler* nor any of the putative cortical dysgenesis gene products are known, one may conclude that the actual

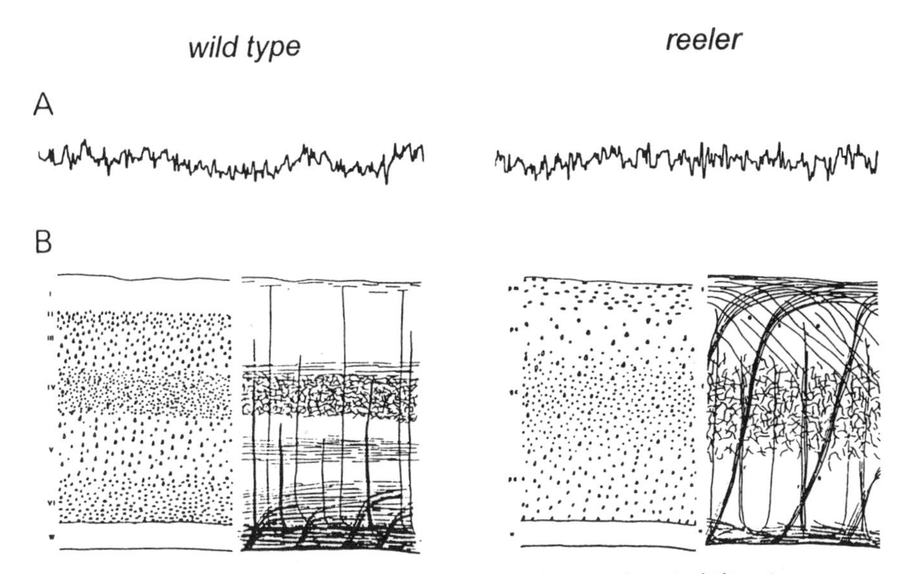

FIGURE 15-1. Single gene migration defects and cortical excitability. Spontaneous electrocorticographic activity **(A)** in awake adult *reeler* mutant shows a pattern identical to the wild-type control mouse, despite a severe defect in cortical neuronal migration **(B)** characterized by an inverted lamination with preserved synaptic targetting. **A** from Noebels, J. L., unpublished data; **B** from Caviness (1976).

position of cell somas within the cortex is less relevant than the defects in membrane excitability and synaptogenesis that may accompany these early developmental errors.

Although the *reeler* mutant suggests that cortical cytoarchitectonics and network excitability may be altered independently, synaptic connectivity is strongly dependent on early firing patterns. Both the fine terminal structure and number of cortical synapses are dramatically altered by abnormal inputs during early development in primary sensory systems (Constantine-Patton et al., 1990; Shatz, 1990), and somewhat later in associative networks, such as the hippocampus (Qiao and Noebels, 1993), demonstrating that membrane excitability properties contribute to synaptogenesis and connectivity at various stages of CNS maturation.

Developmental Control of Intrinsic Membrane Excitability

Central neurons have distinct developmental timetables for the acquisition of specific ion channels, receptors, and transporter molecules. A major challenge has been to determine how the appearance of these molecules is dictated by intrinsic preprogrammed signals interacting with developmental cues. For example, functional subunit diversity of NMDA receptors arises in migrating granule cells of the cerebellar cortex (Farrant et al., 1994), and may depend on positional as well as chronological events. *In vitro* experiments suggest that external signaling between neighboring cells regulates the delayed expression of three specific GABA receptor subunits (α1,β2, γ2) in the cerebellum (Beattie and Siegel, 1993). Alterations in the expression levels of these genes could produce unique ligand-gated ion channels with dissimilar single channel properties (Angelotti et al., 1993). Neuron-neuron interactions between wild type and *weaver* mutant granule cells can rescue inherited defects in neuronal differentiation (Gao et al., 1992). Glial interactions also contribute to neuronal maturation. A particularly clear example of nonsynaptic glial control over neuronal excitability is illustrated by the axonal plasticity seen in the *shiverer* mutation, which reveals that a deletion of the myelin basic protein gene (expressed only in oligodendrocytes) results in a secondary upregulation of type II sodium channels in the plasmalemma of hypomyelinated central axons (Noebels et al., 1991; Westenbroek et al., 1992b).

Since many ion channels and neurotransmitter receptors are heteromultimeric proteins, regulated expression of subunit genes is a fundamental feature of normal development. The first clear example of this process was provided by the ontogenetic changes in cholinergic receptor gating currents at the motor endplate (Sakmann and Brenner, 1978) due to γ to ϵ subunit switching of genes for the nicotinic acetylcholinergic receptor (Mishina et al., 1986). Ion channels behave similarly, as illustrated by the embryonic switch from TTX-insensitive to TTX-sensitive sodium channels in the adult muscle fiber membrane due to the inverse temporal regulation of two sodium channel genes (Kallen et al., 1990). In the brain, glutamate receptors undergo a different type of molecular metamorphosis; alternative splicing of the transcript results in two distinct developmental isoforms with differing functional properties (Monyer et al., 1991), one of which predominates in the adult GluR AMPA subtype (Burnashev et al., 1992a). The glycine receptor also has fetal and adult isoforms (Becker et al., 1988), and developmental switching alters channel kinetics; incorporation of the adult isoform markedly shortens the mean channel open time, and the resulting fast IPSCs permanently alter signaling in central glycinergic pathways (Takahashi et al., 1992). The *spastic* mouse provides an example of an inherited defect in the developmental program controlling glycine receptor gene switching. Homozygous *spastic* mice develop motor spasticity at 2 weeks of age due to a reduction in central glycine receptors. Normal levels of the fetal isoform are present in the adult, but the adult strychnine-sensitive isoform is sharply reduced, indicating that the mutation directly interferes with the age-dependent regulation of the gene (Becker et al., 1992). A phenotypically related mutant, *spasmodic*, shows a mutation in the α1 glycine receptor subunit (Ryan et al., 1994). Since many other excitability molecule superfamilies are likely to undergo developmental substitution, inherited subunit switching defects may constitute a distinct category of early-onset neurological excitability disorders.

GENETIC STRATEGIES IN CEREBRAL EXCITABILITY ANALYSIS

Several strategies have been pursued to analyze the role of individual genes in modifying cortical excitability (Fig. 15-2). Given the complexity of reciprocal interactions between gene control and neuronal pathophysiology, knowledge of a single identified mutant gene product itself is often not adequate to predict the final behavior of the network, and the compensatory regulation of other genes at subsequent intervening steps must be identified to understand fully the phenotypic expression of a mutant gene error. This search for cellular pleiomorphisms is particularly critical in the analysis of the developing central nervous system, since delayed effects of gene lesions secondarily alter excitability in entirely unpredictable ways.

Candidate Genes

Central nervous system genes are now being isolated and mapped onto mouse and human chromosomes. Once sequenced, their anatomical expression at different ontogenetic stages can be defined using *in situ* hybridization techniques, and the spatial distribution in cellular membrane domains, co-assembly into functional homomers or heteromers, and colocalization with other gene products can be evaluated with subunit-specific monoclonal antibodies. Expression data of this kind in wild type brain are now available for three sodium channel subtypes (Mandel, 1992)—N, L, and T type calcium channels (Soong et al., 1993; Westenbroek et al., 1990, 1992a)—and various potassium channel subtypes, (Sheng et al., 1993; H. Wang et al., 1993).

Along with the familiar genes, partial sequences from cDNA library screening, known as expressed sequence tags (ESTs), form the starting point for the cloning of novel cortical genes. An alternate approach involves the identification of intragenic rearrangements whose phenotypic expression is likely to target the brain. Since a small number of interesting neurological disorders are now known to be caused by triplet repeat mutations, a search for this specific genomic lesion may be an efficient strategy to identify cerebral excitability genes. One example is the set of gene mutations characterized by the abnormal amplification of triplet repeat sequences. Interestingly, all members of this class so far identified express

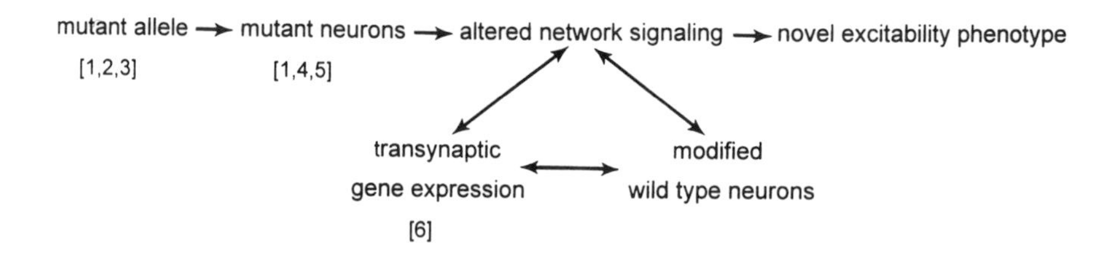

1. candidate genes
2. targetted mutagenesis
3. spontaneous mutations
4. genetic ablation
5. gene transfer
6. activity-induced genes

FIGURE 15-2. Experimental strategies in the neurogenetic analysis of cerebral excitability. Single gene involvement in specific neuronal populations, firing patterns, or neurological disease phenotypes can be evaluated at various points of entry during the development of cortical circuits. Candidate genes can defined by linkage mapping, screening of cDNA libraries after isolation by subtraction hybridization, anatomical expression, or known function.

neurodegenerative CNS phenotypes (Martin, 1993), possibly because a prolonged accumulation of cytotoxic metabolites or a lack of DNA repair in non-replicating central neurons can proceed without being masked by successive generations of younger unaffected cells. Since excitotoxicity is a candidate mechanism for cell death in these disorders, any primary gene defect that enhances repetitive firing and depolarization could also predispose toward neuronal loss. This suggests that the same class of membrane errors that cause abnormal synchronization and epilepsy when expressed in neocortical or hippocampal neurons might lead to hereditary tremor, ataxias, or other movement disorders when expressed in cerebellar or subcortical motor circuits. These neurodegenerative diseases will provide highly instructive clues regarding gene control of cerebral excitability once their mechanisms are better understood.

Several examples of single locus mutations affecting the cerebral cortex arising from amplification of repetitive gene sequences have been identified. In the case of the fragile X mental retardation syndrome, the lesion is generated from the steady expansion of a triplet CGG repeat in the 5′ untranslated region of the FMR-1 gene (Caskey et al., 1992), raised from the normal copy number of fewer than 50 to levels in excess of 2,000. Although the nature of the translational processing defect is not yet defined, high levels of FMR-1 expression are found in cerebral cortex (Hinds et al., 1993). Human cDNA library screening at high stringency has resulted in the isolation of nine genes with triplet repeats that may be candidate neurological disease genes, and some estimate that over 100 more remain (Ross et al., 1993). One identified clone was that for myotonin protein kinase, the defective gene in myotonic dystrophy that shows hundreds of CTG repeats in the 3′ untranslated region, a rare site for disease mutations, compared with only 5 to 30 in unaffected controls. The gene is also expressed in brain and shows strong homology to a cAMP-dependent protein kinase (Brook, 1992). Northern blot analysis revealed that eight other novel cDNAs with length polymorphisms were also expressed in various brain regions, including neocortex (Li et al., 1993).

Membrane hyperexcitability phenotypes can also arise from highly conserved intragenic base pair rearrangements that alter the physiology of the peptide translation product. For example, two independent mutations in the human muscle sodium channel α subunit cause phenotypically similar syndromes of periodic paralysis characterized by repetitive myotonic burst discharges and depolarization block. Both mutant alleles disrupt inactivation of sodium currents without altering single-channel conductance (Cannon and Strittmater, 1993; Cummins et al., 1993). The mutations each involve substitution of a single nucleotide, and in each case the mutated residue sits in the membrane-spanning segment near the cytoplasmic face. The effect on gating behavior is highly sensitive to increases in extracellular potassium, and therefore the mutant channel displays activity-dependence with the added potential for escalating positive feedback. A second interesting example of inherited repetitive firing is a point mutation involving a phenylalanine to cysteine substitution in the gene for a skeletal muscle chloride channel identified in families with recessive and dominant myotonia (Koch et al., 1992). The recessive form can be explained by a total loss of function, as seen in ADR mice (Steinmeyer et al., 1991), whereas a dominant negative phenotype could result from the combination of native subunits with a dysfunctional mutant gene subunit in a homomultimeric channel, as seen in a Drosophila *shaker* mutant (Gisselmann et al., 1989). Mutations of this nature have not yet been described in CNS neurons, but could play a key role in the expression of paroxysmal hyperexcitability phenotypes. These examples are reminders that our primitive understanding of the "function" of any particular neural gene remains incomplete until the phenotypic effects of gene overexpression and permutation are also analyzed.

Modification of Gene Expression by Induced Mutagenesis

Targeted mutagenesis of genes expressed in oocytes has enabled the direct prediction of structure-function relationships of ion channels from their gene sequences. When the excitability of the neocortex, rather than the permeability of a single channel, is at issue, however, the whole brain is the only appropriate expression system. *In vivo* expression of mutant transgenes is thus a logical first step in defining a cortical excitability gene (Rossant, 1990), but two obstacles remain: (1) the time interval required before the neurological phenotype can be assessed and (2) the large number of allelic variants to be screened before a desired phenotype is found.

Defined deletions are a powerful strategy when used to modify a precise cellular phenotype. However, a significant disadvantage of null mutations in mouse brain is that a substantial fraction produce clinically silent neurological phenotypes. In these cases, it is presumed that the locus is not truly redundant, but rather is a member of a group of genes overlapping in function. Although apparently not essential, these loci may still play a conditional role in neuronal excitability, if those conditions can be serendipitously discovered. Two clear examples of latent neurological excitability phenotypes are (1) the gene mutation in the ryanodine receptor in muscle responsible for malignant hyperthermia that prolongs calcium entry when triggered by exposure to specific anesthetics (MacLennan and Phillips, 1992) and (2) the point mutation in the α6 $GABA_A$ receptor subunit, which converts the normally insensitive isoform of a cerebellar granule cell-specific $GABA_A$ receptor to an anomalous isoform that is highly sensitive to benzodiazepines (Korpi and Seeburg, 1993).

These examples also emphasize that neurological phenotypes may often arise from mutations causing a net gain of function, either by conformational changes that augment the active state of the molecule or by overexpression of the wild-type gene product. Overexpression can also negatively interfere with function by altering the subunit ratios of multimeric protein subtypes to produce a dominant negative molecular phenotype. So far, about one-half of the genetic diseases discovered involve reduced levels of expression of a native gene product, whereas the other half involve abnormal function. It is becoming increasingly clear that specific excitability disease phenotypes show genetic heterogeneity, being replicable by several types of mutations in different genes, each transmitted in several possible modes of inheritance. The key point is that gene deletion mutants do not define the function of a gene, but illustrate what the remainder of the brain can do without it.

Transgenic Deletion Chimeras. One approach to eliminating a target gene is to combine two existing chromosome deletion mutations that overlap at a desired locus. For example, a brain-specific gene thought to encode a secretogranin was mapped to the mouse chromosome 9 between the *dilute* and *short ear* loci and was deleted by the appropriate cross (Kingsley et al., 1990). Despite obvious drawbacks—the inability to eliminate single genes and the requirement for fortuitous mapping of the target gene to a region where deletion mutants are already available—the method can be used to exclude participation of a given gene in a specific neurological phenotype. A more flexible approach is the ability to reliably produce germ line chimeras generated by homologous recombination in embryonic stem cells in order to delete defined gene sequences. One relevant example is the CaMKII-deficient mutant mouse (Silva et al., 1992a), demonstrating the developmental influence of this molecular signal on cerebral excitability.

Spontaneous Cortical Excitability Mutations and Epileptic Phenotypes

Electrocorticography is a simple assay to screen mouse mutants rapidly for interesting cortical excitability phenotypes. Abnormal gene-linked EEG patterns in a mutant compared to its co-isogenic wild-type control littermate provide unambiguous identification of an excitability gene and a singular opportunity to trace its cellular expression in a reproducible developmental system (Noebels, 1979). Among 110 spontaneous mouse mutant genotypes surveyed to date, phenotypic EEG variants fall into two excitability categories: continuous (or nearly so) rhythmic alterations and intermittent synchronous discharges.

The recessive *mocha* mutation provided the first evidence that the frequency of background neocortical rhythms is a heritable trait (Noebels and Sidman, 1989). The spontaneous EEG activity in awake *mocha* (*mh/mh*) homozygotes displays a virtually exclusive pattern of constant, high-voltage, bilaterally synchronous, 6 to 7 Hz rhythms unaffected by ongoing motor behavior (Fig. 15-3). The pattern is not seen in either (+/*mh*) or (+/+) mice, which both show low-voltage, desynchronized cortical patterns. The persistent theta rhythm cannot be desynchronized, even transiently, by any modality of sensory stimulation. This gene-linked neocortical hypersynchrony is remarkable for the fact that neocortical theta rhythms have not been observed in other neurological mutant loci or inbred mouse strains surveyed to date; it provides unambiguous evidence that the predominant frequency of spontaneous brain rhythms can be expressed as a hereditary trait under the control of a single recessive locus. The *mocha* mutant constitutes a critical experimental model to ex-

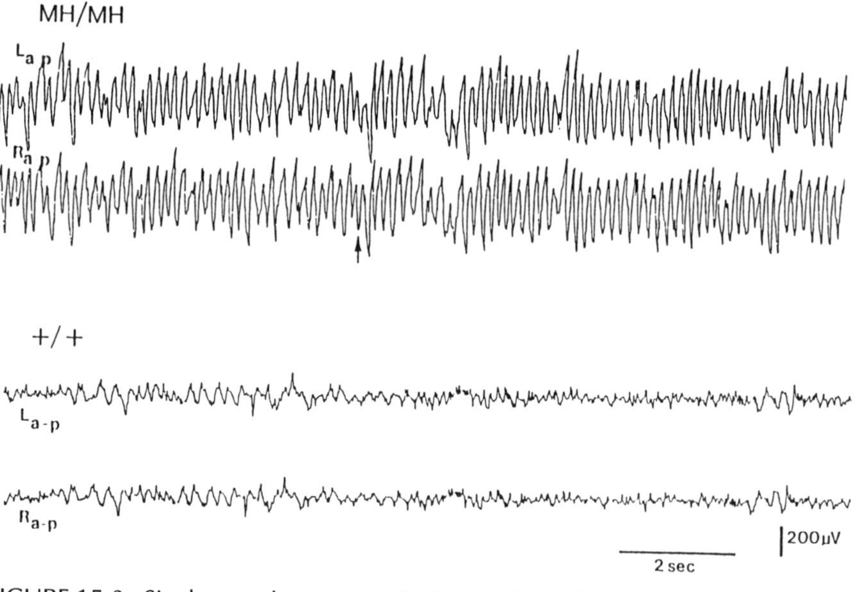

FIGURE 15-3. Single gene locus control of cortical synchronization. The recessive *mocha* locus produces a pattern of constant theta rhythms in the mutant neocortex that cannot be interrupted by sensory stimulation (arrow, air puff). High-amplitude theta rhytmicity is not seen in EEG recordings from wild-type mice (from Noebels and Sidman, 1989).

plore the relationship between cortical synchronization and higher-level information processing in the developing mammalian neocortex.

A more common mutant EEG phenotype is the episodic interruption of normal background desynchronized activity by generalized hypersynchronous spike-wave discharges, similar to patterns seen in human generalized absence or "petit mal" epilepsy. Epilepsy is an unstable behavior of cortical circuits characterized without exception by abnormal neuronal bursting and hypersynchronization, and *epilepsy genes* can therefore be defined as mutant alleles that favor the spontaneous expression of this cortical excitability phenotype. Electrophysiological survey of known single locus mutations in the mouse has revealed that epilepsy genes actually represent a special subset of neurological loci, since only a fraction of mutant mice with CNS defects show EEG abnormalities or spontaneous seizures as an element of the neurological phenotype. This observation made clear at an early stage of the neurogenetic analysis of epilepsy that seizure disorders, despite their high incidence, are neither inevitable nor nonspecific neuropathological outcomes of most inherited CNS disorders, but rather reflect a specific trait of cortical network instability (Noebels, 1979).

Five single locus epileptic mutants—*tottering, lethargic, mocha2j, ducky, and stargazer*—have been identified that share a specific pattern of non-convulsive spike-wave cortical discharges (Fig. 15-4; Noebels, 1984a). Neocortical hypersynchronization in the *tottering* (*tg*, chromosome 8) mutant is controlled by a gene-linked proliferation of noradrenergic locus coeruleus axon terminals in target regions of the *tg* brain, and neonatal correction of the inherited hyperinnervation with a selective neurotoxin prevents the expression of epilepsy (Noebels, 1984b). Intracellular microelectrode studies in brain slices demonstrate that *tg* mutant neurons show a latent voltage-dependent prolongation of burst firing that is unmasked only under depolarizing conditions and appears at the developmental onset of seizures (Helekar and Noebels, 1991, 1992). Excess activation of the noradrenergic beta receptor facilitates the neuronal bursting *in vitro* and may account for one, but not all (Helekar and Noebels, 1994) of the intervening cellular mechanisms underlying the mutant excitability defect.

The *stargazer* mutant (*stg*, chromosome 15) shows a more severe, yet otherwise identical, cortical spike-wave seizure disorder. Unlike the *tg* mouse, no noradrenergic abnormalities have been identified in *stg* brain, nor can they be prevented by neonatal NE depletion, indicating that (1) the two mutant genes produce spike-wave seizures through different intervening neuromodulatory defects and (2) the aberrant LC axon overgrowth in *tg* is not secondarily induced by seizures, but is closely linked to the mutant locus (Qiao and Noebels, 1991). In contrast, seizure-induced hippocampal mossy fiber sprouting is a striking finding in adult *stg/stg* mutants and is far more intense than in *tg/tg* mutants, consistent with the nearly twofold higher rate of seizures in the *stargazer* mice (Qiao and Noebels, 1993). Reactive axon reorganization and neosynaptogenesis have not been previously described in petit mal epilepsy and may contribute to abnormal signaling in the epileptic brain. Interestingly, spike-wave seizures in *stargazer* mice are not blocked by inhibitors of $GABA_B$ receptors, as they are in a third mutant mouse, *lethargic (lh,* chromosome 2; Hosford et al., 1992; Qiao and Noebels, 1992). *In vitro* studies of *stargazer* neurons reveal an abnormal bursting pattern that is distinct from *tottering* neurons, and recent evidence indicates that the neocortex shows specific signs of synchronous network hyperexcitability (Keegan and Noebels, 1993).

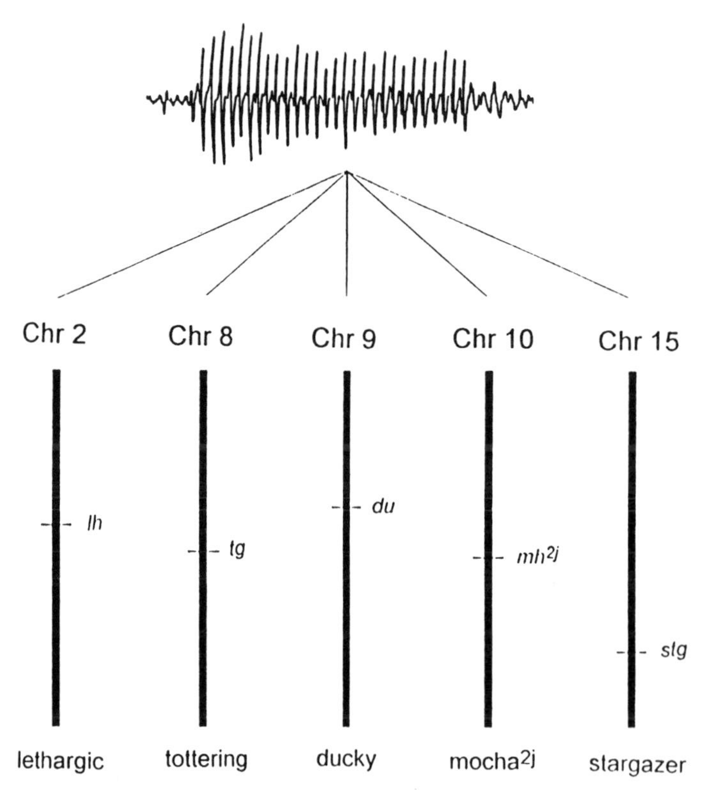

FIGURE 15-4. Electrocortical survey of mapped single locus neurological mutations in mice demonstrates genetic heterogeneity of the spike-wave epilepsy phenotype. Five recessive mutations all give rise to the spontaneous expression of cortical spike-wave synchronous discharges, with each accompanied by a behavioral arrest seizure. Analysis of the mutants reveals distinct intervening excitability defects underlying the same pattern of inherited epilepsy.

Comparative analysis of the single locus epileptic mouse mutants has revealed important parameters of inherited epileptogenesis that were not evident from human genetic studies. First, gene mutations at different loci produce similar abnormal cortical synchronization phenotypes, demonstrating absolute genetic heterogeneity of the spike-wave EEG trait. Second, the gene mutations act on distinct intervening cellular mechanisms, accounting for the differences in seizure severity and pharmacology between the mutants. Third, some cellular elements of the epileptic phenotype may be secondary, activity-induced changes in membrane excitability and synaptic connectivity that follow the onset of seizures. These elements of gene- and seizure-linked plasticity will become more clearly defined as the pathological analysis moves further toward the molecular defect in the individual mutants.

Targeted Genetic Ablation of CNS Cells

Selective ablation of an entire cell population expressing a specific gene is another method of dissecting excitability in the CNS. One targeting strategy involves the use of a lethal gene construct that is widely delivered but selectively expressed under the control of a cell type-specific promoter. Regulated expression of the transgene leads to cell death, and the method has been useful in developmental studies (Evans, 1989). In one study, a lethal transgene incorporating an attenuated copy of the diphtheria DT-A gene, tox 176, was linked to a myelin basic protein promoter sequence. Subsequent expression of the DT-A transgene during fetal and postnatal development led to selective ablation of oligodendrocytes, diffuse brain dysmyelination, tremors, ataxia, and severe seizures, mimicking the *shiverer* mutant phenotype (Bernstein and Breitman, 1989).

A modification of this strategy involves killing cell groups that have been engineered to express an enzyme capable of converting an exogenously applied photosensitive dye into a cytotoxin. Selection can then take place by exposing cells in a specific region to a phototoxic wavelength of light (Nirenberg and Cepko, 1993). Transgenic mouse lines incorporating a construct with a β-galactosidase gene linked to the promoter region of the somatostatin or VIP neuropeptide genes were used to create specific classes of vulnerable cortical neurons. When the neocortex was subsequently photoactivated *in vivo* after superfusion with the fluorescein analog, there was selective killing of somatostatin- and VIP-containing cells in all cortical laminae. A major advantage of this technique is the ability to select the region and defer the time of death until a designated postnatal age. This advantage may soon be extended to other transgenes by constructs that include drug-responsive promoter elements.

Gene Transfer into CNS Circuits

Methods that deliver a single gene to central neurons may be the ultimate key to clarifying the role of candidate excitability molecules. Genes can be selectively transferred into specific circuits within the central nervous system by DNA microinjection, packaging in liposomes, replication-defective viral vectors, and grafts of cells derived from other genotypes. Replication-defective strains of herpes simplex virus (Fink et al., 1992) and adenovirus (Le Gal La Salle et al., 1993) are promising vectors for introducing novel patterns of gene expression into the CNS without cytolysis for periods of several months. Genes can be transferred into neurons *in situ* or in cell culture before brain grafting (*ex vivo*). An alternate method of gene delivery is to graft undifferentiated, embryonic neuronal or glial progenitor cells expressing a desired genotype (Groves et al., 1993; Snyder et al., 1992). One example of this technology on a larger scale involves the transfer of an avian epileptic phenotype by subtotal lower brainstem grafts from the primordia of genetically hyperexcitable mutants into wild-type strains (Teillet et al., 1991).

Activity-Dependent Gene Expression in the Brain

A critical category of excitability genes includes those in which transcription is dynamically linked to patterns of afferent signaling or non-synaptic cell-cell interactions. These genes have been found by serial determinations of mRNA levels using Northern blot and *in situ* hybridization (Gall et al., 1991) or by screening brain cDNA libraries using mRNA probes derived either from the neocortex of untreated rats or those that had undergone metrazol-induced seizures (Qian et al., 1993).

Immediate Early Genes. Significant emphasis has been placed on immediate early gene (IEG) expression, in particular the proto-oncogenes *fos* and *jun* that initially defined the members of this group (Curran and Morgan, 1991; Sheng and Greenberg, 1990; Sudol et al., 1993). A transgenic mouse line bearing a *lacZ -fos* promoter fusion construct has been crossed with various neurological mutant mice, revealing *fos* expression patterns that overlap with dying cells; in wild-type mice, *fos* expression is evident in cells fated for elimination by programmed cell death during terminal differentiation (Smeyne et al., 1993). Excitotoxic death after injury is also a potent stimulus for *fos* expression, and IEG expression is also reliably induced by some but not all patterns of abnormal synchronous stimulation. For example, convulsive epileptiform discharges induced by kainic acid stimulate intense *c-Fos* expression (Gall et al., 1991), whereas the non-convulsive generalized spike-wave pattern of spontaneous seizures in the mutant mouse *stargazer* does not (Nahm and Noebels, 1993). Since both models show a striking pattern of seizure-induced synaptic reorganization (Qiao and Noebels, 1993), *fos* expression does not appear necessary for this type of plasticity. Transgenic *c-Fos* null mice demonstrate no overt CNS pathology (Wang et al., 1992). Experimental stimulation of these deletion mutants should help determine whether the AP-1 and CREB sites that bind the *fos-jun* dimer play a critical role in activity-dependent neuronal plasticity.

Synaptic Transduction Mechanisms. Regulation of immediate early gene expression in neurons by synaptic and voltage-dependent membrane activity is mediated by distinct second-messenger signaling pathways, and intracellular calcium ion levels play an important intervening role while still permitting selective patterns of gene activation. For example, Ca^{2+} influx through the NMDA-gated and the L-type voltage-gated calcium channels differentially activate gene transcription. CaMKII, a Ca^{2+}-calmodulin-dependent protein kinase, is activated by calcium entry through either NMDA or L-type channels; however, only the L-type channel activity requires CaMKII for signal transduction to the nucleus. The NMDA channel activity interacts with regulatory elements for immediate early genes that are apparently not critical for L-type channel-induced *fos* expression (Bading et al., 1993). A similar distinction separates responses triggered by NMDA or AMPA receptor subtypes (Lerea and McNamara, 1993). These data provide a general example of how single membrane receptors manage to maintain distinct pathways through the cytoplasm to the nucleus. Other ways of preserving the integrity of membrane-nucleus regulatory steps may include significant differences in the spatial localization, timing, and the magnitude of calcium entry through different pore types, as well as the selective participation of subsequent coupled reactions. For example, differential subcellular patterns of Ca^{2+} propagation waves activated by muscarinic receptor subtypes have recently been identified by dynamic imaging of calcium-sensitive fluorescent dyes. M_2 muscarinic receptors are coupled to pertussis toxin-sensitive G proteins and weakly activate PI hydrolysis. M_3 receptors potently stimulate PI hydrolysis and activate a pertussis toxin-insensitive G protein pathway. Confocal imaging of intracellular Ca^{2+} activity in oocytes reveals that simultaneous activation of both receptors stimulates periodic bursting patterns of cytoplasmic Ca^{2+} waves. In contrast, selective M_2 activation increased the number of focal sites of Ca^{2+} release, whereas selective M_3 activation evoked a large propagating subplasmalemmal wave (Lechleiter et al., 1991). Since local concentrations of free calcium ions can remain compartmentalized by dendritic spines and the presence of other organelles, these phenomena hint at the potential for complex patterns of Ca^{2+}-mediated gene expression elicited by synaptic signaling and may underlie the ability of specifically positioned synaptic terminals to compete for the ability to trigger nuclear transcription responses resulting in persistent increases in synaptic strength (Bliss and Collingridge, 1993).

Temporal Patterns of Gene Expression. After intense synaptic activation, a steadily increasing number of genes have been identified with transcriptional activity that is altered at different latencies and for variable periods of time. In general, a post-stimulus interval of 1 to 24 hours encompasses most of the two- to tenfold increases in basal expression of neuronal and glial genes. Their function falls into several teleological categories, such as homeostatic, restorative, or enhanced differentiation. However, it is unclear where the functional overlap between these responses lies, and the major distinction in many cases may be simply one of degree. In the first category are genes responding to altered synaptic or metabolic demands where the compensatory changes remain generally consistent with the native molecular phenotype of the cell (Gall et al., 1990; Tsaur et al., 1992). The second and third categories include a range of molecules involved in preventing or repairing potential damage by virtue of their ability to buffer intracellular calcium and/or promote growth (Dugich-Djordjevic et al., 1992; Merlio et al., 1993), as well as newly expressed genes that can radically alter the phenotype of the cell. A clear example of the latter genes is found in the novel expression of neuropeptide Y in dentate granule cells after kainic acid stimulation of the hippocampus (Bellman et al., 1991), a finding also observed in the epileptic mutant mouse *stargazer* (Chafetz, Nahm, and Noebels, unpublished data). The distribution of this neuropeptide in a pattern confined to the intra-axonal vesicle population suggests that the neo-NPY is part of the releasable transmitter pool, and thus, the neurotransmitter phenotype of the granule cells has been converted. This degree of plasticity in the differentiated state raises fundamental new questions regarding the changing biology of neural networks in chronic excitability disorders of the central nervous system.

CONCLUSION

Like the "reazione nera" employed by Cajal to define the connectivity of neural circuits, genetic probes are now redrawing the molecular anatomy of the neocortex and, with it, the functional architecture of the brain. This new framework brings enormous coherence to the way we will view neural plasticity in the future. For example, cortical maps of ion channel gene expression predict burst-firing properties in specific neuronal subsets. Spatial and temporal maps of NMDA receptor and CaMK II expression define specific sites and developmental stages where one subtype of long-term synaptic plasticity can occur and simultaneously exclude that type of plasticity from consideration in circuits where and when it is not expressed. The ability to test these hypotheses directly using genetically engineered nervous systems provides a sharp increase in our ability to appreciate the significance of molecular diversity in the brain.

The true dimensions of the biological potential for plasticity in cortical circuits have not yet been gauged fully, and most of the questions that must be immediately confronted are of the most basic nature. What is the functional significance of low basal levels of mRNA expression of rare transcripts in neurons, and how high must they rise to become relevant? How is the expression of functionally related genes linked during development? Can they be uncoupled? Inherited diseases in both humans and mice continue to provide direct clues into the identity of unexpected genetic control mechanisms, and raise critical new challenges. When do compensatory or activity-induced changes become deleterious, and can they be blocked? How can we recognize "benign" DNA polymorphisms from those that could initiate conditional disease phenotypes? Advances in our understanding of the levels of gene control over excitability in the cerebral cortex will create a neuropathological framework to classify and stage the degree of plasticity in CNS diseases at the molecular phenotypic level and will contribute importantly to our future ability to reverse damaging patterns of gene activity in neurological disorders.

Acknowledgments. I would like to thank Santosh Helekar, Walter Nahm, and Xiaoxi Qiao for their contributions to this research which has been supported by NIH NS29709, Epilepsy Research Center, Mental Retardation Research Center, and the Bluebird Circle Foundation for Pediatric Neurology.

16

Development of cortical excitation and inhibition

RUDOLF A. DEISZ and HEIKO J. LUHMANN

The sequence of developmental modifications in allo- and neocortical areas is similar in most mammals (Marín-Padilla, 1992); however, the exact time course of these alterations is species and area dependent. The following description of age-dependent changes in the structure and function of the cortical excitatory and inhibitory systems covers experimental observations obtained predominantly from the neocortex of one species, the rat, in which most of the electrophysiological *in vitro* studies were performed. To allow a correlation between the different anatomical and physiological parameters and to minimize the influence of layer-specific differences due to the "inside first-outside last" pattern of neocortical development, data from layers II/III of the rat neocortex are presented when available. To highlight certain aspects of excitatory and inhibitory transmission, we resort to findings from other cortical areas, structures, and species as well.

DEVELOPMENT OF CORTICAL EXCITATION

A wealth of neurochemical, electrophysiological, and molecular biological data have firmly established the role of L-glutamate as an excitatory neurotransmitter in the central nervous system (see Mayer and Westbrook, 1987; Monaghan et al., 1989; Nicoll et al., 1990; and Chapter 5 for reviews). Among the ionotropic excitatory amino acid (EAA) receptors, the N-methyl-D-aspartate (NMDA) receptor is of particular interest. The participation of NMDA receptors in synaptic transmission seems to be a common feature of central neurons; it has been demonstrated in neocortex (Thomson, 1986), hippocampus (Collingridge et al., 1988b; Hablitz and Langmoen, 1986), striatum (Cherubini et al., 1988), and nucleus accumbens (Pennartz et al., 1991).

EAA receptors and the NMDA receptor in particular have been implicated in neuronal plasticity (for review see McDonald and Johnston, 1990). NMDA receptor binding exhibits peaks during hippocampal development, both in rat (postnatal day (P) 6 to 8; Tremblay et al., 1988) and human hippocampi (24 to 27 weeks of gestation, Represa et al., 1989b). NMDA receptor antagonists impair spatial learning (Morris et al., 1986) and prevent ocular dominance shifts after monocular deprivation (Kleinschmidt et al., 1987; Rauschecker and Hahn, 1987). NMDA receptors also play a crucial role in long-term potentiation (LTP; see Collingridge and Bliss, 1987; Bliss and Collingridge, 1993, and Chapter 7). Moreover, the outgrowth of neurites of fetal hippocampal neurons is regulated by EAA receptors (Mattson et al., 1988). Finally, excitotoxic effects through EAA receptors (for review see Choi, 1988) may cause cell death and collateral elimination, processes that determine the postnatal development of cortical circuitry.

Because there have been fairly exhaustive reviews (McDonald and Johnston, 1990; Vaughn, 1989) we focus on recent data from the rat neocortex. Aspects of EAA transmission related to aging, such as the progressive decrease of glutamate binding during adolescence in humans (Kornhuber et al., 1988), or decreases in NMDA receptor-mediated depolarizations and long-term potentiation in the aged rat cortex (Baskys et al., 1990b), have been omitted for brevity.

Glutamate Levels and Receptor Densities

Postnatal development of excitatory transmission is characterized by a transient overproduction of EAA binding sites and a complete pattern of tissue levels of glutamate. The regional specificity and the temporal differences among species relate to periods of peak synaptic plasticity and consolidation of synaptic networks (McDonald and Johnston, 1990). Shortly after birth, on postnatal day 2 (P2), glutamate levels in the visual cortex are low, about 60% of that of adults. During the first postnatal week glutamate concentration slowly increases, with a peak at P25. Between P25 and P90 the glutamate level gradually decreases and remains constant thereafter (Erdö and Wolf, 1990a). Despite the monotonous increase in overall tissue concentration, the K^+-induced glutamate release from synaptosomes exhibits an initial fall from parturition until about P7, followed by a transient increase between P8 and P14, before a steady increase toward adult levels proceeds (Collard et al., 1993).

Binding studies of radiolabeled EAA agonists have revealed marked changes in the time course, distribution, and density of binding sites during postnatal development. Binding to NMDA receptors was detectable on P1, but the adult pattern was not reached until P7 (Insel et al., 1990). Between P14 and P28 NMDA receptor binding levels were well above the adult levels; the peak in cortical binding was attained at P28 when the values were about 150% (outer laminae) of the matched adult levels (Insel et al., 1990). The density of AMPA binding sites progressively increased until P21, reaching 150% of the adult level. Compared to NMDA binding sites, the overshoot in AMPA binding sites was much briefer during development (Insel et al., 1990). Also kainate binding sites exhibited a transient overshoot during development with a peak at P21, the overshoot being much more pronounced in the inner cortical layers (Miller et al., 1990).

Binding studies on cortical homogenates also revealed distinct developmental changes. Binding to NMDA receptors (determined with tritiated CPP) was low on P2 and increased rapidly from P14 to P25. Between P25 and P40 the increase slowed and reached a peak on P40 (Erdö and Wolf, 1990b). From P40 up to 1 year NMDA binding decreased (Erdö and Wolf, 1990b). The decrease of NMDA binding sites continued up to 29 months after birth (Tamaru et al., 1991). Binding to kainate and AMPA receptors is perinatally approximately the same as in adults. During the first postnatal days both kainate and AMPA binding increases, reaching a maximum on P6. From P6 on, kainate and AMPA binding gradually declines to reach adult levels at P28 (Erdö and Wolf, 1990b). During adolescence AMPA binding remains constant, in contrast to NMDA binding (Miyoshi et al., 1991).

A transient increase in excitatory transmission was also inferred in the kitten visual cortex (area 17 and 18) using the high-affinity uptake of D-aspartate as a marker of excitatory transmission. During postnatal weeks 5 to 10 the activity of D-aspartate uptake was about threefold higher that in adults (Fosse et al., 1989). This peak of D-aspartate uptake corresponds to the period of maximal dendritic branching and synapse density attained during postnatal weeks 5 to 8 (Cragg, 1975; Winfield, 1981). The increase in high-affinity uptake for D-aspartate occurred even when the animals were reared in the dark (Fosse et al., 1989).

An interesting developmental change is the perinatal change in cerebral concentrations of kynurenic acid, a broad EAA receptor antagonist. In fetal rat brain kynurenic acid levels were prenatally fivefold those in adults and then declined rapidly at P1, reaching adult concentrations at 7 days after birth (Beal et al., 1992). Such a rapid decrease in kynurenic acid would unleash excitatory mechanism shortly after parturition.

Molecular Biology of Excitatory Amino Acid Receptors

During the last few years, cloning of complementary DNAs encoding EAA receptor proteins and expression in an *in vitro* system, such as the *Xenopus* oocyte, have revealed that each of the main EAA receptor classes represent a family of closely related receptors (see Chapter 5).

During postnatal development *in situ* hybridization studies revealed marked regional and temporal differences in the expression of the mRNAs of AMPA receptor subunits. In particular, the expression of flip/flop splice variants of GluR1 to 4 (or GluRA to D) subunits exhibit pronounced alterations during development in the neocortex (Monyer et al., 1991). GluR1 flip mRNAs are expressed throughout postnatal development; the GluR3 flip mRNA, however, seems to increase during postnatal development. The mRNA for the GluR1 flop splice variant is expressed comparatively late, whereas GluR3 flop mRNA is virtually absent during postnatal development (Monyer et al., 1991). Clear developmental changes exist in the expression of NMDA receptor subunits (Monyer et al., 1994; Watanabe et al., 1992) and metabotropic glutamate receptors (Ryo et al., 1993; Shirasaki et al., 1993). In addition, the alteration in the kinetics of NMDA receptor-mediated components of synaptic responses is consistent with temporal differences in the expression of NMDA receptor subunits and their splice variants during maturation.

Properties of Excitatory Transmission

In most cortical neurons of the adult rat, orthodromic stimulation with low intensities evokes EAA receptor-mediated excitatory postsynaptic potentials (EPSPs; Jones and Baughman, 1988; Sutor and Hablitz, 1989a). A prominent feature of these EPSPs in the adult neocortex is their enhancement with depolarizations (Deisz et al., 1991; Hwa and Avoli, 1992a; Stafstrom et al., 1985; Sutor and Hablitz, 1989b; Thomson, 1986; Thomson et al., 1985). Such an effect cannot be explained solely by the properties of kainate or AMPA receptors, which should yield decreasing synaptic responses with depolarization. The participation of NMDA receptors in the generation of EPSPs in the neocortex (Jones and Baughman, 1988; Sutor and Hablitz, 1989b; Thomson, 1986; Thomson et al., 1985) partly accounts for the increase in amplitude and duration of synaptic responses.

The contribution of NMDA receptors, however, seems to comprise only a small fraction of EPSPs in the visual cortex of rats (P22 to P40). Application of the NMDA antagonist AP5 reduced the amplitude of early EPSPs only to 84% (Jones and Baughman, 1988). The areas of EPSPs in neurons from the somatosensory cortex (layer II/III) of young adults (150 to 180 g) were reduced by less than 20% by AP5 (Deisz et al., 1991). The NMDA component is expectedly enhanced at depolarized levels and constitutes up to 50% of the EPSP of layer V neurons in the visual cortex (Jones and Baughman, 1988). In layer II/III neurons of somatosensory cortex, AP5 caused only about a 15% change of the EPSP area at depolarized levels (Deisz et al., 1991). The weak effect of AP5 on EPSP enhancement in the somatosensory cortex indicates a more pronounced contribution of NMDA receptors to EPSPs of neurons in the visual (Jones and Baughman, 1988) and frontal cortex (Sutor and Hablitz, 1988a). A precise comparison of these data, however, is hampered by the different levels of Mg^{2+} used in the different laboratories.

During postnatal development marked alterations in the EPSPs occur. Between P11 and P20 a transient appearance of late, polysynaptic EPSPs, mediated by NMDA receptors, has been described (Luhmann and Prince, 1990b). Compared to adult neocortex (Jones and Baughman, 1988; Deisz et al., 1991; Sutor and Hablitz 1988a), these polysynaptic components are much more pronounced. Part of the transient nature was attributed to a delayed maturation of inhibition with respect to the establishment of excitatory connections (Luhmann and Prince, 1990b). The enhancement of excitatory transmission by bicuculline methiodide at P6 to P8, indicating an early presence of inhibition (Burgard and Hablitz, 1993), is not in conflict with the notion of Luhmann and Prince (1990b). The early inhibition probably dampens excitation in a tonic fashion, but a spatiotemporally well-tuned inhibition necessary to rein in the developing excitation lags behind (see below). However, the transient appearance of late, polysynaptic EPSPs does not imply the absence of NMDA receptors at earlier stages. Data from fetal and postnatal rat neocortex indicate that functional NMDA receptors are already expressed on neurons of the cortical plate (E 18, LoTurco et al., 1991). These NMDA receptor channels exhibit the appropriate pharmacology and voltage dependence in neocortex (but see Ben-Ari et al., 1988). Spontaneous currents are antagonized by D-AP5 and blocked by Mg^{2+} in a voltage-dependent fashion (LoTurco and Kriegstein, 1991). Interestingly younger cells in the ventricular zone lack NMDA receptor channels (LoTurco et al., 1991), yet the subplate receives excitatory input during embryonic development in the kitten (Friauf and Shatz, 1991). These NMDA receptor channels participate also in evoked synaptic transmission at least from P3 onward (Burgard and Hablitz, 1993), indicating an adequate postsynaptic location that is accessible to the released transmitter.

Several possibilities considered below may account for the transient enhancement of EPSPs (Luhmannn and Prince, 1990b). Little is known about the functional changes of NMDA receptors during postnatal development in the neocortex. In the superior colliculus, however, the NMDA receptor-mediated synaptic currents are several times longer than in mature neurons, indicating altered properties during development (Hestrin, 1992a). Rearing in the dark delays the progressive shortening of NMDA receptor-mediated EPSCs (Carmignoto and Vicini, 1992) and the decrease of NMDA components of the visual response between 3 and 6 weeks postnatally (Fox et al., 1992). To what extent alterations in release or uptake mechanisms for glutamate (Fosse et al., 1989) contribute to the change in time course remains to be established. The difference in time course of NMDA receptor-mediated currents may be related to any of the known modulatory mechanisms of the NMDA receptor. However, the modulations of NMDA receptors in the adult neocortex by Mg^{2+} (Thomson et al., 1985) and glycine (Thomson et al., 1989) have also been demonstrated during early development (LoTurco et al., 1991). Developmental changes in the expression of NMDA receptor subtypes with different sensitivity to protein kinase C (Durand et al., 1992; see below) may participate in the hyperexcitability during development. In addition, the feedback control of glutamate release by presynaptic L-AP4 receptors, shown in cultured hippocampal neurons (Forsythe and Clements, 1990), might enable repetitive EPSPs, if less efficient in the immature neocortex.

Modulation of EPSPs by Intrinsic Currents in Adult and Neonatal Neocortex

In adult rat neurons, the areas of EPSPs elicited just below threshold stimulus intensities are increased about fourfold compared with the resting potential, and the insensitivity to AP5 renders a contribution of NMDA receptors unlikely (Deisz et al., 1991). The underlying recruitment of voltage-dependent conductances would serve as an AND gate linking synaptic activation with postsynaptic activity. Alterations in neuronal properties (McCormick and Prince, 1987) suggest a postnatal maturation of voltage-dependent currents, which in turn may affect the voltage dependence of EPSPs; hence, the characteristics of this type of AND gate.

The existence of a persistent Na^{+} current in neurons of the adult hippocampus and neocortex is well established (Connors et al., 1982; Hotson et al., 1979; Spain et al., 1987; Stafstrom et al., 1984). Early experiments suggested that this current contributes to the enhancement of EPSPs by depolarization (Stafstrom et al., 1985). Moreover, a transient low-threshold Ca^{2+} (T) current (Friedman and Gutnick, 1987; Sutor and Zieglgänsberger, 1987) has been described in neocortical neurons; this current participates in the EPSP enhancement as well (Deisz et al., 1991). The EPSP enhancement at depolarized levels exhibits a pronounced time dependence: EPSPs evoked early (100 msec) after the onset of the depolarization are enhanced fourfold, whereas those evoked late (more than 200 msec) are only about augmented twofold (Deisz et al., 1991). Intracellular application of the lidocaine derivate QX 314 reduces the EPSP enhancement as reported by many investigators (Hwa and Avoli, 1992a; Sutor and Hablitz, 1988b), but the enhancement of EPSPs evoked early persists, and the ratio of EPSP areas evoked early versus those evoked late is unchanged, remaining near 1.8 ("time ratio"; Deisz et al., 1991). The voltage- and time-dependent augmentation of EPSPs persists in the presence of AP5 (Deisz et al., 1991), rendering a NMDA receptor contribution unlikely. Simulation of EPSPs by brief current pulses reveals a time- and voltage-dependent enhancement of the resulting voltage transients, which is quantitatively similar to that of EPSPs. The abolition of time-dependent enhancement by Co^{2+} indicates an involvement of the T current in the time-dependent component of enhancement (Deisz et al., 1991). In addition to the enhancement with depolarization, EPSPs are also attenuated and shortened at hyperpolarized levels by an I_h type of current (Deisz and Zieglgänsberger, unpublished observation).

The T current selectively enhances temporally closely related EPSPs (Deisz et al., 1991) and was proposed to mediate the induction of LTP in the kitten visual cortex (Komatsu and Iwakiri, 1992). Participation of the T current in EPSP modification was also inferred from the observation that depolarizing pulses in the presence of D-AP5—without synaptic stimulation—resulted in a transient enhancement of EPSPs (Kullmann et al., 1992). The entry of Ca^{2+} following synaptic activity, resistant to antagonism of NMDA receptors and prevented by hyperpolarization (Miyakawa et al., 1992), suggests that voltage-dependent Ca^{2+} conductances are also recruited during synaptic activity in hippocampal neurons.

The clear delineation of two components of EPSP enhancement—a time-invariant enhancement by a QX 314-sensitive persistent Na^+ current and the time-dependent T current-mediated component—allows an estimate of their relative contributions during development. At P4 to P10 the linear, or outwardly rectifying, current voltage relationship indicates that both the persistent Na^+ and the I_h are not fully developed. During this period, EPSPs are not shortened by hyperpolarization (Deisz and Zieglgänsberger, unpublished observations), indicating that in the absence of I_h the EPSPs are less sensitive to voltage in the hyperpolarizing direction (Hwa and Avoli, 1992a). The voltage-dependent enhancement with depolarization is much smaller in neurons before P12 and resembles that of QX 314-treated neurons, yet the adult time ratio is detectable as early as P7 (Deisz and Zieglgänsberger, unpublished observations). During the postnatal development the voltage-dependent enhancement of EPSPs gradually increases (Deisz and Zieglgänsberger, unpublished observations), corresponding to the pronounced increase of the persistent Na^+ current (Alzheimer et al., 1993a). The constant time ratio attributable to the T current suggests that it contributes early to EPSP enhancement and is fairly constant during the postnatal period. This view accords with the constancy of T currents exhibiting similar amplitudes, kinetics, and pharmacology in neonate (2 weeks) and aged (44 weeks) rat hippocampal CA1 neurons (Takahashi et al., 1989). Moreover, increases in internal Ca^{2+} in synaptoneurosomes at P8 have been taken as evidence for an early coupling of AMPA receptors to voltage-dependent Ca^{2+} channels (Benavides et al., 1988).

Recent evidence indicates a Ca^{2+} influx through AMPA receptor channels (Keller et al., 1992); nevertheless, the interaction of EPSPs with voltage-activated intrinsic currents constitutes not only a source of Ca^{2+} influx but also serves as an AND gate. Other sources of current have attracted much less attention and should be briefly addressed here; for instance, the Na^+/K^+ ATPase. To our knowledge no data are available from neocortical neurons, but in hippocampal pyramidal cells, the electrogenic sodium pump increases in efficacy during postnatal development (Fukuda and Prince, 1992). Moreover, it was shown that Na^+/K^+ ATPase delimits the glutamate-induced depolarization (Fukuda and Prince, 1992), and therefore it may dampen depolarizations during periods of excessive excitation. In addition, Ca^{2+} entry through NMDA receptor channels activates a Ca^{2+}-dependent K^+ conductance in postnatal hippocampal (Zorumski et al., 1989) and neocortical neurons (Mistry and Hablitz, 1990), which limits depolarizations during NMDA receptor activation. On the other hand, Ca^{2+}-dependent K^+ conductance is reduced by NMDA receptor activation by a protein-kinase mediated-mechanism (Baskys et al., 1990a). Attenuation of either repolarizing mechanism or delayed development of the K^+ conductance with respect to the NMDA receptor increase would favor NMDA receptor-mediated depolarization and might contribute to the transient increase of NMDA components (Luhmann and Prince, 1990b).

Intracellular Cascades

Several cellular cascades—the cyclic nucleotides, phospholipase A2, arachidonic acid pathways, and phosphoinositide (PI) metabolism—are known to be linked to EAA receptors. EAA receptors stimulate phospholipase A2 to release arachidonic acid. The pronounced depression of glutamate uptake by arachidonic acid may enhance excitatory transmission, and this sequence has been implicated in LTP (Barbour et al., 1989).

Inositol phospholipid metabolism activated by EAA receptors (for review see Smart, 1989) is of particular interest from a developmental point of view. EAA-stimulated PI breakdown is maximal in the immature brain (P6) and correlates with the critical period for synaptic modifications of the visual system (for review see McDonald and Johnston, 1990). As mentioned above, a cDNA from rat cerebellum encoding a metabotropic glutamate receptor with the appropriate features has been cloned (Masu et al., 1991). The deduced amino acid sequence indicates that this receptor contains seven membrane-spanning domains, although the amino acid sequence and structural architecture of this receptor differ from conventional G-protein-coupled receptors. Expression of the receptor in *Xenopus* oocytes yields glutamate-, quisqualate-, and trans-ACPD-(but not NMDA or kainate) induced inositol 1,4,5 triphosphate (IP_3) formation and intracellular Ca^{2+} mobilization (Masu et al., 1991). The divergent pharmacological features of the receptors mediating phosphoinositide hydrolysis among different types of neurons (see Smart, 1989 for review) suggest that this receptor may be the first of another family. Electrophysiological evidence indicates that metabotropic glutamate receptors participate in excitatory transmission in cultured hippocampal neurons. In the presence of EAA antagonists, glutamate decreased potassium conductances, and a synaptic component persisted with properties similar to the responses induced by agonists of metabotropic glutamate receptors (Charpak and Gähwiler, 1991; Charpak et al., 1990). Metabotropic glutamate receptors have also been implicated in long-term depression (see Miller, 1991 for a brief review).

The ability of the other branch of the PI cascade, diacylglycerol (DG), to activate protein kinase C (PKC) is well known. PKC (measured by phorbol ester binding) has a marked peak during postnatal development (Kumar and Schliebs, 1992). Intracellular injection of PKC selectively enhances the response to NMDA (78% increase in NMDA currents at −80 mV), an effect that was attributed to its reductoin of the voltage-dependent block of NMDA receptor channels by Mg^{2+} (Chen and Huang, 1992).

In this context, it is tempting to speculate about the origin of the hyperexcitability during postnatal development. We feel that from the data reviewed, complemented by a few reasonable assumptions, a tentative view of postnatal hyperexcitability emerges. During the transient peak in the expression of EAA receptors, including NMDA receptors, a proportionate fraction of metabotropic glutamate receptors may be expressed. This assumption seems reasonable since PI metabolism exhibits a pronounced peak during postnatal development around the same time when PKC is at its maximum level. The peak of PI metabolism and DG production would coincide with the peak of PKC. Considering the peak of

NMDA receptor expression and the enhancement of NMDA receptor responses by PKC, this scheme may readily account for the postnatal hyperexcitability and seems particularly attractive when possible developmental changes in the expression of splice variants exhibiting different modulation by PKC (Durand et al., 1992) are considered.

Mobilization of intracellular Ca^{2+} is, however, not confined to metabotropic glutamate receptors. NMDA receptors also induce a late phase of elevation of intracellular Ca^{2+} in cultured hippocampal neurons (6 to 12 days in vitro, DIV). This late increase occurs in the absence of extracellular Ca^{2+} and seems to be selectively induced by NMDA receptor activation. Neither quisqualate, kainate, nor depolarization by 50 mM K^{+} mimicked this response. The late elevation of intracellular Ca^{2+} involves PKC as evidenced by the reduction by sphingosine and H7 (Harada et al., 1992). Glutamate-induced elevation of intracellular Ca^{2+}, which is reduced by removal of extracellular Ca^{2+} and APV, is enhanced in amplitude and time course in cortical neurons in culture between 2 and 8 DIV (Wahl et al., 1989). A more detailed investigation of EAA receptor-mediated Ca^{2+} signaling and of the consequences of increased Ca^{2+} binding proteins is lacking during development (Solbach and Celio, 1991). Comparison of glutamate-induced excitotoxicity and Ca^{2+} signaling of human and rat neocortical neurons revealed marked differences (Mattson et al., 1991).

Neuronal activity rapidly activates several immediate early genes that code for transcription factors. Age-dependent alterations in the expression of the immediate early genes have been reported in primary cortical cultures. Young cultures (3 to 7 days in vitro) exhibited low staining for *Jun-B*, *c-Jun*, and *c-Fos* compared to mature cultures (21 to 25 days in vitro). The basal immunoreactivity to *Jun-B* and *c-Fos*, but not *c-Jun*, is reduced by tetrodotoxin or NMDA receptor antagonists, whereas picrotoxin markedly increases the percentage of neurons displaying immunoreactivity to *c-Fos*, *c-Jun*, *Jun-B*, and *Zif 268* (Murphy et al., 1991). The resting BDNF and NGF mRNA synthesis of hippocampal neurons, induced presumably by spontaneous activity, can be decreased by kynurenic acid but not by MK-801 (Zafra et al., 1990). The rapid transcriptional regulation of immediate early genes and neurotrophic factors of central neurons during neuronal activity may be relevant to neuronal stabilization and synapse maintenance, particularly during periods of enhanced excitatory transmission. The AMPA- and NMDA-induced changes in cytoskeleton were age dependent, appearing between DIV 4 and 8 and reached a maximum between DIV 10 and 14 (Bigot et al., 1991). Application of EAA agonists induced the transition from a diffuse arrangement of microtubule-associated protein (MAP2) and tubulin into filaments. These effects could be distinguished from neurotoxic effects and were age dependent: in cortical neurons at DIV 14 NMDA was neurotoxic with few changes in MAP2 positive cellular morphology. However, at earlier stages, NMDA induced significant cytoskeletal rearrangement without obvious cellular damage (Bigot et al., 1991).

DEVELOPMENT OF GABAERGIC INHIBITION

Although glycine (Ito and Cherubini, 1991) and various neuropeptides (Colmers et al., 1988; Johnston, 1988; Parnavelas et al., 1988a) may have an inhibitory action on cortical cells, especially during early development (see Chapter 14), the most powerful inhibitory transmitter in the mammalian central nervous system is γ-aminobutyric acid (GABA). Therefore the following review covers mainly the ontogenesis of the cortical GABAergic system.

Anatomical Changes and Molecular Biology

The GABA-synthesizing enzyme, glutamic acid decarboxylase (GAD), can be demonstrated in the fetal rat brain already on embryonic day (E) 14 (van Eden et al., 1989), 8 days before birth. At birth, approximately 10% of all neocortical neurons are GABA-accumulating local circuit neurons, and this ratio does not change significantly during later development until adulthood (filled squares in Fig. 16-1; Wolff et al., 1984). During the first 3 postnatal weeks, interneurons show a marked increase in the size of their cell bodies and in the complexity of their dendritic tree (Miller, 1986; Parnavelas and Uylings, 1980). This morphological differentiation is accompanied by a steady rise in GAD activity (open circles in Fig. 16-1; McDonald et al., 1987). Axons arising from inhibitory neurons are first evident as short processes on P3 and increase in complexity and extent in the two subsequent weeks (Miller, 1986). This growth in axonal arborization correlates with the formation of inhibitory synapses on pyramidal cells. Symmetrical axosomatic synapses are first evident on P6 (Miller and Peters, 1981), and their density increases during the following 2 weeks (filled circles in Fig. 16-1; Blue and Parnavelas, 1983; Miller, 1986). These data indicate that GABAergic neurons are present well before the inhibitory synaptic network is established. In addition, $GABA_A$ and $GABA_B$ receptors are evident in the early prenatal period, and $GABA_A$ receptor density increases dramatically during postnatal development (Shaw et al., 1991). The functional role of GABA during early ontogenesis, at a time when inhibitory neurons and GABA receptors are present but inhibitory synaptic connections are still lacking, remains to be determined.

Molecular biological techniques have provided the basis for the structural characterization of the GABA receptor (see Chapters 5 and 6). Five subunits, coded by a family of at least 15 genes, have been described for the $GABA_A$ receptor. According to the strength of the *in situ* hybridization signal, the α 1-3, β 2,3, and γ 2 subunits probably play an important role in cortical GABAergic inhibition (Persohn et al., 1992). The expression of $GABA_A$ receptor subunits exhibits marked alterations during embryonic and postnatal development. In both early and postnatal cortex the mRNAs for the α 2,3 and 5 and the β 3 subunits exhibit a pronounced expression. Subsequently the cortical expression of these four genes decreases, and in the adult expression of α 1,4, β 2, and δ subunits dominates (Laurie et al., 1992). A PKC-mediated downregulation of α 1, β 2, and γ 2S subunits (Kellenberger et al., 1992), if present during postsynaptic development, might cause a transient decrease in GABAergic function.

Physiological Changes

The prominent action of GABA in the adult neocortex is inhibitory (Krnjevic, 1974). On the postsynaptic site, GABA activates $GABA_A$- and $GABA_B$-type receptors associated with an increase in Cl^- and K^+ conduc-

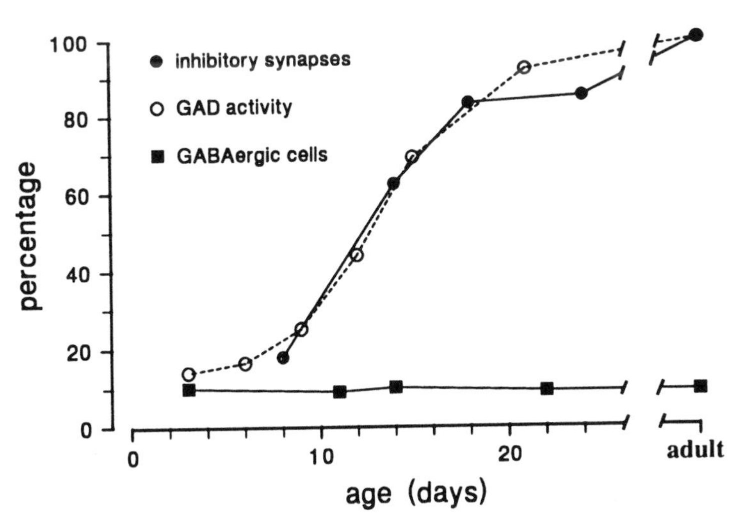

FIGURE 16-1. Postnatal maturation of inhibitory synapses (filled circles, data from Miller, 1986) and GAD activity (open circles, data from McDonald et al., 1987) in layers II/III of rat visual cortex. Data are expressed as percentage of adult (>P28) values. The proportion of GABA-accumulating neurons in the rat visual cortex (filled squares, data from Wolff et al., 1984) is expressed as a percentage of all neurons in layers I to VI.

tance, respectively (see Chapter 6). In addition to this postsynaptic action, GABA activates presynaptic $GABA_B$ receptors, thereby modulating its own release (Deisz and Prince, 1989). Inhibitory postsynaptic potentials (IPSPs) in mature allo- and neocortical areas are constituted of a $GABA_A$ receptor-mediated Cl^--dependent fast (f-) IPSP and a $GABA_B$ receptor-mediated K^+-dependent long-lasting (l-) IPSP (Howe et al., 1987a; Thompson et al., 1988). Whether distinct sets of presynaptic fibers arising from two different populations of GABAergic neurons selectively activate the $GABA_A$- or $GABA_B$-type receptor, as proposed for the amygdala (Sugita et al., 1992), remains to be determined in the cortex. The characteristic sequence of a hyperpolarizing f-IPSP followed by l-IPSP can be demonstrated in almost every regular-spiking cell located in the supragranular layers of the adult neocortex (P38 cell in Fig. 16-2A, B; >P28 columns in Fig. 16-2C).

However, the potency of this inhibitory input is age, cell, and layer dependent. Bursting cells in layer V (Chagnac-Amitai and Connors, 1989; Chagnac-Amitai et al., 1990; see Chapter 9) and some of the fast-spiking cells (McCormick et al., 1985) receive only a weak inhibitory input. In accordance with the gradual maturation of axosomatic symmetrical synapses in layers II/III of the rat cortex, orthodromically evoked IPSPs are not evident during the first postnatal week (Kriegstein et al., 1987; Luhmann and Prince, 1991a; P8 cell in Figs. 16-2A and B). Less than 4% of the neurons recorded in young cortex (P4 to P10) show a hyperpolarizing IPSP (P4 to P10 columns in Fig. 16-2C). In juvenile cortex (P11 to P16), most cells reveal a biphasic IPSP, but at this age the peak conductance of the f- and l-IPSP is still significantly smaller than in mature (≥P28) neurons (P14 cell in Figs. 16-2A–2D). Comparable developmental changes have been described in the cat visual cortex (Komatsu and Iwakiri, 1991). In adult cortex and independent of the species analyzed (*rat*: Connors et al., 1988; Luhmann and Prince, 1991a; *cat*: Connors et al., 1988; Komatsu and Iwakiri, 1991; *human*: McCormick, 1989b), the f- and l-IPSP are associated with an increase in membrane conductance by 60 to 90 nS and 10 to 20 nS, respectively, indicating similar intracortical inhibitory mechanisms in these species.

In contrast to the relatively late development of orthodromically evoked IPSPs, spontaneous $GABA_A$ receptor mediated activity is already present in the neonatal cortex (Luhmann and Prince, 1991a). Whether this tonic GABAergic activity during early postnatal development exerts a trophic influence and plays a role in the formation or modification of synaptic connections (Wolff, 1981) remains to be analyzed. The depolarizing action of GABA in the immature hippocampus (Ben-Ari et al., 1989; Cherubini et al., 1991) and cortex (Yuste and Katz, 1991) supports the hypothesis of a trophic action of GABA in early ontogenesis (see below). A depolar-

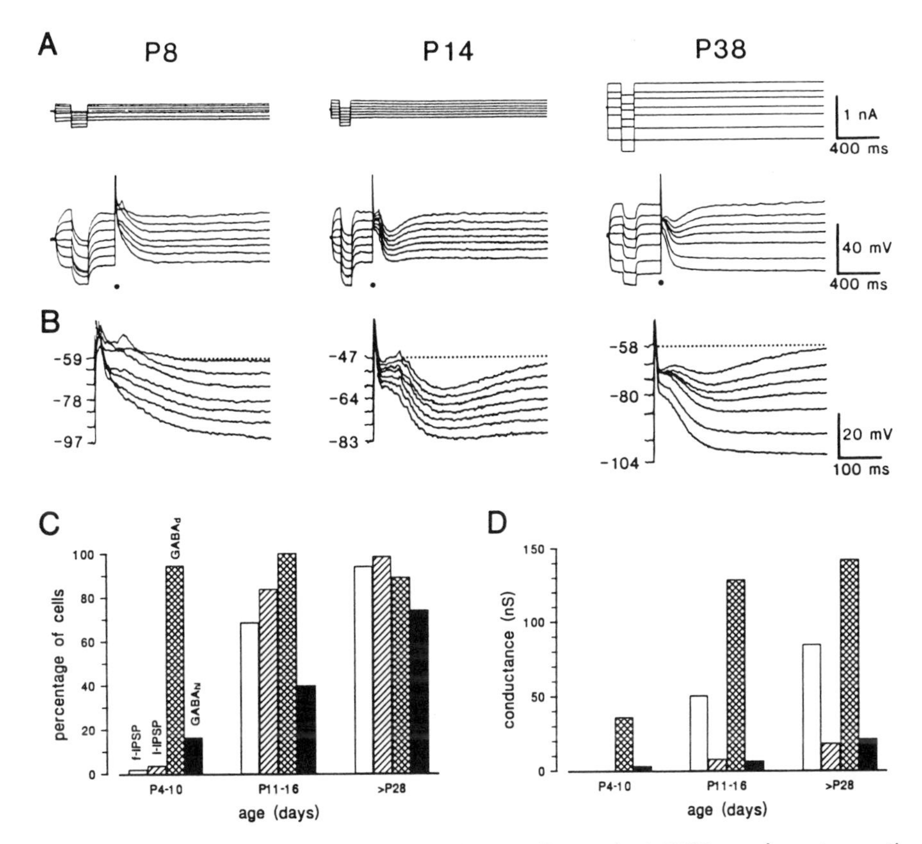

FIGURE 16-2. Development of orthodromically evoked IPSPs and postsynaptic GABA responses in layers II/III of rat neocortex (data from Luhmann and Prince, 1991). **A**. Intracellular recordings from a P8 (left), P14 (middle), and P38 cell (right). Current protocols are shown in the upper traces; voltage responses to suprathreshold stimulation of the underlying white matter (dot) are displayed below. **B**. Same recordings as in (**A**), but at faster sweep and higher gain. Note the presence of hyperpolarizing biphasic IPSP in P38 cell. **C**. Developmental changes in the percentage of cells showing a f-IPSP (open bars) and a l-IPSP (shaded) to synaptic stimulation of the afferent pathway in neocortical slices obtained from P4 to P10 (left), P11 to P16 (middle), and >P28 (right) rats. Maturation of GABA responsiveness is demonstrated by the percentage of cells with a $GABA_d$ (cross-hatched) and a $GABA_{hl}$ (filled) response to application of GABA to the soma. **D**. Developmental changes in the peak conductances of the IPSPs and GABA responses. Note similarities in the peak conductance of the l-IPSP and the $GABA_B$ receptor-mediated $GABA_{hl}$ response in the juvenile and adult age group. Same symbols as in **(C)**.

izing effect can be reliably observed when GABA is locally applied to the soma or dendrite of young neocortical neurons. At this age, hyperpolarizing GABA responses can rarely be observed (Luhmann and Prince, 1991a). In contrast to immature cells, adult neurons generally respond to application of GABA to the soma with an initial hyperpolarizing fast ($GABA_{hf}$, reversal potential $E=-76$ mV) response, followed by a depolarizing ($GABA_d$, $E= -54$ mV) and a hyperpolarizing late ($GABA_{hl}$, $E=-80$ mV) component (Connors et al., 1988; Luhmann and Prince, 1991a; Scharfman and Sarvey, 1987, 1988). Both the $GABA_{hf}$ and the $GABA_d$ responses are mediated by the $GABA_A$ receptor, whereas the $GABA_{hl}$ component requires activation of the $GABA_B$ receptor and can be also elicited by baclofen.

Two $GABA_A$ responses with significantly different reversal potentials have been demonstrated in the hippocampus (Djorup et al., 1981; Thalmann et al., 1981) and in the neocortex (Connors et al., 1988; Scharfman and Sarvey, 1985), but the exact mechanism underlying these distinct $GABA_A$ components is still unknown. Whereas the $GABA_{hf}$ response has been shown to be predominantly restricted to the soma, the $GABA_d$ component can be reliably evoked at dendritic regions. A contribution of Na^+ ions to the $GABA_d$ response (Andersen et al., 1980), concurrent activation of glia cells and neurons (Bormann and Kettenmann, 1988), an involvement of another receptor (Alger and Nicoll, 1982b), a difference in the $GABA_A$ receptor subunit composition (Shaw et al., 1991), and activation of a bicarbonate conductance (Kaila and Voipio, 1987; Kaila et al., 1989) have been discussed as possible mechanisms for the existence of a biphasic $GABA_A$ response. Another likely hypothesis requires the existence of an intracellular Cl^- gradient with relatively high Cl^- concentrations in the dendritic processes and lower Cl^- concentrations in the soma. An active Cl^--extruding process has been initially described in spinal motoneurons (Lux, 1971) and was subsequently demonstrated in the hippocampus (Misgeld et al., 1986; Thompson and Gähwiler, 1989b) and neocortex (Thompson et al., 1988). A Cl^- outward transport at the soma would lower the intracellular Cl^- concentration in the perikaryon and shift the reversal potential toward more negative values. This hypothesis has been recently confirmed in cultured hippocampal neurons using a Cl^--sensitive fluorescent probe (Hara et al., 1992; Inoue et al., 1991). These data indicate that the Cl^- concentration in the dendrites is in the range of 35 mM ($E_{Cl} \approx 35$ mV) and in the soma is approximately 6.6 mM ($E_{Cl} \approx 79$ mV). Interestingly, younger cultured pyramidal cells show a larger Cl^- concentration in the soma of about 11 mM ($E_{Cl} \approx 66$ mV; Hara et al., 1992), suggesting that the efficacy of the Cl^- pump is age dependent. This hypothesis has been also proposed by Zhang et al. (1991) and by Luhmann and Prince (1991) for the developing hippocampus and neocortex, respectively.

A developmental increase in the efficacy of a somatic Cl^- extrusion would be also in accordance with the age-dependent modifications in the neocortical GABA responsiveness. Depolarizing GABA responses can be reliably obtained from the soma and dendrites of immature neurons (Luhmann and Prince, 1991), indicating that functional $GABA_A$ receptors are present at this early age (cross-hatched columns in Fig. 16-2C). During postnatal development the somatic Cl^--extruding mechanism becomes more efficient, and hyperpolarizing $GABA_A$ components are evident at the soma. This developmental change in the $GABA_A$ responsiveness is accompanied by a shift in the reversal potential of the Cl^--dependent f-IPSP from −55 mV in juvenile to −70 mV in adult neurons. However, the concurrent activation of long-lasting NMDA receptor-mediated EPSPs in immature neocortex (Burgard and Hablitz, 1993; Luhmann and Prince, 1990a) also contributes to the depolarizing action of the f-IPSP at this age.

The functional maturation of the $GABA_B$ receptor is reflected in the gradual postnatal increase in the $GABA_{hl}$ (filled columns in Fig. 16-2C) and baclofen responses. The proportion of neurons showing a $GABA_B$ response increases from approximately 20% in young to about 70% in adult cortex. In agreement with the age-dependent rise in the conductance of the l-IPSP, the peak conductance of the $GABA_{hl}$ response increases significantly from 3 nS in young to 21 nS in adult neurons (Fig. 16-2D), probably reflecting an increase in the density of functional postsynaptic $GABA_B$ receptors.

Functional Aspects

Experimental evidence indicates that GABA may play different functional roles during early and late development of the cortex. In the perinatal period, when Cl^--extruding mechanisms are still functionally immature, GABA mainly exerts a depolarizing action (Cherubini et al., 1991). This depolarization may reach the threshold for the activation of the low-threshold Ca^{2+} channel and induce a rise in the intracellular Ca^{2+} concentration (Yuste and Katz, 1991), which may be a prerequisite for neuronal growth and plastic modifications during early ontogenesis. Release of GABA from neuronal perikarya, dendrites, and GABAergic terminals and spontaneous $GABA_A$ receptor-mediated activity precede the consolidation of inhibitory synapses, suggesting a non-neurotransmitter trophic function of GABA during early cortical development (Wolff, 1981; Wolff et al., 1984).

In the adult cortex, GABA acts as a classical neurotransmitter with an inhibitory function. Intracortical GABAergic inhibition considerably influences, or even controls, the functional properties of single cells. For example, in the visual cortex of the cat, certain receptive field properties are modulated by inhibitory mechanisms. Orientation selectivity (Sillito, 1975; Tsumoto et al., 1979), direction specificity (Eysel et al., 1988; Sillito, 1977), and the property of end inhibition (Bolz and Gilbert, 1986; Sillito and Versiani, 1977) are strongly influenced by intracortical inhibition. In accordance with the gradual maturation of the GABAergic system, these functional properties also undergo dramatic changes during postnatal development (Frégnac and Imbert, 1984). The role of the different inhibitory interneurons in mediating these and other functional aspects of intracortical information processing is largely unknown. At least seven types of GABAergic local circuit neurons can be distinguished on the basis of their distinct axonal arborization pattern (see Chapter 8), but little is presently known of their intrinsic membrane properties, synaptic input or output, and postnatal development.

Pathophysiological Aspects

GABA plays an important role in controlling normal and pathophysiological cortical excitability (Ribak, 1991). Already minor decreases in the efficacy of intracortical inhibition lead to the expression of propagating epileptiform activity (see Chapter 9). In addition, the GABAergic system reacts very sensitively to external disturbances, such as hypoxia (Krnjevic et al., 1966; Luhmann and Heinemann, 1992; Rosen and Morris, 1993) or local lesions (Cornish and Wheal, 1989). Such minor impairments in inhibitory function could shift the cortical network into a hyperexcitable state and induce long-term structural modifications, e.g., by axonal sprouting (Sutula et al., 1988). A hyperexcitable state is transiently expressed in immature hippocampus (Swann et al., 1991) and neocortex (Burgard and Hablitz, 1993; Luhmann and Prince, 1990a) and may be essential for activity-dependent synaptic modifications during early development. As a tentative mechanism we propose that a transient peak of PKC during development mediates a temporary minimum of inhibition together with a maximum of NMDA-mediated excitability. This hyperexcitability also lowers the threshold for the induction and manifestation of epileptiform activity during early developmental periods (Aicardi, 1986; also see Chapter 17).

17

Cortical epileptogenesis in the developing human brain

PETER KELLAWAY

Traditionally human focal epilepsy has been thought to arise from an area of *cortical damage*, and the models of focal epilepsy used in research have been based on this concept. However, although the concept holds true for the adult brain, focal epilepsy is more common in children who have no evidence of such a lesion. This chapter gives a description and analysis of two types of focal epileptogenesis that are unique to the developing brain and for which there is no historical, clinical, or laboratory evidence of a structural brain lesion. In the first type, the focal epileptogenesis has its origin in a genetically determined cellular defect. In the second, the focal epileptic process is a result either of deprivation or chronic distortion of sensory input during a critical period of brain development.

About 56% of all cases of epilepsy has an onset by the age of 15 years (Fig. 17-1A), and these two distinct mechanisms of epileptogenesis are of importance because seizures of focal origin are the most common type of epilepsy in children and in about 60% of these children the epilepsy is *not* associated with structural brain lesions. These children are neurologically intact and have no history of disease or injury to the brain, and their electroencephalograms are essentially normal except for the presence of foci of spike discharge.

The distribution of foci according to locus in 1,500 children with spike foci of all etiologies are shown in Table 17-1. The distribution is in accord with the epileptogenicity of the various cortical areas as demonstrated by afterdischarge studies using electrical stimulation. About 72% of the foci are localized in the centro-temporal region, 16% are in the frontal region, and 12% in the occipital region. The same table shows the percentage of

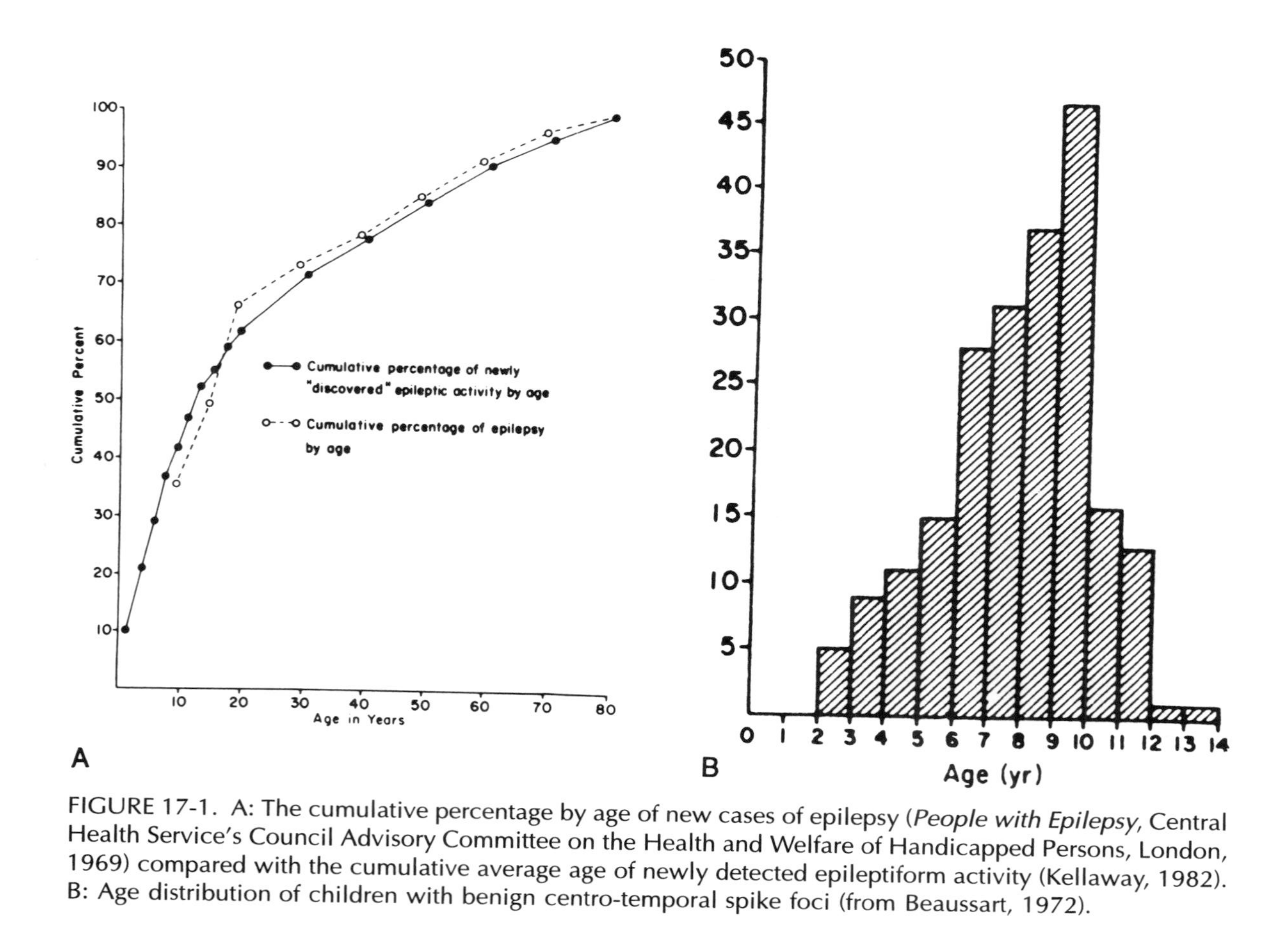

FIGURE 17-1. A: The cumulative percentage by age of new cases of epilepsy (*People with Epilepsy*, Central Health Service's Council Advisory Committee on the Health and Welfare of Handicapped Persons, London, 1969) compared with the cumulative average age of newly detected epileptiform activity (Kellaway, 1982). B: Age distribution of children with benign centro-temporal spike foci (from Beaussart, 1972).

spike foci for which etiology cannot be established. At least 60% to 70% of children with spike foci in the central and temporal regions and about 50% of children with frontal and occipital foci do not have historical or laboratory evidence of injury or disease of the brain (Kellaway, 1981). For this reason and because these spike foci and the associated clinical seizures cease to exist after a short period of time, clinicians have referred to these entities as benign focal spikes and benign focal epilepsy of children (Lüders et al., 1987).

FOCAL EPILEPTOGENESIS OF GENETIC ORIGIN

The focal spikes that occur in children who have no evidence of a palpable brain lesion have characteristics that are quite stereotypic. The spikes are typically quite high in voltage, they occur frequently (especially in sleep), they have a characteristic configuration, and there is an after-going slow wave. There is no relationship between the amount of spike activity and the number of clinical seizures, and similar foci occur in a significant number of children who do not have clinical seizures (Cavazzuti et al., 1980; Lerman and Kivity-Ephraim, 1981; Smith and Kellaway, 1964b).

These spike foci occur within a circumscribed time period during development. The age distribution of individuals with this type of spike focus in the centro-temporal region is shown in Figure 17-1B. The peak age incidence is between 7 and 10 years (Beaussart, 1972); the spike foci are rare before the age of 3 years and are not present after the age of about 17 years. Occipital spikes of this type show a similar circumscribed range of age occurrence, but with an earlier peak incidence (Kellaway, 1981).

Figure 17-2A shows a typical example of this type of spike focus. The voltage at the scalp may exceed 200 mV, spike rates of up to 70 spikes per minute are common, and the spikes often occur in bursts lasting several seconds. Although the interictal spikes exhibit a degree of amplitude and waveform variation, the average characteristics of the spike, in terms of

TABLE 17-1. Distribution of Foci in 1,500 Children

	Distribution of Spike Foci (%)	No Apparent Etiology (%)
Temporal	41	60
Central	31	70
Frontal	16	54
Occipital	12	68

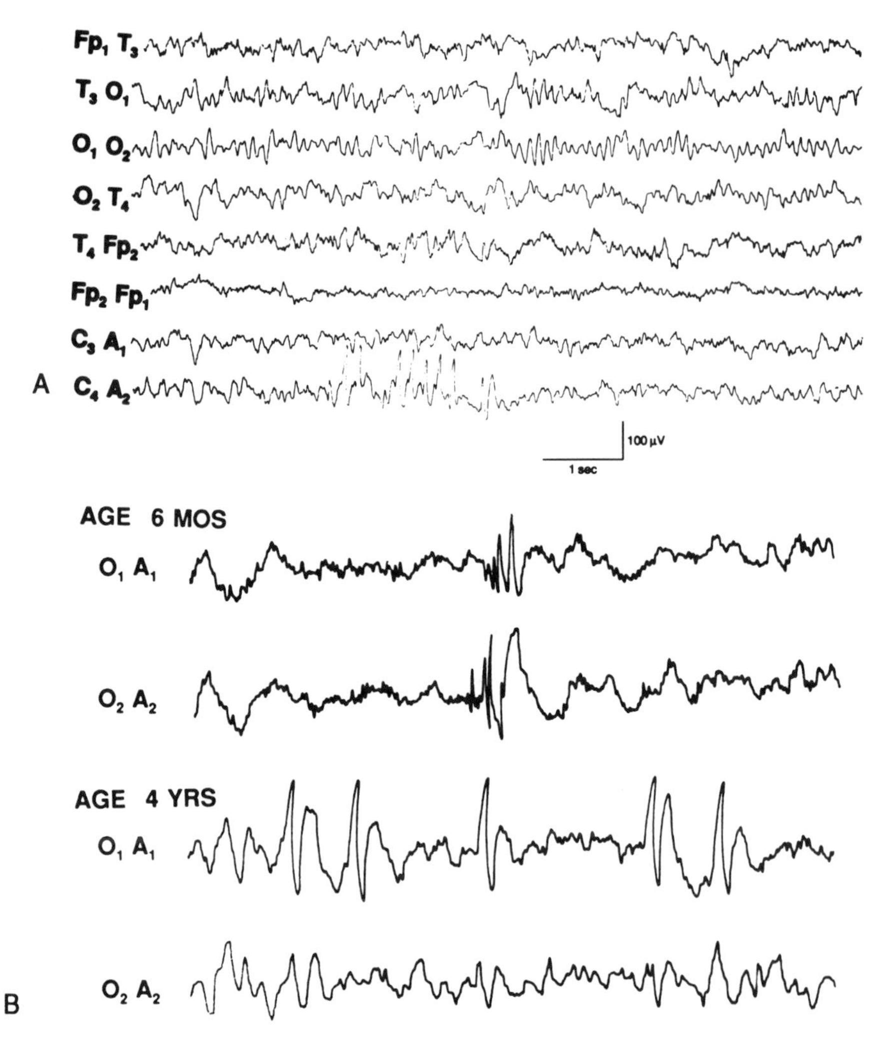

FIGURE 17-2. A: High-voltage benign rolandic focal spikes in the right central (C4) region, ear to scalp recording. The spikes have a maximum voltage of about 140 mV and often occurred in short bursts as shown. B: Changes in the configuration and amplitude of occipital spikes in a child blind since birth as a consequence of retinopathy of prematurity: At age 6 months sporadic low-voltage, very sharp, short-duration spikes were recorded in the left and right occipital regions independently. At age 4 years 6 months, the spikes were more prevalent both awake and asleep, higher in voltage and longer in duration; had blunter peaks; and occurred more frequently on the left side.

morphology, are quite characteristic in this type of epilepsy (Frost et al., 1992) and differ from those of focal epilepsy associated with structural brain lesions.

The field of benign centro-temporal spikes in most children is characterized by a horizontal dipole with maximum negativity in the centrotemporal region and a positivity in the frontal region (Blume, 1982; Gregory and Wong, 1984; Gutierrez et al., 1990). The considerable distance between the points of maximal negativity and positivity is quite unique, and it is unlikely that this is a true dipole in the usual sense of reflecting the two ends of a local generator. The precise mechanism by which this bipolar field is generated has not been established. It may be that the positivity is generated at a distance by a thalamocortical volley secondary to the corticothalamic volley arising at the site of the cortical focus or that both are generated by a volley from a subcortical (thalamic) source.

"Migration" of Foci

Unlike the spike foci associated with structural cortical lesions, which remain localized to a relatively circumscribed area, the non-lesional spike focus may appear to move from the area in which it was first localized to disparate areas of the same or even the opposite hemisphere. This phenomenon was first described by Gibbs and Gibbs in 1953, who believed that what they called the migration of spike foci was age related (Gibbs et al., 1954). However, Trojaborg (1966), in a serial study of 242 children, showed that frequent transient changes in spike focus location may occur within weeks or even days, and, more recently, long-term monitoring studies have shown that new and disparate foci may appear within a period of a few hours with the onset of sleep (Kellaway et al., unpublished data).

The Nature of the Epileptogenesis

Although clinicians have given this type of seizure disorder the name Benign Focal Epilepsy of Childhood, the shifting sites of focal discharge and the existence of a benign multifocal epilepsy (Moure and Kellaway, 1980) suggest that this type of cortical epileptogenesis is consequent to a widely distributed propensity of the cortex to become focally epileptogenic under certain conditions. The factors that determine when and if a focus of spike activity is to be activated at a particular cortical location have yet to be determined. However, in terms of what is known from animal studies, three factors that may play a role are (1) the different degrees of excitability of various cortical areas; for example, high excitability of the temporal and central cortex compared to the frontal and occipital cortex; (2)

differences in the maturation rate of various cortical regions; and (3) neurotransmitter activity at various times and in various states, e.g., circadian and ultradian modulation (Kellaway and Frost, 1983).

Etiology and Pathogenesis

If, as the evidence indicates, this type of epileptogenesis is not due to an acquired structural lesion, how is it engendered? There is reasonably good evidence that benign focal epilepsy with centro-temporal spikes is genetic. A family history of epilepsy has been reported to occur in from 30% (Bray and Wiser, 1965) to 41% (Degen et al., 1988) of cases studied. A summary of the various studies related to the question is shown in Table 17-2. The results, tested against different genetic hypotheses, indicate that an autosomal dominant gene with a particular age penetrance is responsible for the EEG trait (see Heijbel et al., 1975).

In this regard, it is of interest that a high percentage of children who have centro-temporal spikes also have generalized 3-Hz spike-and-wave bursts (Ambrosetto and Gobbi, 1975; Ambrosetto and Viovanardi, 1975; Beaussart, 1972; Bray and Wiser, 1965; Kato, 1977; Lerman and Kivity, 1975). Absence seizures may indeed be the only, or the predominant, clinical seizure type. An autosomal gene, with age-dependent penetrance, is also believed to be responsible for this EEG trait (Serratosa et al., 1990). Another observation that relates to this association concerns an EEG finding common to both generalized 3-Hz spike-and-wave epilepsy and epilepsy with centro-temporal spikes. When the 3-Hz generalized spike-and-wave bursts or the centro-temporal spikes have been abolished or reduced to low levels by anticonvulsant drug therapy, focal spikes may appear *de novo* in the midline central region. This finding suggests that these midline spikes are an electrical signature of a primary mechanism common to both

TABLE 17-2. Genetics of Epilepsy with "Benign" Centro-Temporal Spike Foci

Reference		Family (%)	Parents (%)	Siblings (%)
Bray and Wiser	1965	30	19	36
Blom et al.	1972	40		
Heijbel et al.	1975			36
Heijbel et al.	1976		27	34
Lerman and Apter	1981		25	
Kajitani et al.	1981	34		
Degen et al.	1988	41		
Kajitani et al.	1980	Three pairs of identical twins		

types of epilepsy and constitute their minimal electrical expression (Kellaway, 1991).

The concept that an epilepsy, with both focal and generalized expression, may derive from a gene-determined cellular defect is not new, but only recently have there been appropriate animal studies providing evidence as to the probable mechanisms involved. Genetic studies of mutant mouse strains have demonstrated that single genes can exert major effects on cellular excitability within discrete central neural pathways (Noebels, 1982, 1984a; Noebels and Sidman, 1979). Neuronal excitability studies have provided evidence for anatomic and biochemical heterogeneity in different parts of the brain, in different parts of the cortex, in different types of neurons, and even in different areas of the cell membrane (Llinás, 1981). Genetic variability may be present at any of these levels. The fact that these mouse strains derive from single locus mutations suggests that the complex behavioral phenotypes are expressions of intermediate phenotypic abnormalities. The fact that a single locus mutation can elicit a complex phenotypic pattern suggests that critical abnormalities in genetically coded protein structure can give rise to a cascade of intermediate phenotypes, resulting in either generalized or focal epilepsy or in both. Also, the cellular expression of single genes may be reflected in developmentally determined alterations in neuronal function within selected CNS cell groups.

The Nature of the Gene-Determined Cellular Defect

The cellular changes responsible for epileptogenesis in benign focal epilepsy have yet to be discovered, and at present it is possible only to speculate, on the basis of animal studies, what the gene-determined cellular defect may be. Two possible mechanisms are described below.

1. *Possible structural-functional changes*: An example is the hyperinnervation of the forebrain by locus ceruleus projections, which is the basis of the generalized epilepsy of the tottering mouse (Noebels and Sidman, 1979). The increased epileptogenicity in this mouse mutant is a consequence of a noradrenergic decrease in hyperpolarization of cortical neurons (Helekar and Noebels, 1991).

2. *Microdysgenesis*: Cortical microdysgeneses have been described in patients with primary generalized epilepsy for many years (Meencke and Janz, 1984; Veith and Wicke, 1968). More recent studies have detailed the following alterations of brain architecture: (1) increase in dystopic neurons in the stratum moleculare, the white matter, the hippocampus, and the cerebellar cortex; (2) an indistinct boundary between cortex and subcortical white matter and between laminae 1 and 2 of the cortex; and (3)

inappropriate columnar arrangement of cortical neurons (Meencke, 1983; Meencke and Janz, 1985). Similar findings have been reported in focal temporal epilepsy (Hardiman et al., 1988). The suggestion is that there may be an accompanying faulty development of synaptic connections that would lead to a state of cortical hyperexcitability.

Whatever the cellular defect may prove to be, the mechanisms that determine the timing and the location of spike foci in this type of epilepsy remain a challenge for future basic research.

CORTICAL EPILEPTOGENESIS CONSEQUENT TO DEPRIVATION OR DISTORTION OF VISUAL INPUT EARLY IN LIFE

Another type of cortical epileptogenesis unique to the developing brain is a result of deprivation or distortion of visual input during a critical period of brain development. It has been shown that destruction of the retina due to retrolental fibroplasia (Kellaway et al., 1955), congenital glaucoma, enucleation of one or both eyes, or optic atrophy occurring in infancy may be associated with the development of spike foci in the occipital regions unilaterally or bilaterally (Kellaway, 1975). Similar spike foci also develop as a result of form deprivation, as in congenital cataract (Smith and Kellaway, 1964a) or suppression amblyopia consequent to strabismus (Stillerman et al., 1952). They also develop in children with congenital failure of the eyes to develop (microphthalmia or anophthalmia).

The pattern of development of the occipital spikes is quite stereotyped and is similar for these various conditions of deprivation or distortion of afferent input: there is a variable period of delay, rarely less than 6 months, before the spikes appear in the EEG. Initially the spikes are very brief, very sharp in waveform, and low in voltage, and they may be evident only in sleep. With increasing age they become higher in voltage and have a longer duration and a blunt waveform (Fig. 17-2B). The inference is that this progression is the result of a gradual increase in the size of the neuronal pool participating in each synchronized discharge. After the age of about 15 years, this succession of changes reverses, and in early adult life the spike may again be very low in voltage, sharp in waveform, and of brief duration, or they may be absent. About 33% of children who have occipital spike foci consequent to deprivation or distortion of visual input have clinical seizures at the time their occipital spike foci are first demonstrated (Smith and Kellaway, 1964a). Long-term studies of a significant number of cases to determine seizure incidence have not been done. Furthermore, the seizures associated with occipital spike foci are often visual sensory, and such seizures may not be apparent in infants.

Mechanism of Generation

The genesis of the occipital spikes and the changes that take place in their characteristics with the passage of time and increasing age most likely are determined by the emergence of new, complex, multisynaptic circuitry. The intricate connectivity that enables the exquisite sensitivity and organization required for stereopsis (Gilbert, 1983) develops over a finite time (Carey et al., 1973; Cragg, 1975). The development of normal binocular connections requires synchronous and *coherent* input from the two eyes during a critical period of brain development (Barlow et al., 1967; Nikara et al., 1968), and spatial alignment is essential (Hubel and Wiesel, 1965; Shlaer, 1971). In the normal adult animal, each visual cortical neuron receives excitatory input from each eye, and if each neuron is to respond appropriately, the two retinal images must lie in an exact spatial relationship. Occlusion of each eye alternately, day by day, produces a progressive shift in eye dominance until only monocularly excitable neurons are present in the visual cortex. This finding has been interpreted as evidence that lateral geniculate cell axons from adjacent laminae compete with one another during development for synaptic surfaces of binocular neurons. The deprived cells are conceived of as being at a disadvantage in the competition (Rose and Blakemore, 1974).

The period of life in which the brain is maximally susceptible to chronic structural and functional defects coincides with a period of most rapid neuronal growth (Cragg, 1975; Hubel and Wiesel, 1970). The period of rapid growth in humans has not been established precisely, but based on studies of the postnatal development of the human lateral geniculate, maximum cell growth occurs during the first 6 to 12 months, and at least 2 years are required for all cells to reach adult size. In the human newborn infant, the primary sensory-receiving areas of the cortex have less arborization and fewer interconnections between cells than most other regions. The number of neurons and the density of their interconnections increase for at least 2 years after term birth (Conel, 1939, 1941, 1947). Clearly these conditions would favor, in the absence of appropriate afferent input, proliferation of recurrent axon collaterals and a net increase in excitatory synaptic drive.

In the kitten, undercutting the cortex results in axodendritic proliferation (Purpura and Housepian, 1961). Spontaneous spike activity that arises as a result of this undercutting may be caused by hyperexcitability of the cortical neurons innervated by the recurrent axon collaterals that have been shown to effect synaptic relationships with pyramidal neurons via the cell body, as well as with basilar dendrites and proximal surfaces of apical dendrites. This reorganization of synaptic circuitry of the undercut cortex would lead to an overall increase in net excitatory drive on pyramidal neurons. However, the development of hypersensitivity after undercutting of cerebral cortex may be a multifactoral process, some aspects of which have been demonstrated in recent studies by Prince and Tseng (1993).

Undercut cerebral cortex in adult animals also shows a slowly developing increase in excitability to the point that spontaneous epileptiform discharge occurs (Echlin, 1959; Grafstein and Sastry, 1957; Sharpless and Halpern, 1962). A critical feature of this phenomenon may be deafferentation of stellate cells (Rutledge, 1969). These cells exert an inhibitory action on pyramidal neurons.

Since glutamic acid decarboxylase (GAD), the GABA-synthesizing enzyme, has been localized in aspiny and sparsely spiny stellate neurons (Ribak, 1978), these neurons are probably responsible for GABA inhibition in the cerebral cortex. Quantitative ultrastructional analysis has revealed a preferential loss of GABAergic synapses in epileptic foci produced by alumina gel in monkey neocortex (Ribak et al., 1982). This loss is caused by terminal degeneration. Other types of terminals in this model are much less affected (Ribak et al., 1979). Cortical epileptogenic foci, both in the monkey and in humans, show reduced GAD activity, GABA concentration, and GABA-receptor binding (Bakay and Harris, 1981; Lloyd et al., 1981). Markers for other transmitters are little changed. The decreased inhibitory function could, in itself, provide a sufficient mechanism for the generation of spikes (Ribak, 1985; Ribak and Reiffenstein, 1982; Ribak et al., 1979).

Similarly, a decrease in inhibitory influence may be present in the occipital cortex of young children with optic atrophy or enucleation early in life. The resulting transneuronal degeneration probably results in deafferentation of stellate cells. A rich pattern of afferent termination is present in layer IV of the cortex where the stellate cells are most abundant, and in visual cortex these neurons are normally rich in GAD (Ribak, 1978).

The occipital spike foci that develop in young children as a consequence of light or form deprivation may be explained on the basis of functional deafferentation at an early critical point of brain development; a statistically significant diminution in the mean number of dendritic spines on apical shafts of layer V pyramidal cells occurs in the visual cortex of mice raised in total darkness, and the stellate cells of layer IV have fewer and shorter dendrites than those of normal animals (Valverde, 1967). In this regard, it is of interest that the development of spontaneous epileptiform activity in chronically isolated cortex can be prevented by daily application of subconvulsive electrical stimulation to the isolated slab (Rutledge et al., 1967).

The development of occipital spike foci in young children with strabismus and unilateral argamblyopia may be explained on a similar basis. Surgically induced strabismus in the cat, performed within an early period of life, produces in the visual cortex a population of exclusively monocular neurons (Hubel and Wiesel, 1965), and small vertical misalignments of about 10° effected with prisms have similar consequences. Amblyopia has been produced in two infant monkeys as a consequent of surgically induced esotropia (von Noorden and Dowling, 1970). In one monkey who had light perception only in the deviated eye, 2 of 47 visual cortical neurons tested could be influenced by the deviated eye. Of the neurons sampled, 55% could not be driven by visual stimulation of either eye. The second monkey, whose acuity in the deviated eye improved with time to 20–300, had more neurons (39 of 68) driven from the deviated eye. Three of these neurons were binocular responsive and were highly dominated by the deviated eye. Thus, although the deviated eye was not deprived of light and pattern stimulation, the net morphological effect and the degree of visual impairment were similar to those seen in monkeys with monocular deprivation due to lid closure during an early critical period (von Noorden, 1973, 1985). The common factor in the two situations was the dissimilarity of input to the two eyes.

In summary, it seems probable that the mechanism by which occipital spikes are generated—as a delayed consequence of deafferentation, light or form deprivation, or disparity of binocular input—is unitary: developmental changes in cortical circuitry seem to be the sole factor common to all three situations. Structural-functional alterations are conceived on the basis of the experimental evidence to result in an increased synchronous excitatory drive and a concomitant decrease in inhibitory input.

V

The vulnerable cortical neuron

18

Glutamate receptors and neuronal death: a beginning

DENNIS W. CHOI

Like the authors of several other chapters in this volume, I began my present line of work in the basement quarters of the Stanford Neurology Department, pulling electrodes and going on raft trips. However, my foray into epilepsy-related electrophysiology never gained the momentum typical of investigators in David Prince's basement. Instead, I became progressively more intrigued with the phenomenon of neuronal cell death.

The 1980s was a time of rapid advances in excitatory amino acid pharmacology and physiology, which permitted specific exploration of John Olney's "excitotoxicity" notion (Olney, 1986)—the idea that the neurotoxicity of endogenous transmitter glutamate could be responsible for some of the neuronal loss associated with disease states. Key experiments emerging from the laboratories of Nils Diemer (Benveniste et al., 1984), Brian Meldrum (Simon et al., 1984), and Steve Rothman (1984) provided strong support for this idea in the specific setting of brain hypoxia-ischemia. Early work with kainate injections into brain in McGeers' laboratory (McGeer and McGeer, 1976) and that of Robert Schwarcz (Coyle and Schwarcz, 1976) raised the interesting speculation that excitotoxicity might be involved in the pathogenesis of Huntington's disease.

My own conversion to Olney's excitotoxicity idea occurred when I found that very brief (2- to 3-minute) exposures to glutamate, at levels found in the ischemic brain, could destroy cultured cortical neurons. As an electrophysiologist accustomed to nursing a single dish through a long day's study, I had to adjust to the destruction scale inherent in toxicity experiments—going through a whole rack of cultures in a single experiment.

Work done in this system, as well as key studies from several other laboratories, has suggested that excitotoxicity is mediated primarily by activation of ion channel-linked receptors. Glutamate activates three major families of ionophore-linked receptors classified by their preferred agonists: N-methyl-D-aspartate (NMDA), kainate, and α-amino-3-hydroxy-5-methyl-4-isoxazolepropionic acid (AMPA; Hollmann and Heinemann, 1994; Watkins et al., 1990). The channels gated by all three receptor subtypes are permeable to both Na^+ and K^+. Channels gated by NMDA receptors, as well as a small subset of channels gated by AMPA or kainate receptors, additionally possess high permeability to Ca^{2+} (MacDermott et al., 1986).

These different glutamate receptor subtypes do not participate equally in the excitotoxic death of cultured cortical neurons (Choi, 1992). Antagonist experiments suggest that most of the neuronal death with brief (minutes) intense glutamate exposure requires NMDA receptor activation. Death can be almost completely blocked by selective blockade of NMDA receptors, although predictably glutamate-induced excitation and acute neuronal swelling are only partially attenuated by the same blockade. In contrast, AMPA/kainate receptors mediate widespread neuronal death only when exposure time is extended for hours. Thus glutamate receptor-mediated neuronal injury on cultured cortical neurons occurs in two main patterns: (1) rapidly triggered excitotoxicity induced by the brief intense stimulation of large numbers of NMDA receptors, and (2) slowly triggered excitotoxicity induced by the prolonged stimulation of AMPA/kainate receptors (or the low level stimulation of NMDA receptors).

Rapidly triggered NMDA receptor-mediated excitotoxic death is critically dependent on the presence of extracellular Ca^{2+}, which is consistent with the idea that rapidly triggered toxicity is initiated by excessive Ca^{2+} influx through the Ca^{2+}-permeable NMDA receptor-gated channel. Slowly triggered AMPA/kainate receptor-mediated excitotoxicity may also be initiated by excessive Ca^{2+} influx. Most channels gated by AMPA or kainate receptors have limited Ca^{2+} permeability, but Ca^{2+} influx can occur as an indirect consequence of Na^+/Ca^{2+} exchanger. In addition, recent studies have indicated that some AMPA receptors may gate channels permeable to Ca^{2+} (Iino et al., 1990), possibly reflecting a molecular composition lacking the Glu-R2 (Glu-RB) subunit (Hume et al, 1991; Verdoorn et al., 1991, see Chapter 5). Recent study suggests that a minority of cortical neurons may possess large numbers of AMPA receptor-gated Ca^{2+} channels and, perhaps as a direct result, are especially vulnerable to damage induced by kainate (presumably mediated by AMPA receptor overactivation; Turetsky et al., 1992). A subpopulation of cerebellar neurons seems to exhibit a similar correlation between expression of AMPA receptor-gated Ca^{2+} channels and heightened vulnerability to damage by kainate (Brorson et al., 1994).

Considerable effort is now ongoing to develop antagonists of specific ionophore-linked glutamate receptors, as well as drugs capable of attenuating the downstream sequelae of receptor activation underlying excitotoxic injury. Viewed broadly as a field, the study of excitotoxicity as well as other processes underlying CNS injury has gathered substantial momentum over the past decade. This momentum has replaced the pessimism and sense of inevitability that permeated consideration of brain or spinal cord injury in the past. The vulnerability of the CNS to damage by acute insults or neurodegenerative conditions is best viewed as the result of multiple specific processes, many related to the cell-cell signaling events special to the brain. Delineation of these processes is likely to facilitate the development of neuroprotective interventions. Investigators in the field have gained confidence that their efforts will lead to effective clinical therapies before long.

19

Measurement of cortical neurotransmitter receptors with radioligand binding: insights into the mechanisms of kindling-induced epilepsy

JOHN E. KRAUS and JAMES O. MCNAMARA

Understanding the pathophysiology of partial epilepsy in molecular terms will likely facilitate development of new and more effective therapies. One approach is to analyze the mechanisms underlying the hyperexcitability in an animal model in the hope that analogous mechanisms are operative in some forms of the human condition. This thinking led us to embark on studies of the kindling model of epilepsy in the late 1970s.

The goal of elucidating the molecular and cellular mechanisms underlying the hyperexcitability of kindling is a daunting one. A host of questions arise. Which part(s) of the brain contain the hyperexcitability underlying kindling *in vivo*? Within those part(s), which population(s) of neurons are responsible? Within those neurons, what is the nature of the hyperexcitability? Is it due to an alteration of synaptic function, of intrinsic properties of the neurons, or other nonsynaptic factors? Given the complexity of these diverse issues, one wonders what might be a logical starting point for such investigations.

We used the measurement of neurotransmitter receptors with radioligand binding techniques as a tool to begin investigation of this model. In this chapter we briefly describe the kindling model, outline the rationale for use of radioligand binding as a tool, consider the advantages and disadvantages of this approach, summarize findings of a host of such studies in the kindling model, and review a few such findings in particular as a means of illustrating the value of such approaches. Although our discussion of radioligand binding in this chapter is from a historical perspective (i.e., its impact on the discovery of anatomical and functional correlates in the kindling model), it is important to note that these techniques continue to be essential in the investigation of the molecular basis of epileptic hyperexcitability.

KINDLING MODEL OF EPILEPSY

Kindling is an animal model of complex partial epilepsy in which periodic application of an initially subconvulsive stimulus eventually results in intense limbic and clonic motor seizures (Goddard et al., 1969; McNamara et al., 1992, 1993). The stimulus is usually a small current (100 to 1000 μA) delivered through an electrode that has been stereotactically implanted into a specific brain region, most often the amygdala. The initial stimulus elicits little change in behavior or electrical brain activity. Additional stimulations result in focal afterdischarge (AD) or electrical seizure recorded with an electroencephalogram (EEG). Initially, the AD is not accompanied by overt behavioral seizures. However, subsequent stimulations induce progressive lengthening and propagation of the AD along with behavioral seizures that increase in intensity over the course of kindling. The following classes of behavioral seizures have been described (Racine, 1972): class 1, facial clonus; class 2, head nodding; class 3, contralateral forelimb clonus; class 4, rearing; and class 5, rearing and falling. An animal is termed "kindled" once the enhanced sensitivity to electrical stimulation has developed (as evidenced by class 4 and 5 seizures). The abnormal neuronal excitability induced by kindling persists for the life of the animal, and spontaneous seizures (i.e., not stimulus induced) may develop after hundreds of electrical stimulations.

The kindled seizures are analogous to complex partial seizures (CPSs) in humans in three important ways: (1) similar EEG activity is recorded from intracerebral electrodes in limbic structures during the seizures; (2) similar behaviors are observed; and (3) kindled seizures and CPSs have similar sensitivities to conventional anticonvulsants (McNamara et al., 1993). Kindling-like processes have been suggested by some to lead to epileptic disorders in humans (see McNamara, 1994). For example, a brain lesion (e.g., glioma, hamaratoma, tumor, trauma) could lead to focal hyperexcitability, which in turn could initiate focal afterdischarges, resulting in a reduction in the seizure threshold and subsequent "kindling" of synaptically connected structures; evidence from human studies supporting such a mechanism has been summarized by McNamara (1994). It seems plausible that insight into the mechanisms of kindling may shed light on some forms of partial epilepsy in humans.

RATIONALE FOR THE STUDY OF NEUROTRANSMITTER RECEPTORS AND FOR RADIOLIGAND BINDING APPROACHES

Modifying the efficacy of chemical synaptic transmission is an attractive way in which the hyperexcitability of kindling could arise. Receptors constitute a critical control point in the chain of events underlying chemical synaptic transmission. A particularly informative example of this principle lies in the striking phenotype, the startle syndrome, accompanying a point mutation of a subunit of the glycine receptor that results in decreased affinity for the transmitter (Shiang et al., 1993). Therefore, study of neurotransmitter receptors in the kindled brain seems to be a reasonable way to initiate the investigation of synaptic function.

Radioligand binding offers a simple, direct, and easily quantitative way to measure receptors. Information on the interaction of ligands with specific receptor types can be readily obtained through the use of radiolabeled ligands that have high selectivity for defined receptors. Determination of receptor density, expressed as a B_{max} value or the maximum amount of ligand bound normalized to protein content, and the affinity of the receptor for the ligand, expressed as a K_D value or the concentration of ligand at which half maximal binding is observed, may be obtained using standard receptor binding assays under equilibrium conditions. Investigation of receptors has been facilitated by the commercial availability of high-specific activity radiolabeled compounds that selectively label pharmacologically defined receptor subtypes.

Adapting radioligand binding to histologic sections, when properly done, facilitates pinpointing the cellular locale of alterations of receptor binding. The excellent quantitative measures combined with the improved anatomical resolution facilitate the development of hypotheses about specific populations of synapses on identified cells that might undergo modifications of efficacy.

These advantages notwithstanding, these techniques have several limitations. We assume for the purposes of this discussion that the proper controls are implemented to assure, based upon use of appropriate pharmacologic compounds, that the binding site in fact represents a receptor. Even so, it is not clear whether the binding sites measured are located at synapses, at extra-synaptic sites in the cell membrane, or even at intracellular membranes (e.g. endoplasmic reticulum); these possibilities cannot be differentiated with light microscopic resolution available from standard radiohistochemical approaches. Likewise, localization of receptor alterations to pre- or postsynaptic regions is not directly possible; however, correlative electrophysiological and pharmacological studies can provide clues. Moreover, the molecular nature of the binding site is not directly addressed by radioligand binding studies. The measured site could actually be a high-affinity state that represents a desensitized form that is not actually functional. Therefore, the *functional significance* of an alteration identified in a radioligand binding assay remains to be established. Since a receptor is only one component of a complex cascade mediating a synaptic event, the presence of a modification of receptor binding site must be viewed as a rationale to proceed with study of synaptic physiology. Finally, important mechanisms of receptor modulation (e.g. phosphorylation) that might occur in a kindled brain could easily escape detection in a radioligand binding assay, e.g., the phosphorylated receptor might undergo dephosphorylation during membrane preparation and elude detection by these methods.

Despite these limitations, the rapidity with which information of extraordinary molecular specificity can be obtained in multiple neuronal populations in multiple brain regions underscores the value of this technique as a starting point for investigation. Modifications can be used to develop a host of hypotheses that can then be tested with other techniques.

THE HIPPOCAMPUS: A MODEL FOR STUDY OF MECHANISMS OF KINDLING

Anatomical delineation of structures containing hyperexcitability is necessary for a meaningful analysis of the mechanisms of the hyperexcitability of kindling. Work by Dr. Oswald Steward and his colleagues suggested that the enhanced propensity to express seizures in the kindled animal was due to hyperexcitability intrinsic to sites spatially remote from the stimulated structure (Messenheimer et al., 1979). Identification of at least one such site would provide a model in which the mechanisms could be analyzed, and this information could guide subsequent studies to determine whether analogous mechanisms are operative elsewhere in the brain.

Dr. McNamara and his colleagues first outlined these two criteria necessary to implicate discrete neuronal populations in the hyperexcitability: (1) destruction or deafferentation of a specific population(s) of neurons must dramatically modify kindling in the intact animal, a finding necessary to link mechanistic studies of specific neuronal populations *in vitro* with the model *in situ;* and (2) abnormal hyperexcitability must be demonstrated intrinsic to these same neurons or their synapses in a kindled brain, which is possible only by *in vitro* electrophysiological study of slices of such neurons removed from the brain following kindling. The question of which of many brain structures might contain the hyperexcitability remained unanswered.

The central role of the hippocampus in temporal lobe epilepsy in humans led Dr. McNamara and his colleagues to suspect that it was pivotally involved in the hyperexcitability of the kindled brain. Results of radioligand binding studies of muscarinic cholinergic, and benzodiazepine receptors provided molecular evidence for a modification intrinsic to discrete populations of neurons in the hippocampus. In animals undergoing kindling by amygdala stimulation, reductions of muscarinic cholinergic receptor binding were found in membranes isolated from the hippocampus (Byrne et al., 1980). Increased density of benzodiazepine receptor binding was also found in hippocampal membranes following amygdala kindling (McNamara et al., 1980). In each instance, these alterations of receptor binding were subsequently localized at least in part to a discrete neuronal population, the dentate granule cells (Savage et al., 1983; Valdes et al., 1982).

Together, these findings provided a strong rationale for study of the hippocampus itself. Dr. McNamara and his colleagues exploited highly selective lesion techniques to begin to address the role of the hippocampus in kindling. Using intrahippocampal microinjections of colchicine, a quite selective neurotoxin for the dentate granule cells, they found that elimination of the granule cells was associated with marked slowing of the rate of kindling development (Dasheiff and McNamara, 1982; Frush et al., 1986). They subsequently found that microknife cuts of the principal excitatory afferents of the granule cells and hippocampus, the entorhinal afferents, profoundly slowed the rate and in some instances prevented kindling development (Savage et al., 1985). Anatomic specificity was demonstrated with both of these experimental approaches, since injections of colchicine into neocortex or knife cuts of neocortex failed to modify kindling development. The similar outcomes emerging from these distinct lesion techniques that removed different components of the hippocampal network led to the conclusion that the hippocampus facilitates formation of kindling *in situ*.

These experiments led to the first *in vitro* electrophysiological studies of hippocampal slices isolated from electrically kindled animals, performed in collaboration with Drs. Greg King and Raymond Dingledine (King et al., 1985). In slices isolated either one day or one month after the last kindled seizure, hyperexcitability was identified in all three principal neuronal populations of the hippocampus: the dentate granule cells, CA2/3 pyramidal cells, and CA1 pyramidal cells. In particular the immediate recipients of a powerful excitatory input from the granule cells, the CA2/3 pyramidal cells, were found to exhibit increased burst firing in slices isolated from kindled animals. The importance of these findings was threefold. First, they clearly implicated the hippocampus as one structure of the kindled brain containing abnormal hyperexcitability. Second, they provided a clue to the mechanism of the hyperexcitability of the kindled brain, since the increased propensity of CA2/3 neurons to exhibit burst firing would be expected to promote the initiation and/or propagation of kindled seizures. Together with the striking effects of lesions of discrete hippocampal neuronal populations on kindling *in situ*, these findings strengthened the rationale for study of the hippocampus in search for the underlying mechanisms. Third, they documented the feasibility of detecting differences between control and kindled animals in electrophysiologic studies of hippocampal slices *ex vivo*, a necessary step for analysis of the mechanisms of hyperexcitability, thereby setting the stage for similar approaches in many other laboratories.

IDENTIFICATION OF SEIZURE-INDUCED ENHANCEMENT OF INHIBITORY SYNAPTIC FUNCTION IN THE KINDLED BRAIN

As outlined above, modification of chemical synaptic function might contribute to the hyperexcitability of a kindled brain. Quite simplistically, the hyperexcitability could be due to enhanced function of excitatory synapses, impaired function of inhibitory synapses, or a combination of the two. Although many additional possibilities exist, regulation of synaptic function constituted a logical starting point for investigation.

Some of the early studies using radioligand binding techniques centered on the $GABA_A$ receptor and its allosteric components. Increased numbers of $GABA_A$ and benzodiazepine receptors were identified in hippocampal membranes isolated from kindled animals and were subsequently localized by quantitative radiohistochemical methods to the dentate granule cells of hippocampus (Shin et al., 1985b; Valdes et al., 1982). The question arose as to whether this biochemical alteration exhibited a functional correlate. The direction of this change, an increase, together with the cellular localization led to the hypothesis that GABA-mediated synaptic inhibition of the granule cells would be increased in slices from the kindled hippocampus. Subsequent electrophysiological studies demonstrated an increase of paired pulse inhibition of granule cells with a time course paralleling the receptor regulation (King et al., 1985), suggesting that the increased expression of $GABA_A$ receptors in the granule cells contributed to enhanced synaptic inhibition. Otis et al. (1994) identified a near-doubling of the number of synaptic $GABA_A$ channels in dentate granule cells from hippocampal slices of kindled animals with a time course matching that of the radioligand binding studies. Interestingly, Otis et al. (1994) argue that, since postsynaptic $GABA_A$ receptors are likely to be saturated by the

GABA concentrations in the synaptic cleft, increasing the number of receptors may be a particularly effective method to augment synaptic inhibition in the mammalian brain.

Several lessons can be learned from these studies. First, the results of both the synaptic physiology and whole cell recordings strongly support the likelihood that the alterations in receptor content identified by radioligand binding assays are *functionally significant.* Therefore, binding studies can provide a valuable framework for the technically more time consuming analyses of synaptic physiology and single channel properties. Additionally, the electrophysiological studies support our original hypothesis that alterations in $GABA_A$ receptors were postsynaptic. Second, in parallel with these studies of the $GABA_A$ receptor, Dr. McNamara and his colleagues have successfully demonstrated the multiplicity of neurotransmitter receptors (summarized in Table 19-1) undergoing regulation in the hippocampus of kindled animals (Byrne et al., 1980; Crain et al., 1987; McNamara et al., 1980; Savage et al., 1984; Shin et al., 1985b). A recurrent theme emerged in each of these studies: the induction of the regulation required multiple seizures, the regulation returned to normal within a week or so after cessation of seizures, and the direction of the regulation would be expected to *decrease* neuronal excitability. This pattern was identical to that found with the $GABA_A$ receptor (Shin et al., 1985b). These findings are consistent with the idea that repeated seizures trigger a long-lasting adaptive response intrinsic to discrete neuronal populations aimed at dampening the increased excitability of a kindled brain and raising seizure threshold. Finally, understanding the molecular details of the brain's adaptive responses should suggest improved methods of pharmacologically regulating neuronal excitability with more specific and effective drugs. Stated differently, identifying the molecular nature of the way in which the brain itself attempts to limit excitability may lead to the discovery of drugs that selectively target $GABA_A$ (or other) receptor subtypes especially important in control of limbic seizures.

ENHANCED EXCITATORY SYNAPTIC TRANSMISSION: A PUTATIVE MECHANISM OF HYPEREXCITABILITY IN KINDLING

Synapses using glutamate and related excitatory amino acids represent the principal excitatory synapses in the central nervous system. Enhanced function of such synapses is one potential mechanism underlying the enduring hyperexcitability of a kindled brain. Upon release from a presynaptic terminal, glutamate can effect a diversity of biological responses by binding to and activating at least four different receptor subtypes. In principle, enhanced function of synapses using one or more of these receptors could contribute to the hyperexcitability.

If enhanced function of one of these subpopulations of glutamatergic synapses were responsible for the hyperexcitability, one expectation is that antagonists of this receptor subtype would inhibit a kindled seizure. To

TABLE 19-1. Receptor Alterations in Limbic Structures of Kindled Rats

Receptor Type	Direction of Change	Magnitude of Change (%)	Time Course*	Tentative Anatomical Locale	Predicted Effects on Synaptic Transmission†
Muscarinic, cholinergic[1,2,3]	↓	15–30	15h ALS; ND 7d ALS	Dentate granule cells	↓
β-Adrenergic[1,4]	↓	30	3d ALS; ND 7d ALS	Amygdala, neocortex	↓
α_2-Adrenergic[5]	↓	20	1d ALS; ND 2wk ALS	Pyriform cortex, amygdala	↓
μ Opioid[6]	↓	17–32	24h ALS; ND 7d ALS	Interneurons in CA1, CA2, CA3	↓
δ Opioid[6]	↓	11–17	24h-7d ALS; ND 28d ALS	Interneurons in dentate gyrus; CA1, CA3	↓
$GABA_A$/benzodiazepine[7,8,9]	↑	22–40	24h ALS; ND 28d ALS	Molecular, granule layers of dentate gyrus	↓
Kainate[10,11]	↓	20–60	24h ALS; ND 28d ALS	Stratum lucidum of CA3, molecular layer of dentate gyrus	↓
AMPA[11]	↓	20–35	24h ALS; ND 28d ALS	Throughout neocortex, hippocampus	↓
NMDA[12]	↑	185	28d ALS	CA3	↑

*Abbreviations: ALS: after last seizure; ND: not detected.

†Assuming a postsynaptic localization of receptors. For predictions on effects of receptor alterations on kindling development/permanence, see individual papers. [1]Byrne et al. (1980); [2]Savage and McNamara (1982); [3]Savage et al. (1983); [4]McNamara et al. (1978); [5]Chen et al. (1990); [6]Crain et al. (1987); [7]McNamara et al (1980); [8]Valdes et al. (1982); [9]Shin et al. (1985); [10]Savage et al. (1984); [11]Okazaki et al. (1990); [12]Kraus et al. (1994).

test this idea, Dr. McNamara and his colleagues determined whether antagonists of the NMDA subtype of glutamate receptor inhibited kindled seizures. They discovered that an NMDA receptor antagonist did indeed inhibit kindled seizures (McNamara et al., 1988), although the animals exhibited several unwanted behavioral effects at effective doses, including ataxia and hypotonia. Similarities in the nature of the unwanted effects obtained with multiple structurally distinct NMDA antagonists suggested that the unwanted effects were mediated by antagonism of NMDA receptors that normally regulate those functions. Taken together, the importance of these findings was threefold: (1) they underscored enhanced function of these synapses as a potential mechanism underlying the hyperexcitability of kindling; (2) they raised the possibility that drugs that inhibited activation of NMDA receptors may effectively inhibit temporal lobe seizures; and (3) the presence of the unwanted effects emphasized the potential pitfalls of antagonizing all NMDA receptors and underscored the need to somehow develop specificity among NMDA antagonists if such drugs were to be useful anticonvulsants clinically.

The presence of an enhanced NMDA component of synaptic function in the kindled brain emerged from the work of Drs. Istvan Mody and Uwe Heinemann (1987). Using hippocampal slices isolated from kindled animals, these investigators demonstrated the presence of an enhanced NMDA component of the EPSP of the entorhinal-granule cell synapse of kindled animals. This finding strengthened the likelihood that enhanced function of this subpopulation of synapses contributed to the hyperexcitability of kindling. Part of this enhanced NMDA component may be due to a modification intrinsic to the NMDA receptor itself since single channel recordings of NMDA receptors in granule cells acutely isolated from slices of kindled animals revealed increases in mean open time and reduced sensitivity to Mg^{2+} block (Köhr et al., 1993).

Radioligand binding assays have led to the discovery of another modification of the NMDA receptor in the kindled brain; namely, in the CA3 region of the kindled hippocampus. Biochemical studies from Dr. McNamara's laboratory demonstrated that hippocampal neurons of kindled animals exhibit an increased sensitivity to NMDA, since NMDA more potently inhibited carbachol-stimulated phosphoinositide hydrolysis in hippocampal slices isolated one day or one month after the last kindled seizure (Morrisett et al., 1989). This increased sensitivity to NMDA was subsequently localized to the CA3 pyramidal cells, which were five times more sensitive to NMDA-evoked depolarizations in slices isolated from kindled animals (28 days after the last kindled seizure) compared to controls (Martin et al., 1992). A significant but smaller increase was evident in slices isolated one day after the last kindled seizure.

One possible mechanism to explain the enhanced sensitivity of hippocampal slices to NMDA demonstrated in our biochemical and electrophysiological studies is an increase in receptor reserve. To test this possibility, binding assays were performed using structurally distinct NMDA receptor-selective radioligands to estimate receptor density in membranes isolated from whole hippocampus of kindled animals killed 28 days after the last kindled seizure. A 47% increase in the binding of the competitive NMDA receptor antagonist [^{3}H]CPP (3-[(±)-2-(carboxypiperazin-4-yl)]propyl-1-phosphonic acid) was found; similar increases were also found for [^{3}H]glycine (↑42%), [^{3}H]TCP (*N*-(1-[thienyl]cyclohexyl) piperidine); (↑28%);(Yeh et al., 1989) and NMDA-displaceable L-[^{3}H]glutamate (↑36%); (Kraus et al., 1994). Importantly, the affinity of the NMDA receptor for [^{3}H]CPP and [^{3}H]glutamate was reduced, which likely explains our failure to detect an increase in the specific binding of L-[^{3}H]glutamate to NMDA receptors in CA3 or any other region of hippocampus using radiohistochemical techniques (Okazaki et al., 1989). Increases in the binding of L-[^{3}H]glutamate, possibly to NMDA receptors, have also been demonstrated in postsynaptic density fractions isolated from cortex of kindled rats 2 weeks after the last kindled seizure (Wu et al., 1990), as well as in hippocampal membranes isolated from pentyleneterazole-kindled rats (Schröder et al., 1993).

To determine the anatomical locale and time course of the kindling-induced changes in NMDA receptor binding, membranes were prepared from microdissected hippocampal regions fascia dentata, CA3, and CA1 for use in equilibrium binding assays (Kraus et al., 1994). Membranes were prepared from animals killed either 28 days or 24 hours after the last kindled seizure. At 28 days, we found that the number of binding sites for [^{3}H]CPP was markedly increased in membranes isolated from region CA3 of kindled animals, with a B_{max} approximately 2.8 times higher than that of controls. Additionally, the affinity of the NMDA receptor for this ligand was lower in region CA3 of kindled animals, as reflected by a three-fold increase in K_D. Changes in B_{max} and K_D were restricted to CA3, with no differences in [^{3}H]CPP binding detected in membranes from either fascia dentata or CA1. Surprisingly, no differences between kindled and control animals in the binding of a second competitive NMDA receptor antagonist, [^{3}H]CGS-19755 (1-(cis-2-carboxylpiperidine-4-yl)methyl-1-phosphonate), were found. In contrast to the threefold increase in B_{max} detected at 28 days, we found smaller increases in B_{max} (↑33%) and K_D for [^{3}H]CPP in region CA3 of kindled animals at 24 hours. In accord with findings at 28 days, no change in the B_{max} for [^{3}H]CPP was detectable in membranes from either fascia dentata or CA1 or in [^{3}H]CGS-19755 in any region. These results indicate that kindling induces the expression of an NMDA receptor in CA3 specifically recognized by the NMDA receptor antagonist CPP.

Experimental evidence suggests that this kindling-induced NMDA receptor is functional. We found that CPP was less potent in inhibiting glutamate-stimulated [^{3}H]TCP binding—a measure of NMDA receptor activation (Bonhaus and McNamara, 1989)—in whole hippocampal membranes of kindled animals isolated 28 days after the last kindled seizure (Kraus et al., 1994), a result in accord with the lower affinity of the NMDA receptor for CPP shown in the binding data. There were no changes in the potency of CGS-19755. Recent electrophysiological evidence utilizing the grease-gap method has demonstrated a reduction in the potency of CPP in antagonizing NMDA-evoked depolarizations (Nadler et al., 1994), paralleling the reduced affinity of CPP in binding studies.

The biochemical data derived from the radioligand binding studies demonstrate that kindling induces a marked increase in the expression of an NMDA receptor ($NMDAR_K$) that is novel insofar as it is recognized by one competitive antagonist ([^{3}H]CPP) but not by another ([^{3}H]CGS-19755). The expression is long lasting and, within the hippocampus, is confined to region CA3. The direction (an increase), time course (present at both one day and one month, but greater at one month after the last kindled seizure), and anatomical locale (CA3 region but not CA1) of the changes in [^{3}H]CPP binding parallel the enhanced sensitivity of CA3 pyramidal cells of kindled animals to NMDA. Such a correlation suggests that $NMDAR_K$ may underlie the long-lasting increased potency of NMDA-evoked depolarization of CA3 pyramidal cells of kindled animals. Also, the localization of this receptor is likely postsynaptic, especially in light of the strong correlation with the enhanced sensitivity of region CA3 to NMDA and the lack of direct evidence of presynaptic NMDA receptors. Together with the findings of Köhr et al. (1993) in the dentate, the time course and direction of these modifications of the NMDA receptor in the CA3 region are consistent with the idea that this receptor contributes to the hyperexcitability of the kindled brain.

Several lessons emerge from these studies of the NMDA receptor with radioligand binding techniques. First, recognition of the NMDA receptor regulation in the CA3 region of the kindled hippocampus by one but not by another competitive antagonist (the differences in the structure of these antagonists are subtle) underscores the extraordinary specificity of the regulation. This provides hope that compounds can be identified that selectively block the form of NMDA receptor expressed in the kindled CA3 region, but have minimal or no unwanted effects due to block of other NMDA receptors. Second, these techniques failed to detect alterations of the NMDA receptor in the fascia dentata membranes, raising the possibility that findings by Köhr et al. (1993) may be due to a post-translational modification that escapes detection with these methods. Finally, these findings underscore the need to move forward with the powerful techniques available from molecular neurobiology, including the diversity of probes for distinct genes (Kraus et al., 1994) and antibodies for subtypes of NMDA receptors.

CONCLUSIONS

Radioligand binding measures of neurotransmitter receptors have proven to be powerful tools with which to initiate investigations of the kindling model of epilepsy. Results of such studies guided lesion and electrophysiological investigations that implicated the hippocampus as one structure in the kindled brain exhibiting abnormal hyperexcitability. Results from such studies subsequently provided a framework for examination of specific synapses on distinct populations of neurons in the hippocampus of kindled animals, experiments that confirmed the presence of synaptic modifications predicted by the binding assays. More recent studies have identified increased numbers of a novel form of the NMDA receptor in the CA3 region of the kindled hippocampus, a regulation that may contribute to the hyperexcitability of the kindled brain. These findings challenge investigators to elucidate the molecular nature of this novel form of the NMDA receptor and how it affects the synaptic function of these neurons and the excitability of the kindled brain.

20
Pathophysiology of cortical synapses and circuits

PHILIP A. SCHWARTZKROIN

This chapter focuses on the cellular features of the epileptic brain that may be key to an understanding of mechanisms underlying the pathophysiological discharge. The major issues are reviewed in historical context to show how our views have evolved from electrophysiological analyses of epileptogenic cortical tissues. Intrinsic cellular and synaptic contributions to the abnormal activities are examined, and illustrations are drawn from studies of chronic animal models of the epilepsies.

HISTORICAL PERSPECTIVE OF UNDERLYING MECHANISMS

Cellular Epileptiform Activity: A Function of Intrinsic Neuronal Pathology or an Emergent Property of Circuitry Interactions?

One of the earliest issues to arise from the study of cellular activities in epileptic cortex was whether there were identifiable neurons that were intrinsically abnormal—"epileptic neurons"—that drove the remaining population of relatively normal neurons. In the initial intracellular studies of Matsumoto and Ajmone Marsan (1964), the idea of an epileptic neuron evolved from their description of the paroxysmal depolarization shift (PDS) as the characteristic intracellular event of the interictal cortical neuron. Further studies from the Prince laboratories (Prince, 1968b, 1969a,b; Prince and Futamachi, 1970) were consistent with the notion of an epileptic neuron in the acutely induced cortical focus (made epileptic via application of a $GABA_A$ antagonist, such as penicillin). Extracellular recordings from the chronic monkey alumina cream model (Calvin et al.,

1968; Wyler et al., 1975, 1978) described a small subpopulation of neocortical neurons that were "intrinsic bursters," which seemed to recruit relatively normal surrounding cells into the epileptic discharge. This idea of an epileptic neuron implied that one should be able to identify abnormalities in the intrinsic properties of candidate cells, properties that would, at the very least, endow the cell with the capacity for independent burst discharge. Such intrinsic features were sought in cellular studies on epileptic neocortex and hippocampus—both *in vivo* and *in vitro*—with relatively little success. Cells in the epileptic brain, at least during interictal periods, appeared quite normal; no changes were found in such properties as resting potential, input resistance, or discharge patterns in response to depolarization.

The existence of an intrinsically bursting neuron in mammalian brain, however, was supported conceptually by descriptions of "pacemaker" neurons in invertebrates (Johnston, 1976; Wilson and Wachtel, 1974). The "pacemaker" concept and its possible epileptogenic consequences received significant reinforcement when the burst properties of CA3 pyramidal cell neurons were described in hippocampal slices (Wong and Prince, 1978; Wong and Schwartzkroin, 1982) (Fig. 20-1A). These cells, in response to intracellular depolarization, fired in burst patterns quite similar to those seen in cells discharging interictal bursts in epileptogenic hippocampus (Dichter and Spencer, 1969a,b). Further study indicated that the bursts were based on voltage-dependent calcium influx, in dendrites as well as soma, and did not require exaggerated synaptic input as a trigger (Schwartzkroin and Prince, 1978; Wong and Prince, 1978, 1979, 1981). A similar/parallel intrinsic burst-generating neuron has been identified in the deep layers of neocortex (Connors and Gutnick, 1990; Connors et al., 1982; see Chapter 9). Although the burst discharge characteristics of these neocortical cells are not so clearly calcium dependent as the bursts of hippocampal CA3 neurons, studies of both cell types suggested that intrinsic burst discharge mechanisms could underlie the PDS characteristic of the "epileptic neuron." Further, cells with these characteristics were located in cortical regions that played a "trigger" or pacemaker role with respect to the surrounding tissue (Gutnick et al., 1982; Wong and Traub, 1983). Finally, recent studies on chronic models of epileptiform activity have reported changes in intrinsic cell properties that may contribute to epileptogenesis (Franck et al., 1988; Prince and Tseng, 1993).

An opposing—although not mutually exclusive—argument developed in parallel to this concept of the "epileptic neuron." Investigators argued that epileptiform discharge is a result of synaptic mechanisms, that the PDS is a "giant EPSP" (Dichter and Spencer, 1969a,b; Prince, 1968a), and that the large hyperpolarization that generally follows the PDS is an ex-

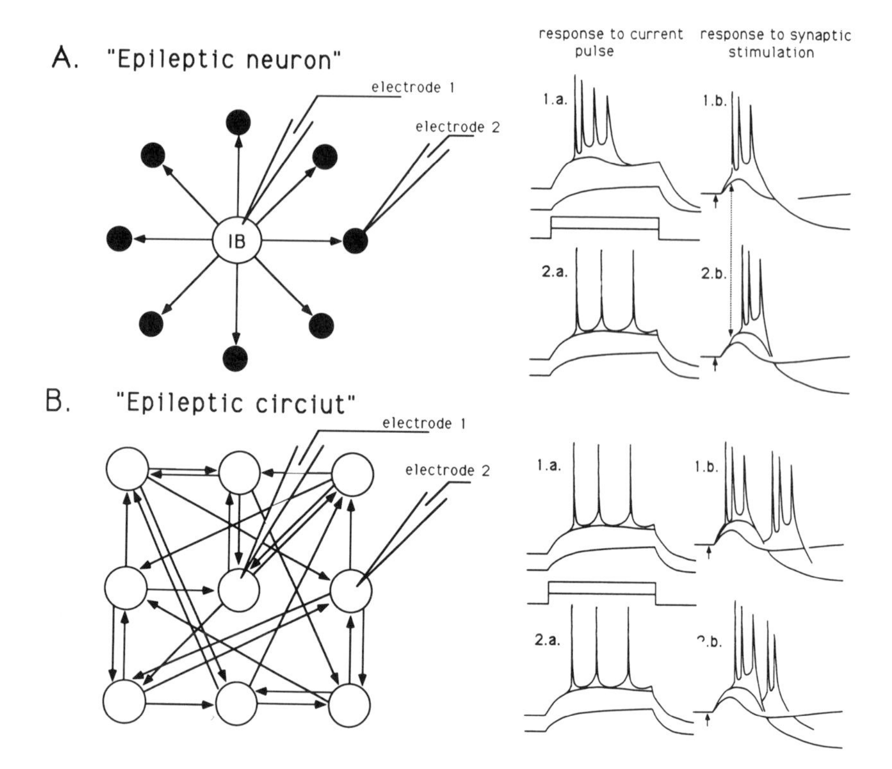

FIGURE 20-1. Diagrammatic representations comparing the epileptic neuron to the epileptic circuit. **A.** The epileptic neuron is an intrinsic burst (IB) generator that drives a population of normal neurons (black circles). Theoretical responses recorded from the IB neuron (electrode 1) and a follower neuron (electrode 2) are shown in traces 1.a,b and 2.a,b, respectively. Intracellular current injection elicits a burst of action potentials from the IB cell (1.a), but only a "normal" train of action potentials from the follower (2.a). Stimulation of an afferent input (short arrows in traces at right) drives an EPSP/(IPSP) in both cells when the input is "subthreshold." Stronger stimulation elicits a burst in the IB cell (1.b), which is followed by a large EPSP or burst in the follower cell (2.b) Note the difference in burst latency in the IB versus the follower cell in response to afferent stimulation. **B.** In the epileptic circuit, a population of relatively normal cells is interconnected via axon collaterals making excitatory synapses. All cells in the circuit respond normally to intracellular current injection (responses at electrodes 1 and 2 in traces 1.a and 2.a, respectively). Afferent stimulation that exceeds threshold drives bursts of action potentials in all cells in the circuit, but the timing and morphology of the burst may vary from cell to cell (cf. 1.b and 2.b), depending on the pathway of excitation in the circuit.

aggerated IPSP. According to this view, the burst discharge in epileptic tissue resulted from "abnormal" synaptic interactions, rather than from intrinsic properties of the neurons, and the PDS phenomenon was an emergent property of the circuitry (Fig. 20-1B). This point of view was strongly supported by Johnston and Brown (1981), who showed that in hippocampal slices treated with penicillin, spontaneous burst discharge observed in a given neuron was not dependent on the cell's membrane potential (as would be expected if the bursts were based on intrinsic membrane properties) and that the PDS underlying the burst behaved like a large synaptic potential. Clearly, there are data that support both sides of this "controversy," and it seems likely that both intrinsic cell properties and circuitry interactions are important contributors to the pathophysiology of the epileptic focus.

Relative Contributions of Enhanced Excitation versus Loss of Inhibition

Given the emphasis on altered synaptic interactions as the underlying basis for epileptogenesis, investigators turned their attention to an evaluation of whether enhanced excitation or reduced inhibition was the key factor. Since many of the initial techniques used to induce experimental epileptogenesis involved acute treatment of cortical tissues with GABA blockers (Purpura et al., 1972a,b), there was early support for the role of loss of inhibition in epileptiform discharge. However, as indicated above, investigators interpreted the depolarization underlying the PDS as a giant EPSP, suggesting that excitatory synaptic connections were significantly enhanced, i.e., either more transmitter was liberated from a given terminal or more terminals released transmitter during the epileptiform event. Paradoxically, although GABA antagonists were used to produce these epileptiform discharges, there appeared to be a giant IPSP following the PDS discharge (Dichter and Spencer, 1969b; Prince, 1968a). More recent studies have shown that this hyperpolarization was due to a calcium-dependent potassium current (Schwartzkroin and Stafstrom, 1980) and/or a $GABA_B$ IPSP (Malouf et al., 1990). Early experiments also showed that surround inhibition was a significant feature of cortical and hippocampal epileptic foci (Dichter and Spencer, 1969a,b; Prince and Wilder, 1967), reinforcing the idea that inhibition could be maintained in and around the epileptic focus. Careful analyses by Dingledine and Gjerstad (1980) and Schwartzkroin and Prince (1980a) showed that the development of epileptiform bursting, in penicillin treated tissue, evolved in parallel with the blockade of GABAergic inhibition in the tissue. These investigators argued that the PDS/depolarization underlying the epileptiform burst was a "release" phenomenon, whereby normally occurring EPSPs *appeared* to grow in amplitude and duration as the shunting/inhibitory action of the IPSP was blocked. Loss of IPSPs as a critical feature of the epileptic brain has been supported in a variety of subsequent studies, including careful titration of GABAergic blockade in neocortex (Chagnac-Amitai and Connors, 1989), demonstration of selected loss of GABAergic neurons and terminals in monkey alumina foci (Ribak et al., 1986), and a variety of observations on the "dormant interneuron" in chronic animal models (to be discussed below).

It should be noted, however, that there are several epileptogenic experimental manipulations that do not block or reduce inhibition, but simply enhance the EPSP amplitudes. For example, tissue treatment with potassium channel blockers (e.g. 4-aminopyridine or tetraethylammonium) results in epileptiform activity under conditions in which inhibition is relatively unaffected (Perreault and Avoli, 1991; Schwartzkroin and Prince, 1980b). Even in recordings from human epileptic tissue, loss of inhibition is not an obvious feature (Avoli and Olivier, 1989; Knowles et al., 1992; but see Chapter 21). Based on the available evidence, it seems clear that either loss of inhibition or facilitation of excitation can lead to epileptiform burst discharge patterns and that each form of abnormal activity must be investigated for its unique underlying mechanisms.

Synchronization

Although a change in excitability is perhaps most obviously a hallmark of epileptogenic tissue, it has long been clear (from the time of Hughlings Jackson, 1931) that synchronization of discharge is at least as important. Indeed, increased excitation without cell synchronization produces no epileptiform activity, whereas even inhibitory events that synchronize the cell population produce electrical activity that looks "epileptiform." This latter point is perhaps most evident in models of absence (or spike-and-wave) epilepsy, which evolves from synchronizing hyperpolarizing (inhibitory) mechanisms (see Chapter 11). If one believes that synchronization is a critical issue, then an important question is how it may be achieved. At least three different mechanisms have been discussed in the literature. The first and most obvious involves conventional depolarizing synaptic interactions. Synchronization of activity via excitatory axon collateral connections was proposed by Dichter and Spencer (1969b), based on their stimulation studies on deafferented hippocampus. Antidromic activation of this population resulted in burst discharges, presumably mediated by local excitatory collateral connections. Synchronization of cell discharge through excitatory interconnections has, more recently, been modeled successfully (Traub and Miles, 1991; Traub et al., 1984) and is supported by observations of excitatory axon collateral connections between cell populations in cortex and hippocampus. Interestingly, in the hippocampal CA3 region, these excitatory connections are significantly enhanced as inhibition is blocked (Miles and Wong, 1987a,b). In normal tissue, excitatory interaction between CA3 cells seems to be shunted significantly by inhibitory synaptic input; as inhibition is reduced, excitatory synchrony is increased dramatically. The strength of synchronization mechanisms and their potential significance in driving neuronal populations have recently been underscored by Michelson and Wong (1991; see Chapter 10), who have shown that, even in the absence of conventional excitatory synaptic transmission, inhibitory neurons may become synchronized via depolarizing synaptic connections.

A second avenue for synchronization is the electrotonic coupling among cells in discharging populations. Studies in both neocortex and hippocampus have revealed significant dye coupling of neurons when Lucifer Yellow was used to label pyramidal cells (Gutnick and Prince, 1981; MacVicar and Dudek, 1980b). Subsequent electron microscopy has shown occasional gap junctions (Dudek et al., 1986), and electrophysiological studies have demonstrated electrotonic coupling directly in a small number of connected pairs in hippocampus (MacVicar and Dudek, 1980a). It is unclear, however, whether electrotonic synaptic interactions contribute significantly to the synchrony of the epileptic focus, especially given the relative infrequency of such coupling compared to conventional chemical synaptic interactions. Modeling studies (Traub and Wong, 1983) suggest that electrotonic synapses may subtly modulate the nature of the discharging focus, but in themselves cannot establish the synchrony. The related electrical phenomena of "ephaptic interaction," based on the flow of current through extracellular space, may play a more significant role in shaping the extent of the epileptic focus (Traub et al., 1985). Investigators have shown that, with the generation of synchronized neuronal discharge, large field effects can influence neuronal discharge propensities (Taylor and Dudek, 1984); these extracellular currents may "recruit" cells at the borders of the focus by contributing to their depolarization, thus enlarging the cell population involved in epileptiform events.

Chronic Models of Epilepsy

As indicated above, most of the early models used for studying mechanisms underlying epileptiform discharge involved the acute application of the $GABA_A$ blockers. Under these conditions, the experimenter knew, a priori, at least one of the basic changes in the discharging tissue. Underlying mechanisms have not been so clear in chronic models. Given that chronic models may provide a better approximation of critical factors in human epileptic brain, an understanding of basic mechanisms in such models seems critical. Until relatively recently, the chronic model that received most attention was that produced in the primate cortex by injection of aluminum hydroxide (Schwartzkroin and Wyler, 1980; Ward, 1972). Although relatively intractable to intracellular investigation, several important observations emerged from extracellular studies of the alumina monkey focus. First, as indicated above, there was apparently a population of abnormal "bursters" characteristic of the alumina focus (but not of comparable normal neocortical tissue); the remaining cells were quite normal, exhibiting burst discharge only as epileptiform activity developed. Second, synchrony of cell discharge and the relationship between cell discharge and the surface EEG were not nearly as "tight" as observed in the acute models on which much of the intracellular analysis had been based. Third, there was little evidence for a stereotyped PDS-like event in the alumina cortex; the bursts were variable in amplitude and duration, and bursts of action potentials were generated from prolonged or enlarged EPSP-like events. Finally, although the mechanisms by which aluminum hydroxide induces epileptiform activity in neocortex remain unexplained, it is clear that long-term (and apparently progressive) alterations in neuronal integrity are associated with the alumina focus in and/or near the focus. In particular, cortical dendrites were found to degenerate (Westrum et al., 1964), with spines and fine dendritic processes most affected. Whether these morphological changes are causally related to the onset and development of epileptiform bursting has not been determined, but the occurrence of gradually occurring structural alterations correlates with the long (several months) incubation time of epileptiform activity in the primate alumina focus. A structural basis for the epileptogenicity is particularly attractive since focal (sometimes multifocal) and secondarily generalized epileptic activity can be maintained for many months or years in this model.

Several other important chronic model systems have been employed to significant advantage. Pumain and colleagues (Pumain, 1981), working with a cobalt focus, have demonstrated dramatic changes in the laminar profile of excitatory amino acid sensitivity of epileptic neocortex. These findings suggest a dramatic broadening of the distribution of NMDA receptors. Tetanus toxin, injected into cortex or hippocampus, also induces a relatively long-term change in excitation, resulting in epileptiform activity in the injection region. In contrast to the cobalt model, recent studies from Jordan and Jefferys (1992) suggest that the tetanus toxin-induced hyperexcitability is due to a loss of inhibitory PSPs; indeed, tetanus toxin has been shown to selectively block the release of inhibitory transmitter

(GABA) from interneurons. Interestingly, these tetanus-toxin effects occur in the apparent absence of any structural damage. Another widely used model involves the injection of kainic acid—systemically, intraventricularly, or directly into sensitive tissue, i.e., hippocampus. Kainic acid induces seizure activity by at least two different mechanisms. One involves the excitation of a broad spectrum of neurons (systemic injections), resulting in generalized seizure activity and damage associated with those generalized seizures (Ben-Ari, 1985). More localized injections of kainate have a relatively specific excitotoxic effect on the CA3 pyramidal cell population (Nadler et al., 1978). The remaining (intact) CA1 cell population becomes hyperexcitable, and animals may generate spontaneous behavioral seizures for many months after the injections. Recent studies have suggested that the development of hyperexcitability in the intact CA1 region of KA-lesioned hippocampus is correlated with a loss of GABAergic inhibition in this region (Fig. 20-2), although the cellular "machinery" for inducing GABA IPSPs seems to be preserved (Franck et al., 1988). These observations, along with similar findings obtained in other chronic models (as outlined below), have given rise to the concept of "dormant inhibitory interneurons."

The most widely used chronic models of epilepsy involve various paradigms of electrical stimulation. Starting with the initial "kindling" model of Goddard (1967), numerous investigators have found that seizure activity can be kindled slowly (McNamara et al., 1985), quickly (Lothman et al., 1985), and with or without major structural damage (Sloviter, 1983 versus Cavazos and Sutula, 1990). In the stimulation model introduced by Sloviter (1983, 1987), specific cell types in the dentate hilus showed selective vulnerability to damage, whereas other cell types (many of them GABAergic) seemed particularly resistant to damage. Kindling studies have reported increases and /or decreases in different receptor populations (Valdes et al., 1982), alterations in receptor sensitivities (Morrisett et al., 1989), changes in inhibitory and/or excitatory synaptic strength (Mody et al., 1988; Tuff et al., 1983), and even modulation of voltage-dependent membrane currents (Mody et al., 1990). The observed alterations vary with the pecularities of the individual model (stimulation protocol), although all models give rise to some form of epileptiform activity. These chronic models of epilepsy have therefore yielded quite a variety of different pictures and have certainly suggested the likelihood of a variety of underlying mechanisms. In particular, however, they have focused attention on the relationship between structural damage (and/or plasticity) and the development of hyperexcitability. These models have also emphasized, however, the possibility that functional changes can occur without obvious structural correlation.

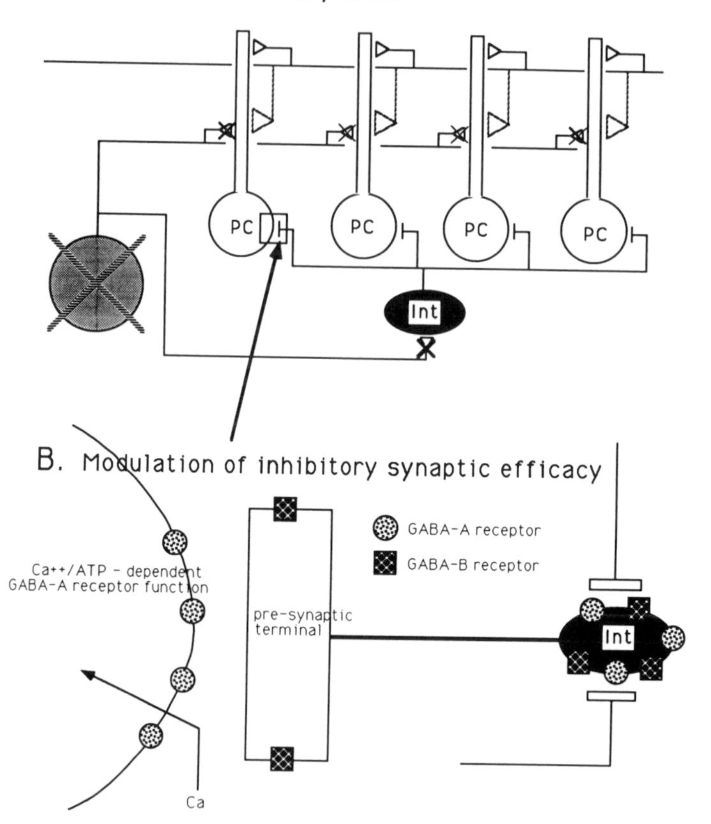

FIGURE 20-2. Inhibition mediated by GABAergic interneurons in hippocampus. **A**. In several animal models of chronic epileptiform activity, it is hypothesized that a significant excitatory drive to the interneuron (Int) population is lost (cell and synapses with X), perhaps because of the selective sensitivity of such excitatory cells to activity- or trauma-induced toxicity. As a result, inhibitory control over the principal cells (PC) is reduced, even though the inhibitory machinery is still present and functional (the interneurons are "dormant"). Although the PCs also lose some excitatory input in this process, it is hypothesized that these synapses are replaced by "sprouting" of excitatory collaterals from other sources. **B**. In addition to the process illustrated in (**A**), modulation of inhibitory efficacy may be achieved through a variety of cellular/molecular mechanisms. For example, the functioning of the $GABA_A$ receptor has been shown to be dependent on intracellular calcium- and ATP-dependent processes. GABA release from presynaptic terminals may be modulated by GABA action at $GABA_B$ receptors on transmitter-releasing terminals or by GABA action at $GABA_A$ and $GABA_B$ receptors on inhibitory interneuron somata.

PLASTICITY OF SYNAPTIC INTERACTIONS AND CONNECTIONS

Changes That May Underlie the Pathophysiology of Hyperexcitability

To understand pathological processes, particularly in the epileptic brain, it has been useful first to understand normal basic underlying mechanisms. In many cases, epileptogenesis means the transition or change of normal neuronal functions to pathological ones, particularly at the synaptic level. Of particular interest in this regard is the phenomenon of long-term potentiation (LTP), a process by which the strength of excitatory synaptic connections between cells can be strengthened depending on the history of activity of that synapse and its neighbors (Bliss and Lomo, 1973; Madison et al., 1991; see Chapter 7). Although it is not the intention of this chapter to provide an overview of LTP mechanisms, it is important to note that at most synapses LTP is dependent on NMDA receptor-mediated mechanisms or at least on receptor activation that allows calcium influx into the cell (Collingridge and Bliss, 1987; Madison et al., 1991; Malenka et al., 1989). The LTP process thus has relevance for epileptogenesis not only as a means of increasing excitatory synaptic strength but also in understanding pathophysiological processes that may be triggered by rises in intracellular calcium (Fig. 20-3). The paradigms by which LTP is initiated are strikingly similar to the stimulation protocols for kindling, and researchers have argued over the question of whether these two phenomena overlap (Sutula and Steward, 1987); indeed, one can view kindling as an extension of long-term potentiation, whereby synaptic interactions are strengthened to a nonphysiological extent. Many recent studies of LTP focus on long-term changes in the release of transmitter and/or in receptor sensitivity or numbers (Bekkers and Stevens, 1990). These issues too have been intensively investigated in models of epilepsy, and particularly in the kindling model (see Chapter 19).

Another major emphasis in the recent epilepsy literature has been on the potential role of sprouting in the development of hyperexcitability (Fig. 20-3). The "sprouting" phenomenon has long been studied in both cortex and hippocampus (Lynch and Cotman, 1975), but the relationships between sprouting and changes in excitability are unclear. In several model systems that are designed to mimic epileptic damage, investigators now recognize sprouting as a potential consequence of the epileptogenic manipulation. In both the kainic acid and the kindling models, loss of hilar neurons is correlated with the sprouting of mossy fiber collaterals into the inner molecular layer of the dentate gyrus (Cavazos et al., 1991; Davenport et al., 1990). A similar change is seen in the human epileptic hippocampus (Sutula et al., 1989). Thus, it is tempting to link sprouting with the development of hyperexcitability. However, Sloviter (1992) has argued that the time course of the excitability change in these models does not correspond to the time course of the sprouting effect and that the two phenomena are independent; indeed, he argues that the sprouting may have a larger effect in reducing hyperexcitability than in producing it. Sprouting of fiber systems other than the mossy fibers also occurs, although its nature has been less well defined. For example, following a kainic acid lesion of CA3 (and the subsequent loss of Schaffer collateral axon terminals in the CA1 region), the density of excitatory synapses in the CA1 dendritic fields is restored (Nadler et al., 1980). The source of the new synapses is unknown, but some remaining fiber system is clearly making "new" synapses onto the CA1 neurons. Similarly, when neocortex

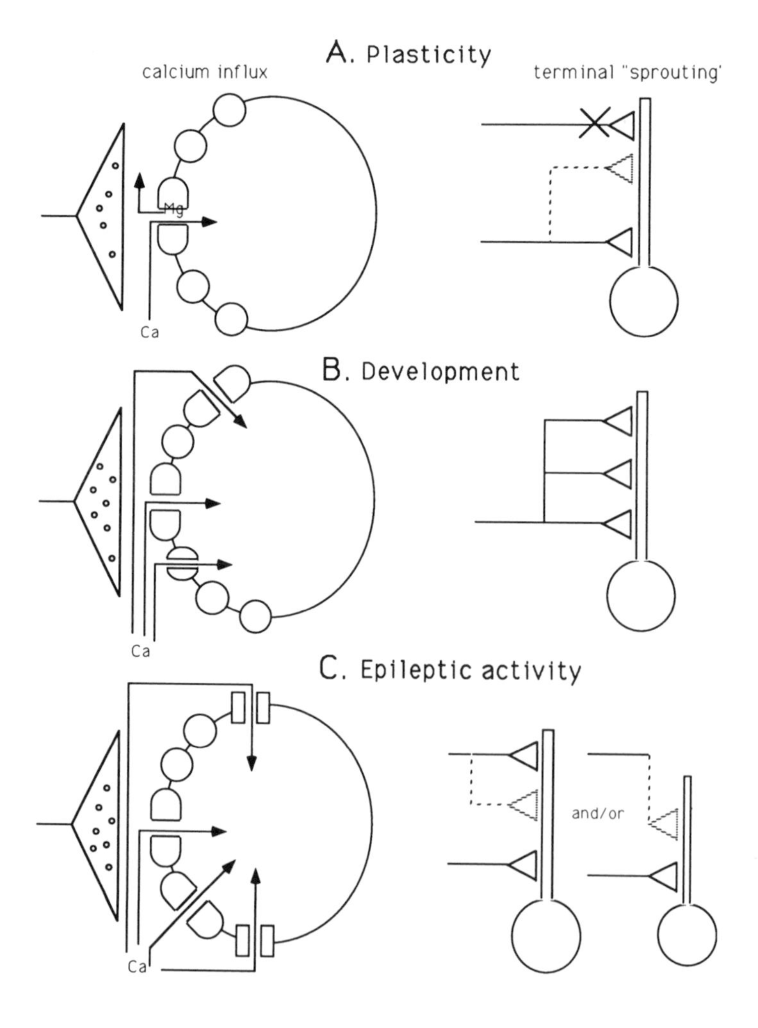

is undercut, to eliminate afferents from subcortical regions, cortical synaptic density initially decreases, but then is restored (and may even become greater than in the normal). Undercut cortex too shows a tendency toward hyperexcitability (Hoffman et al., 1994; Prince and Tseng, 1993; Rutledge, 1969). In these models, the development of hyperexcitability is not clearly correlated with the development of "sprouted" synapses.

Finally, pathological processes seen in epileptic brain often seem to involve changes in ionic homeostasis. It has long been known that large changes in extracellular ion concentrations are correlated with cortical epileptiform discharge (Prince et al., 1978) and that changes in the extracellular ionic environment can trigger epileptiform activity—increasing potassium (Traynelis and Dingledine, 1988), decreasing chloride (Avoli et al., 1990a), or decreasing calcium or magnesium (Konnerth et al., 1986a; Tancredi et al., 1990). Recently, several other interesting "ionic" features of epileptogenic cortex have been described. For example, in slices of hippocampal tissue from kindled rats, hippocampal neurons show a greater sensitivity to potassium (i.e. are more likely to burst synchronously) than

FIGURE 20-3. Parallels in processes underlying normal synaptic plasticity (in mature brain), developmental "sorting" of synaptic inputs, and cell/synaptic processes implicated in hyperexcitability/epileptiform activity. Two processes are illustrated —calcium influx and terminal elaboration/pruning ("sprouting"). **A**. At "plastic" synapses, calcium influx (through NMDA channels) is thought to be key to the change in excitatory synaptic efficacy. NMDA channel function is magnesium/voltage dependent; the membrane must be depolarized to expel the Mg^{2+} blockade before calcium can flow through the channel. At a structural level, terminal sprouting has been implicated as a possible normal mechanism underlying learning and is certainly found when a neighboring afferent input is removed. **B**. During development, calcium influx is thought to play many roles, including the fine tuning of patterns of neuronal connectivity. Here, too, calcium influx through NMDA receptors is critical. However, the immature NMDA receptor seems to be less magnesium/voltage dependent. In addition, calcium may enter via variant forms of non-NMDA receptors. Formation of synapses is often found to be "exuberant" in immature cortex; that is, a given afferent system will form more synapses (and less selectively) than will be characteristic of the mature cortex. "Pruning" of extra/inappropriate synapses is dependent, at least in part, on NMDA/calcium-mediated mechanisms. **C**. Epileptiform activity (i.e., hyperexcitability) is often characterized by cell burst discharges. Underlying the burst is a large membrane depolarization that is due, at least in part, to calcium influx via synaptically mediated (i.e. NMDA receptor) and voltage-gated channels. In the epileptic brain, there is also evidence suggesting that hyperexcitability is associated with terminal/collateral alterations, including sprouting (with or without associated deafferentation). Changes in synapse location must also occur, since there are often alterations in cell morphology, including cell shrinkage and loss of fine dendrites and dendritic spines.

cells from comparable tissue from normal animals (King et al., 1985). In addition, there seems to be an enhanced calcium influx into granule cells from kindled rats (as compared to granule cells from normal tissue), a phenomenon correlated with the development of an "abnormal" NMDA receptor-mediated synaptic component (Mody et al., 1990). Investigators have also suggested that Na^+, K^+-ATPase in kindled tissue is abnormal, perhaps accounting for some of the changes in sensitivity to extracellular potassium (Grisar et al., 1992). Although it is unclear whether the potassium pump in question is localized to neurons or glia, there has long been a suspicion that glial function in epileptic tissue is somehow abnormal (Pollen and Trachtenberg, 1970). Although experimental data have failed to support that hypothesis (Prince, 1976; Ward, 1978), recent imaging studies on cultured glia from human epileptic brain have shown these cells to be quite different from comparable cells from normal brain, generating much more dramatic waves of calcium transmission (Gunel et al., 1991).

Characteristics of Immature Tissue

In recent years, research on normal developmental aspects of cortical and hippocampal physiology has provided potentials insights for understanding abnormal function. Both experimental and clinical investigations have shown that the immature CNS, at least during a certain window of development, tends to be unusually seizure prone (Aicardi, 1986; Moshe et al., 1983; Swann et al., 1988; Tharp, 1987). Thus, it is particularly interesting to assess and compare developmental changes with those processes that might underlie pathologic activities in cortical neurons.

There are several intriguing parallels between those plastic phenomena outlined above for pathological tissue, and "normal" processes occurring during cortical development (Fig. 20-3). For example, during development, the nature and function of various transmitter receptors seem to be quite different from those in the adult, particularly with respect to properties that might facilitate tissue excitability. In immature rat hippocampus, investigators have shown that NMDA-gated channels are transiently increased in density (Tremblay et al., 1988) and have a reduced voltage-dependent blockade and heightened glycine modulation compared to the mature channel (Ben-Ari et al., 1988; Gaiarsa et al., 1990). Additionally, non-NMDA-gated channels, which in the adult generally admit little calcium, in the immature animal consist of subunits that apparently admit much greater levels of calcium (Monyer et al., 1991; see Chapter 5). Similarly, GABA-gated (inhibitory) channels seem to be quite different in the immature neuron. Indeed, the immature cortical $GABA_A$ receptor seems to mediate a primarily depolarizing response, rather than the conventional hyperpolarizing IPSP seen in adults (Ben-Ari et al., 1989; Mueller et al., 1984; Swann et al., 1989). An explanation for this difference is still not clear; the immature GABA depolarization may be based on different chloride concentration gradients and thus, presumably, differential activity of chloride exchange mechanisms (Luhmann and Prince, 1991), or it may be a result of changes in the nature of the $GABA_A$-gated ionophore. Given evidence that the $GABA_A$ receptor composition undergoes developmental changes (Araki et al., 1992), the latter possibility cannot be discounted.

Other experiments have focused on differences in ionic homeostatic mechanisms. Immature hippocampal neurons show much lower levels of Na^+, K^+-ATPase activity than mature cells (Haglund et al., 1985), perhaps explaining the less effective maintenance of extracellular potassium concentration in immature tissue (Haglund and Schwartzkroin, 1990) and the observation that potassium increases in immature epileptic brain are much larger than those seen in the adult (Mutani et al., 1974). Thus, the heightened epileptogenicity of immature tissue may be attributed to a complex set of feedback effects dependent on maintenance of the extracellular potassium environment.

Finally, an analogy can be made between the potential epileptogenetic effects of sprouting and the overexuberant production of axon collaterals characteristic of the immature nervous system. In hippocampus, the excitatory axon collaterals of immature CA3 pyramidal cells are remarkably "overgrown," with these axons apparently associated with a large "excess" of excitatory synapses (Deitch et al., 1990). During development, there is a gradual pruning away of some of these connections to yield a mature level of connectivity. However, in the immature hippocampus, this hyper-interconnectivity is associated with increased excitability and support of long periods of electrical afterdischarge (Swann et al., 1993). Increased cell connectivity is also reflected in another potential contributing factor to the hyperexcitability of immature cortex—the degree of electrotonic coupling seen early in development. Primarily on the basis of intracellular dye injections, investigators have shown greater coupling in the immature animal than in the adult (Connors et al., 1983), providing a potentially potent mechanism for synchronization of neural activity.

Neuron Specific Vulnerability and the Correlation Between Structural and Physiological Pathologies

Many forms of nervous system pathology are characterized by particular patterns of structural abnormality and/or cell loss. Perhaps the cells most sensitive/vulnerable (at least among cortical structures) to seizure-induced damage (Babb et al., 1984a,b; Cavazos and Sutula, 1990; Mouritzen-Dam,

1980), percussive head damage (Lowenstein et al., 1992), hypoxia/ischemia (Franck, 1993), and even excitotoxicity (Mathern et al., 1992) are identifiable subpopulations in the hilus of the dentate gyrus of hippocampus. These vulnerable neurons include mossy cells, as well as an interneuron population that co-localizes somatostatin, NPY, and GAD (Sloviter, 1991b). The mossy cell vulnerability has been examined in animal model systems, and various hypotheses have been developed to explain the sensitivity of these cells (Schwartzkroin et al., 1990; Sloviter, 1987). Absence of adequate intracellular calcium buffering was suggested, since these cells contain no parvalbumin or calbindin (Schwartzkroin et al., 1990); indeed, in an *in vitro* model system, stimulation-induced mossy cell damage can be blocked by intracellular injection of a calcium chelator (Scharfman and Schwartzkroin, 1989). The cells also display intrinsic characteristics that favor depolarizing over hyperpolarizing conductances (Buckmaster et al., 1993) and are inserted in synaptic circuits (including "positive feedback" loops) that favor excitatory over inhibitory PSPs (Buckmaster et al., 1992). If the favored synaptic targets of these excitatory cells are inhibitory interneurons, as hypothesized but not yet proven (Sloviter, 1987), then loss of mossy cells would reduce the inhibitory tone in the dentate and facilitate "maximal dentate activation" (Stringer and Lothman, 1989). Loss of the mossy cells, in any case, seems to give rise to mossy fiber "sprouting" (alluded to above) and has spawned several theories relating this structural damage to excitability. However, hilar cell damage that occurs as a result of other types of experimental "trauma" does not invariably give rise to seizure activity or hyperexcitability; thus, the interaction between structural lesions and functional hyperexcitability remains a critical but unresolved issue.

Changes in Functional Inhibition

Given these uncertainties about the role of cell loss in epileptogenesis, it is especially interesting that recent studies have found that functional inhibition may be lost or significantly decreased in hyperexcitable/epileptiform tissue, but that inhibitory interneurons, GABA receptors, and the remaining machinery needed to produce synaptic inhibition remain intact. Indeed, under special conditions, the apparently lost inhibitory potentials can be evoked (Bekenstein and Lothman, 1993; Franck et al., 1988; Sloviter, 1991 a,b), reinforcing the view that the "lost" inhibition was due to functional, not structural changes (Fig. 20-2A). In stimulation models and in the intraventricular kainic acid model, the "dormant" interneuron seems to be quiescent because of a relative loss of excitatory input. It is proposed that the cell populations that normally excite these interneurons (e.g., the CA3 pyramidal cells in the CA1/kainic acid model) are damaged and/or lost; since excitation of the inhibitory neurons is therefore reduced significantly, the interneurons are less effective in inhibiting their postsynaptic targets.

In contrast, several investigators have also shown that *overuse* of the inhibitory system can result in the loss of functional inhibitory hyperpolarizations. Repetitive stimulation of hippocampal circuits leads to a gradual loss of GABAergic IPSPs (McCarren and Alger, 1985; Wong and Watkins, 1982). This alteration may be due in part to receptor desensitization, although desensitization may not provide the entire explanation (Stelzer et al., 1988). Wong and colleagues have shown that functional GABAergic inhibition is dependent on the level of intracellular calcium (too much calcium leads to receptor dysfunction) and on receptor phosphorylation (the absence of sufficient ATP leads to the loss of the inhibitory effect) (Chen and Wong, 1991). Thus, molecular modulation of the receptor plays an important role in the inhibitory efficacy of GABA release (Fig 20-2B).

A further complication, recently introduced into the GABA literature, deals with the role of "disinhibition," particularly as mediated by $GABA_B$ receptors. Stimulation in hippocampal orthodromic pathways, particularly in the dentate gyrus, results in the release of GABA, which apparently reduces subsequent GABA release from inhibitory interneurons (Mott and Lewis, 1991). This effect has been attributed to GABA interactions with $GABA_B$ receptors on interneuron somata that prevent the interneurons from firing (Misgeld et al., 1989) or GABA interactions with $GABA_B$ receptors on interneuron terminals that reduce the release of GABA (Fig. 20-2B). Whichever the pathway, this mechanism results in a relative increase in excitability in the region in question. Related to this finding is the observation that inhibitory interneurons within the hippocampus contact not only projection cells (pyramidal cells and granule cells) but also other inhibitory interneurons (Lacaille et al., 1987). Thus, depending on which interneuron subpopulation is activated by a particular input, there may be more or less disinhibition. Similarly, GABAergic afferents from the septum have been shown to make selective contact with inhibitory interneurons in the dentate gyrus (Freund and Antal, 1988).

Finally, as mentioned above, GABA release may have depolarizing as well as hyperpolarizing effects, depending on the region of cell membrane affected and on the nature of the postsynaptic receptors. Repetitive stimulation results not only in a decrease of the hyperpolarizing GABAergic component but also in the facilitation and growth of a depolarizing GABAergic component (Grover et al., 1993). In some cases the two events may cancel each other out; alternatively, the depolarizing effect may become so overwhelming as to become functionally excitatory. This GABA-modulated excitation may play a significant role in the synchronous activity of GABAergic neurons, which can be induced in the absence of excitatory synaptic transmission (Aram et al., 1991).

It is striking that so many of these mechanisms in which GABA inhibition is functionally reduced depend on repetitive (or over-) activation of the synaptic circuitry—exactly the condition in the epileptic brain in which cells are firing repetitively. Regardless of how the initial epileptiform activity is initiated, repetitive activation during seizure-like discharge may result in the loss of *functional* inhibition and thus provide "positive" feedback that perpetuates and even increases excitatory activity.

21

Pathophysiology of cortical areas from human epileptics

LEONA M. MASUKAWA

Epilepsy is a clinical disorder that is manifested by abnormal brain electrical discharges and therefore is appropriate for electrophysiological study. Until relatively recently, mechanisms of epilepsy have been largely examined in animal models that have established several possible mechanisms that are described in companion chapters of this book. The use of en bloc temporal lobectomies as an effective treatment of complex partial seizures intractable to drug treatment has permitted physiological studies of human epileptic tissue that were not previously possible. In this chapter emerging results from the application of *in vitro* brain slice technology to surgically resected tissue are discussed against the background of *in vivo* studies of epileptic patients to give an up-to-date view of the pathophysiology of human epilepsy.

IN VIVO STUDIES OF HUMAN EPILEPSY

Electroencephalographic (EEG) recordings from epileptic patients provided the earliest evidence of cortical pathophysiology—"sharp waves" during interictal periods and repetitive spiking accompanying high-frequency oscillations during a seizure (ictal phase). The site of abnormal electrophysiological activity cannot be localized precisely using this recording method because of volume conduction that underlies the EEG signal. The development of surgical therapies for epilepsy led to the use of invasive recordings from the intact brain through depth electrodes to localize epileptic foci more accurately in humans. Chronic depth electrode implants also provided localized monitoring from several areas of the brain simultaneously during interictal and ictal periods (Crandall et al., 1963; Engel, 1983) and allowed both localization and tracking of abnormal activity through the brain. From extracellular recordings in early studies, spontaneous repetitive neuronal burst discharges were observed in the neocortex (200 to 300/sec; Calvin et al., 1973), the hippocampus (136/sec), and the amygdala (15.8/sec; Babb et al., 1973) of epileptic patients. These burst discharges have been viewed as the cellular reflection of epileptiform activity during the interictal phase in humans.

Spontaneous bursts of action potentials (interictal bursts) occurred in characteristic patterns of either "stereotyped" bursts in which a constant high-frequency firing occurs or a "structured" burst in which a pause of relatively constant duration occurs between the first or second spike and subsequent train of spikes (Calvin et al., 1973; Fig. 21-1A). The latter pattern, termed a "long first interval burst," was similar to bursts reported by Calvin et al. (1973) in the alumina treated monkey cortex. Both Babb et al. (1973) and Calvin et al. (1973) reported a decrement in amplitude of the action potentials during the burst discharge in human epileptics. Somewhat surprisingly, cellular burst patterns were observed in the hippocampus contralateral to the seizure focus (nonepileptic side), as well as in the epileptic focus where spike frequencies (short bursts of two to four

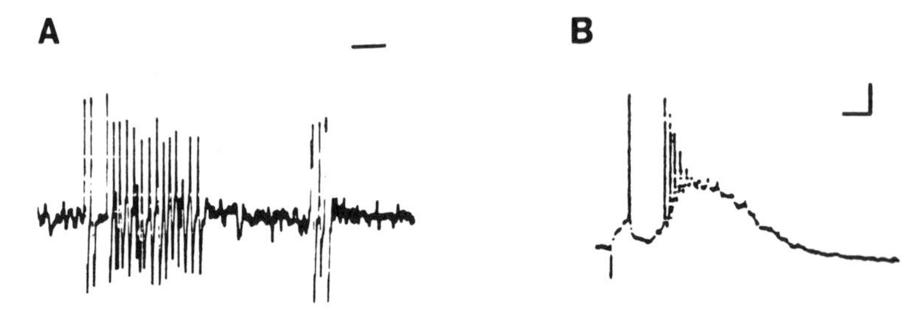

FIGURE 21-1. *In vivo* and *in vitro* neocortical burst responses from human epileptics. **A**. *In vivo* unit records from the cerebral cortex of a patient during a craniotomy for therapeutic surgical excision of epileptogenic areas showed stereotypic burst responses characterized by a high-frequency spike discharge in which the timing was relatively constant between spikes. The structured burst response shown here is marked by a longer interval between the first or second spike and subsequent spikes in the burst. This burst formation is termed the "long first or second interval burst," respectively. Illustrated here is a long second interval burst (from Calvin et al., 1973). **B**. An intracellular *in vitro* brain slice recording from a neocortical neuron in tissue removed for the therapeutic treatment of epilepsy also exhibited a "long first interval burst" (from Strowbridge et al., 1992b). Calibration for (**A**) 10 msec and (**B**) 10 mV and 25 msec.

spikes) occurred at a rate as high as 100 to 300/sec (Isokawa-Ackesson et al., 1987). The observations in human epileptic neocortex and hippocampus of variable frequency and duration of the burst discharge, in both presumably normal and abnormal areas of temporal lobe epileptics, seem on the surface to be contrary to the basic assumption that burst discharges are an electrophysiological sign of abnormality. Burst discharges, however, can be a normal pattern of directly evoked neuronal activity—that is, intrinsic activity (via membrane depolarization) in rodent cortex (layer V) and hippocampus (CA3); (Connors et al., 1982; Kandel and Spencer, 1961a)—although they are not normally synaptically evoked in *in vitro* brain slices. Thus, spontaneous bursting of units may not be a sign of abnormality per se except that it can occur experimentally under unusual conditions, such as during exposure to bicuculline, picrotoxin, 0 Mg^{2+}, or 4-AP, which alter synaptic input (Fig. 21-2). Recruitment of bursts is often thought of as a synaptic synchronization of responses from many neurons. However, burst discharges in neurons of human neocortex were not often correlated in time with each other and were generally not correlated with the occurrence of EEG sharp waves (Calvin et al., 1973; Ishijima et al., 1975; Wyler et al., 1982), indicating that not all neurons were synchronous in their activity. The lack of correlation between field interictal bursts and unit burst may be due to a confined focus of activity that does not involve all neurons in neighboring areas. The separation of individual cell responses and population responses is an important one to make while

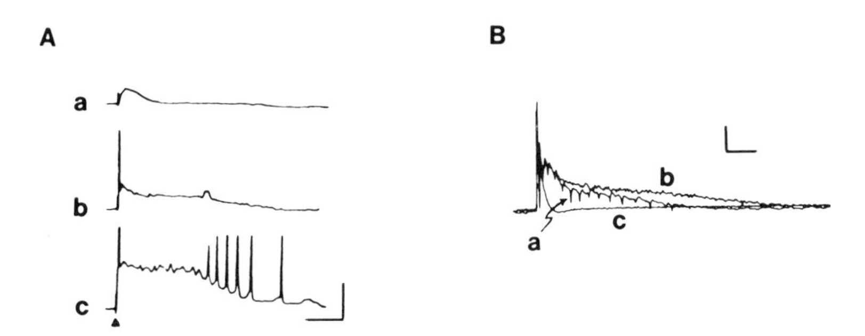

FIGURE 21-2. Involvement of the $GABA_A$ and NMDA receptor during neocortical and dentate granule cell responses recorded intracellularly from epileptic patient brain slice. **A**. Bicuculline (20 μM) prolonged on orthodromically induced EPSP (a and b), ending in a long duration burst response composed of repetitive action potentials (c) in a neocortical neuron (from Avoli and Olivier, 1989). **B**. A burst response that was recorded from a neocortical cell under normal conditions (a) and after exposure to 0.6 mM Mg^{2+} (b) was blocked by 4 μM CPP (c) indicated the dependence of the burst on activation of an NMDA receptor (from Urban et al., 1990).

interpreting physiological data from epileptics. In fact, Calvin et al. (1973) stated: "The observed examples of independence (of bursting) speak strongly for the existence of burst mechanisms which are not merely secondary to massive synchronous recruitment."

Other signs of pathophysiology have been reported in epileptic human cortex during electrical stimulation in addition to spontaneous activity described above. Wyler and Ward (1981) reported that, in response to direct repetitive cortical stimulation (in the range of 1 to 10 Hz), a burst of action potentials lasting up to 30 seconds occurred in epileptic cortex, but not in adjacent normal cortical areas. In most temporal lobe epileptic patients, whose hippocampi are characterized by significant sclerosis, stimulation of entorhinal cortex ipsilateral to the focus elicited an extracellular field response in hippocampus that was simple in configuration (Rutecki et al., 1989). In contrast, patients without hippocampal sclerosis exhibited responses that were complex in configuration. Rutecki et al. (1989) attributed this complex response to the presence of a full complement of neurons in less pathological tissue. Another possible indicator of pathophysiology is the abnormal threshold for an evoked response or afterdischarge in epileptics. It is yet to be resolved whether the threshold for an afterdischarge in the epileptic human hippocampus is altered on the side ipsilateral to the presumed focus. Cherlow et al. (1977) and Bernier et al. (1990) from *in vivo* depth recordings report higher and lower thresholds, respectively, on the focal side. Cherlow et al. (1977) suggested that a higher stimulation intensity may be the result of a decrease in the number of excitable elements in sclerotic tissue. A lower threshold, in contrast, might indicate an increase in the synaptic efficiency or in hyperexcitability.

ANATOMICAL PATHOLOGY

Hippocampal sclerosis (cell loss) has been considered an anatomical hallmark of temporal lobe epileptics when Sommer (see Sano and Malamud, 1953) in 1880 reported drastic reductions in cell numbers in all fields of the epileptic hippocampus and especially in an area of CA1 later referred to as the Sommer sector. A high variability in hippocampal sclerosis from autopsy material from temporal lobe epileptic patients was described subsequently (Margerison and Corsellis, 1966; Mouritzen-Dam, 1980). Recently, several laboratories have confirmed cell loss in hippocampal subfields of temporal lobe tissue resected during the surgical treatment of epilepsy. Kim et al. (1990) used a restricted clinical variable, the presence or absence of a tumor, to categorize the epileptic patient population. They reported that in tumor-associated epileptic temporal lobe, there was a small, nonsignificant loss of cells throughout the hippocampus and dentate gyrus in comparison to autopsy control tissue. Cell densities of both the tumor-associated tissue and controls were significantly greater than "cryptogenic" epileptic temporal lobe in which there was no apparent structural lesion. Further examination of lesion-associated epileptic tissue demonstrated that, when a tumor was present in the hippocampus, cell loss was greater than in normal (autopsy) controls (Fried et al., 1992). This study also demonstrated that cell loss was greater in patients with earlier seizure onset, but the degree of cell loss was not related to the duration of the seizure disorder. Fried et al. (1992) proposed that cell loss may not be a simple consequence of duration of seizure, but rather may be related to the period in development of the first seizure occurrence.

Others have detailed the variation in cell organization in sclerotic tissue. In a subpopulation of patients, Houser (1990) described dispersion of cells in the dentate granule cell layer, resulting in a widening of the layer that was related significantly to the density of dentate granule cells. A decrease in the polymorphic hilar neuronal population was correlated with the widening of the granule cell layer. The dispersion of granule cells into the molecular layer suggested an alteration in the targets of afferent fibers from those that are normally contacted in granule cells. It was suggested that developmental abnormalities were important in the widened granule cell layer and elongation of the granule cell body. Houser (1990) also noted an unusual variant of the granule cell layer exhibited by two discrete layers instead of one. Similar morphological anomalies were also reported by Babb et al. (1991). These observations emphasize the detailed variation in cellular organization in human epileptics.

Babb et al. (1984b), utilizing depth electrode recordings, reported a close link between the degree of hippocampal sclerosis and the site of seizure onset. When anterior hippocampal onset occurred, cell density in the anterior hippocampus was lower than in the posterior hippocampus. When patterns of temporal lobe seizure onset or focal spiking were multifocal, however, there were no significant differences between anterior and posterior cell density in the hippocampus, dentate gyrus, or neocortex. Spencer et al. (1992) reported a significant correlation of ictal spiking with cell loss in the CA1 pyramidal cell population and no significant loss in all other regions, suggesting that electrical abnormality is associated with cell loss in this specific hippocampal field that is usually vulnerable to other insults. Urban et al. (1993) reported in CA1 that abnormal EPSPs characterized by small amplitude and distorted waveform were always observed in the highly sclerotic tissue group. However, the "causal" relationship between cell loss and seizure activity cannot be easily determined from postsurgical data. Although a correlation may exist between regional cell loss and seizures, areas of limited cell loss may play a greater role in epilepsy than do areas of greatest pathology. Areas of greatest cell loss may be indicative of the aftermath of excessive or long-term focal activity.

In addition to cell loss, reorganization of mossy fiber terminals has been demonstrated in the dentate gyrus and CA3 region of epileptics. Mossy fibers from the dentate granule cells, which normally project to hilar neurons and CA3 pyramidal cells, have been shown to innervate the infrapyramidal region of the CA3 neurons and the molecular layer of the dentate granule cells, two areas normally devoid of innervation (Babb et al., 1990; deLanerolle et al., 1989; Houser et al., 1990; Isokawa et al., 1993; Masukawa et al., 1992; Represa et al., 1989a; Sutula et al., 1989) in epileptic patients. Because of the excitatory effects of mossy fiber terminals on CA3 pyramidal neurons (Andersen et al., 1971) and hilar interneurons (Muller and Misgeld 1990; Scharfman et al., 1990), investigators have proposed that the reorganized innervation is also excitatory and may add to the hyperexcitability of the region (Fig. 21-3). Babb et al. (1991) have shown "sprouted" mossy fiber synaptic boutons in the molecular layer of the epileptic human dentate gyrus where they make contacts onto granule cell dendrites.

IN VITRO PHYSIOLOGY OF HUMAN BRAIN SLICES

The technique of *in vitro* brain slices has allowed electrophysiologic examination of individual neurons and circuits in animal models that was previously not possible by *in vivo* approaches. *In vitro* recording was first applied to human brain tissue by Kato et al. (1973) and then specifically

→

FIGURE 21-3. Correlation between mossy fiber reorganization and abnormality of the antidromic response of dentate granule cells from temporal lobe epileptic patients (from Masukawa et al., 1992). **A**. Mossy fiber reorganization varied among patients from no molecular layer Timm stain (a), mild stain (b), to dense stain (c). **B**. Antidromically evoked responses in hippocampal human brain slices also varied for these patients from a single population spike (b-d) to multiple population spikes (e-f). As the number of population spikes increased, the abnormality of the response increased. In the rat dentate gyrus, the antidromic response was a single population spike (a). **C**. The correlation between the graded abnormality of the antidromically evoked response and the graded molecular layer stain density was highly significant ($r_s = 0.86$, $P<0.001$, $n = 13$). The data point surrounded by a square was from a patient who was surgically treated for a tumor, and who did not experience seizures. The ordinate and abscissa are graded values based on the antidromic response and molecular layer Timm stain intensity, respectively. Calibration for (**A**) 200 μm; for (**B**) 2 mV and 6.5 msec.

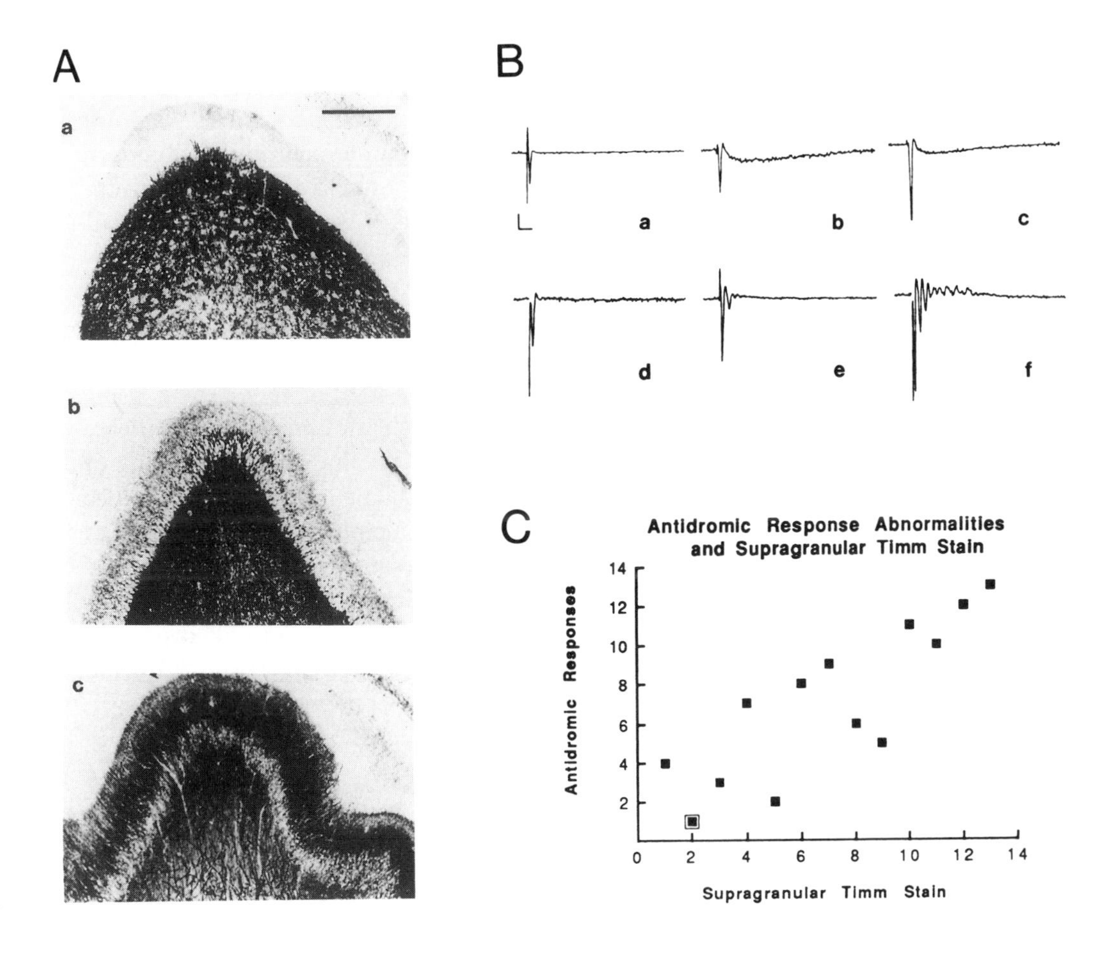
A
a
b
c
B
a
b
c
d
e
f
C
Antidromic Response Abnormalities and Supragranular Timm Stain
Antidromic Responses
Supragranular Timm Stain
0
2
4
6
8
10
12
14

to specimens from epileptic patients by Schwartzkroin and Prince (1976). In neocortical tissue, there have been few reports of significant differences in neuronal electrical characteristics—resting membrane potential, membrane resistance at the resting potential, membrane time constant, or action potential threshold—between cells from "epileptic" tissue (defined as areas that display interictal discharges *in vivo*) and "nonepileptic" regions (Avoli and Olivier, 1989; Foehring and Waters, 1991; Lorenzon and Foehring 1990; Schwartzkroin and Knowles, 1984; Schwartzkroin and Prince, 1976; Tasker et al., 1992; Williamson et al., 1993). In dentate granule cells from tumor-associated epileptic patients, the passive membrane properties and some voltage-dependent conductances were not obviously different from normal guinea pig granule neurons (Williamson et al., 1993), indicating the general normality of cells from nonfocal human tissue. Although there are several similarities between electrical characteristics of human and rodent cortical neurons, some neuronal properties have been found to differ between these two species. Connors et al. (1982) described three types of intrinsic responses of neocortical cells to membrane depolarization in the guinea pig, which included a bursting and nonbursting response. Intrinsic bursting has not been reported in some studies of neocortical human neurons in nonepileptic and epileptic patients (Foehring et al., 1991; Tasker et al., 1992), but is illustrated in a study of epileptic neocortex by Avoli and Olivier (1989). The infrequent occurrence of "intrinsic" bursting and relatively normal membrane conductances in human neocortical cells of epileptics emphasizes the importance of the role of synaptic activity in the generation of epileptiform discharges.

The greater importance of synaptic input in comparison to intrinsic electrical properties in the generation of abnormal activity is illustrated in several studies of hippocampal neurons from epileptics. Knowles et al. (1992) have reported that the intrinsic electrical characteristics of hippocampal neurons from sclerotic/epileptic and tumor/nonepileptic tissue were not significantly different, but that there was a reduction of the IPSP in epileptic tissue. When cells from epileptic patients (presumably nontumor related) were classified on the basis of PSP characteristics, Isokawa et al. (1991) reported that cell input resistance at the resting potential was higher in dentate granule cells with normal EPSP/IPSP responses (40.2 ± 4.33 megaohms) than in cells with prolonged EPSPs and no discernible IPSP, which were presumed to be abnormal (31.8 ± 2.63 megaohms). Strowbridge et al. (1992a) also found a similar relationship for neocortical neurons from the temporal lobe. In this study, neurons were grouped according to the presence or absence of a structural lesion in the involved tissue. In the lesion group, the percentage of neurons exhibiting prolonged EPSPs was 46% compared to 0% in the nonlesion group. The average input resistance in cells from lesion versus nonlesion tissue was 34.6 ± 3.1 megaohms and 49.2 ± 5.7 megaohms, respectively, a relationship similar to that observed by Isokawa et al. (1991). Differences in cell input resistance between normal and epileptic tissue may not be evident in studies when averaging of data from a large number of cells is used in the analysis and variability predominates. Although this explanation is not entirely consistent with the study by Strowbridge et al. (1992a) in which half of the cells from the lesion group would be considered normal based on the EPSP configuration, grouping of the cells according to their location as lesion or nonlesion tissue—as opposed to epileptic versus nonepileptic—may have contributed to the greater consistency of the neuronal population and electrical characteristics.

The heterogeneity of synaptically evoked and spontaneous activity observed in early *in vivo* studies (Babb et al., 1973; Calvin et al., 1973; Ishijima et al., 1975; Wyler and Ward, 1981) has been borne out in studies of brain slices from tissue resected from epileptic patients. The early study by Schwartzkroin and Prince (1976) reported synaptically evoked intracellular responses that varied between single action potentials and burst of spikes. Further intracellular recordings from neocortex again demonstrated a range of evoked responses that included proximal depolarizations and/or bursts of three or more action potentials (Prince and Wong, 1981). Burst responses were present in 21 of 50 neurons from epileptogenic cortex, but were completely absent in recordings from the neocortex of nonepileptogenic patients. Similar intracellular response variability and burst discharges have been a pattern described in subsequent studies (Avoli and Olivier, 1989; Isokawa et al., 1991; Reid and Palovcik, 1989; Schwartzkroin and Haglund, 1986; Schwartzkroin and Knowles, 1984; Schwartzkroin et al., 1983; Strowbridge et al., 1992b), and in field responses (Isokawa et al., 1991; Masukawa et al., 1989, 1992; Uruno et al., 1994; Figs. 21-1B, 21-2, 21-3, and 21-4B). Strowbridge et al. (1992a) found that a greater incidence of prolonged or abnormal EPSPs, including long-lasting bursts of action potentials (paroxysmal depolarizations) and long first interval bursts, occurred in the lesion-associated tissue (Fig. 21-1B). Throughout a group of epileptic patients, synaptically evoked field responses during stimulation of orthodromic or antidromic input vary across a wide spectrum of responses from single to multiple population spikes (Uruno et al., 1994; Fig. 21-4), supporting the non-uniformity of electrical abnormality of resected tissue from epileptic humans.

The lack of spontaneous activity in neocortical and hippocampal human brain slices from epileptic patients has been reported widely (Avoli and Olivier, 1989; Isokawa et al., 1991; Knowles et al., 1992; Schwartzkroin and Haglund, 1986; Schwartzkroin and Prince, 1976; Schwartzkroin et al., 1983; Tasker et al., 1992; Wuarin et al., 1990), except under modified pharmacological or ionic conditions (Avoli et al., 1987, 1988; Wuarin et al., 1990). Schwartzkroin et al. (1983) suggested that this absence of spontaneous discharge may be due to the lack of a sufficiently large population of synchronized neurons. Another possible explanation is that areas extrinsic to the resected tissue may be necessary to initiate spontaneous activity.

In a number of studies, investigators found that burst responses could be evoked by repetitive orthodromic stimulation delivered at 0.5 to 10 Hz. Varying the stimulus frequency altered the effect. Working *in vivo*, Wyler et al. (1982) reported that 5 to 10 Hz stimulation could augment spontaneous action potential frequency particularly in cortex that was suspected as epileptogenic. In slice studies, Avoli and Olivier (1989) reported complex burst discharges in 2 of 19 neocortical cells when synaptically

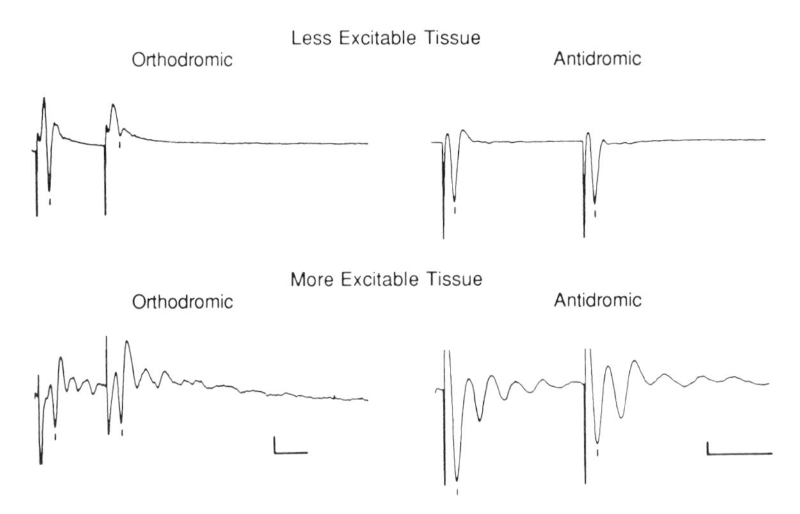

FIGURE 21-4. Responses of the dentate granule cell layer to paired-pulse orthodromic and antidromic stimulation. The interstimulus interval was 20 msec. Responses were divided into a more excitable and less excitable group based on multiple or single orthodromically evoked population spike/s, respectively. Orthodromic paired-pulse stimulation resulted in the inhibition of the population spike of the second response in the less excitable tissue. There was no antidromic paired-pulse interaction in this type of tissue. Orthodromic paired-pulse stimulation resulted in the facilitation of the first spike of the second response in the more excitable tissue. Antidromic paired-pulse stimulation led to inhibition of the first spike of the second response. A single antidromic stimulus elicited multiple spikes, and secondary spikes were not suppressed. These data indicated the range of responses exhibited in epileptic human tissue and that the loss of paired-pulse (feedback) inhibition is accompanied by hilar-induced paired-pulse inhibition (from Uruno et al., 1994). Calibration: 1 mV and 10 msec.

activated by trains of stimuli at 0.5 to 2 Hz. Using extracellular field recording, Masukawa et al. (1989) reported an increase in the number of population spikes elicited from dentate granule cells during 1-Hz trains; these results are reminiscent of rodent brain slice studies (Alger and Teyler, 1979) in which synaptic potentiation occurred during repetitive orthodromic stimulation, but at a significantly higher frequency (10 Hz). In normal rat slices, a 1-Hz frequency produces depression of the field response. These examples of potentiation of excitatory orthodromic responses during repeated stimulation may involve processes that are brought into play during the transition to the ictal state. Reduction of synaptic inhibition or increase in NMDA receptor mediated excitation has been suggested as possible mechanisms that might bring about low-frequency potentiation in tissue from epileptic patients (Masukawa et al., 1989, 1991). If the balance of excitation is altered in epileptic tissue, further compromise of the balance may be induced by repeated stimulation, leading to further imbalance and burst discharges.

Several transmitter systems as well as alterations in the levels of transmitters and receptors have been identified in the human neocortex and hippocampus (see for example de Lanerolle et al., 1989, 1992; Geddes et al., 1990; Houser et al., 1990; Hosford et al., 1991; McCormick 1989a; McCormick and Williamson, 1989; McDonald et al., 1991; Robbins et al., 1991; Strowbridge et al., 1992b; Tremblay et al., 1985). An apparent loss of inhibitory input has been reported whenever burst discharges were observed (Avoli and Olivier, 1989; Haglund et al., 1992; Isokawa et al., 1991; Knowles et al., 1992; Masukawa et al., 1989; Schwartzkroin and Prince, 1976; Schwartzkroin et al., 1983; Strowbridge et al., 1992b; Uruno et al. 1994; Wuarin et al., 1990). However, the complete loss of $GABA_A$-mediated inhibition is not always found in epileptic patient tissue. This is illustrated by the ability of bicuculline to lead to further enhancement of the EPSP amplitude and duration and to induce burst discharge (Avoli and Olivier, 1989; Masukawa et al., 1989; Tasker et al., 1992; Uruno et al., 1994; Wuarin et al., 1990; Fig. 21-2A). Further, Uruno et al. (1994) provide evidence for an increase in inhibitory input originating from hilar stimulation in specimens that also show greatly diminished orthodromically evoked feedback inhibition (Fig. 21-4). Correlations of evoked responses suggest that the loss of feedback inhibition occurred more readily than the loss of feedforward inhibition (Uruno et al., 1994).

Electrophysiological reports substantiate contributions of both non-NMDA and NMDA receptor activation to the prolonged EPSP in neocortex and dentate gyrus from epileptics (Avoli and Olivier, 1989; Avoli et al., 1987, 1989; Hwa and Avoli, 1992; Isokawa and Levesque, 1991; Masukawa et al., 1991; Urban et al., 1990; Wuarin et al., 1990; Fig. 21-

2B). Isokawa and Levesque (1991) reported a correlation between the presence of an NMDA component in the granule cell EPSP, loss of spines, and occurrence of "beaded shafts" on granule cell dendrites. Such correlations support the hypothesis that an increase in NMDA-mediated conductances leads to the eventual destruction of epileptic neurons. Additionally, these data are consistent with the idea that alterations in cell and synaptic structure act together with a greater involvement of the NMDA receptor to enhance hyperexcitability. Receptor binding studies support the likelihood that changes in glutamate receptors might underlie the increase in the EPSP, in addition to the loss in synaptic inhibition (Geddes et al., 1990; Hosford et al., 1991; McDonald et al., 1991).

An abundance of anatomical descriptions of epileptic hippocampal tissue further supports the view that altered synaptic circuitry may result from cell loss and reorganization. Loss of hilar interneurons has been assumed to lead to a decrease in inhibition, whereas reorganization of mossy fibers in the dentate has been proposed to underlie a feedback excitatory function (Tauck and Nadler, 1985). In support of this proposal, Masukawa et al. (1992) reported a significant correlation between the abnormality of antidromic responses recorded from dentate granule cells and the extent of mossy fiber reorganization (Fig. 21-3). Those specimens that exhibited the greatest molecular layer sprouting also responded abnormally to antidromic stimulation with multiple spikes that were dependent on activation of AMPA-type glutamate receptors (Uruno et al., 1994). Reorganization of mossy fibers was also correlated with hilar cell loss, although not strongly (Babb et al., 1991; Houser, 1990; Masukawa et al., 1992). It is still unclear whether seizure activity in humans precedes these anatomical changes or whether the anatomical pathology is a consequence of the seizures, although the latter possibility is more consistent with a highly variable Timm stain pattern and intensity in the molecular layer of the dentate gyrus in a population of epileptic patients (Masukawa et al., 1992). The degree of reorganization and antidromically as well as orthodromically evoked responses were quite variable across epileptic patients (Masukawa et al., 1989, 1992; Uruno et al., 1994) and were perhaps correlated with the site of the focus or the history of seizure activity.

CONCLUSION

Human tissue studies have had a relatively short history. Evoked responses and neuronal electrical characteristics that have been described in neurons from normal human neocortex form an important body of data linking human and rodent electrophysiological studies. The less often observed abnormal or hyperexcitable responses associated with epileptic tissue suggest that epileptogenicity may be highly localized in areas that are not sampled frequently. This later observation also suggests that a majority of cells in epileptic cortex could be, in fact, normal in their electrical behavior and connectivity. The focus may simply be a highly restricted area composed of a relatively low percentage of abnormal neurons that act as recruiters of other neighboring neurons. Generally, epileptic patients whose tissue was examined in these studies experienced complex partial seizures and therefore had long periods of controlled or "quasi-normal" neuronal behavior, which could explain the relative normality of most or some areas of resected tissue. It is only under certain, as yet unknown circumstances that seizures are experienced, and it may be only under those conditions that the balance between synaptic excitation and inhibition is lost. Future experiments should refine our understanding of the interplay between excitation and inhibition within the context of varying neuroanatomy and pathophysiological responses.

22

Clinical strategies for neuronal protection

GREGORY W. ALBERS

Cortical neurons are injured by acute insults, such as ischemia, trauma, and epilepsy, as well as by a variety of chronic neurodegenerative disorders. Although the events that initiate neuronal degeneration vary from disorder to disorder, the fundamental processes that ultimately cause neuronal death are remarkably similar (Choi, 1988). Over the past decade, major advances in molecular and cellular neurobiology have converged to outline a "final common pathway" of neuronal injury, which involves overactivation of excitatory amino acid receptors, excessive intracellular calcium fluxes, subsequent activation of catabolic enzymes, and production of free radicals. These processes are influenced by a variety of neurotransmitters, neuromodulators, alterations in gene expression, and growth factors. Strategies aimed at interfering with these fundamental processes of cell injury are likely to be applicable to a wide variety of neurological disorders. This chapter briefly reviews the scientific rationale behind two emerging neuroprotective strategies and emphasizes the challenges involved with establishing the efficacy of these novel treatments in patients.

MECHANISMS OF CORTICAL NEURONAL INJURY

The pathological elevation of intracellular calcium levels is a common trigger for cell death, not only in the brain but in other tissues as well (Siesjo and Bengtsson, 1989). Neurons, however, are significantly more susceptible to injury than are other cell types. A potential explanation for neuronal susceptibility is provided by the "excitotoxic theory" of neuronal injury

(Rothman and Olney, 1986). This theory posits that neuronal injury is triggered by a toxic extracellular accumulation of excitatory amino acids, specifically glutamate—the major excitatory neurotransmitter in the brain. The excitotoxic theory is compatible with the calcium cytotoxicity hypothesis because excessive extracellular glutamate activates receptors that are coupled to membrane channels with high permeability to calcium. Therefore, any pathological process that leads to an extracellular accumulation of excitatory amino acids could also trigger a rapid intracellular calcium influx.

Although it seems that the largest amount of intracellular calcium accumulation during acute brain injury occurs via glutamate-mediated ion channels, there are several additional sources for intracellular calcium influx, including injury-induced depolarization and activation of voltage-gated calcium channels. In addition, energy depletion can reverse the operation of the sodium/calcium exchanger, resulting in further increases in intracellular calcium. Release from intracellular storage sites and nonspecific leakage through damaged membranes are other sources of injury-induced calcium accumulation. The toxic intracellular calcium levels that result from these and other mechanisms trigger an array of neurochemical changes, including activation of proteases and phospholipases, the eventual generation of free radicals, and lipid peroxidation.

The neuronal injury cascade described above offers multiple sites at which to pharmacologically intervene in what can otherwise be the inevitable progression to neuronal death. This chapter focuses on two new clinical strategies that have rapidly moved from the laboratory to the bedside—excitatory amino acid antagonists and inhibitors of oxygen radical-mediated lipid peroxidation.

EXCITATORY AMINO ACID ANTAGONISTS

Because the excitotoxic hypothesis of neuronal injury suggests that the neuronal injury cascade begins with the excessive activation of glutamate receptors, glutamate antagonists have been extensively studied as potential neuroprotective agents. Glutamate receptor pharmacology is extremely complex and consists of three major families of ionophore-linked receptors, as well as metabotropic receptors that can release calcium from intracellular stores by G-protein-linked activation of inositol-1,4,5-trisphosphate and diacylglycerol (Westbrook, 1993). The inotropic receptors, named for their preferred pharmacological agonists, are N-methyl-D-aspartate (NMDA), kainate, and alpha-amino-3-hydroxy-5-methyl-4-isoxazolepropionic acid (AMPA). Channels gated by these three receptor subtypes are permeable to sodium and potassium. The NMDA receptor also has a high calcium conductance. Early experimental studies focused on the NMDA receptor and employed a wide variety of NMDA antagonists. In addition, there is considerable human experience with several NMDA antagonists, such as dextromethorphan, a commonly used antitussive, and the dissociative anesthetics, ketamine and phencyclidine (PCP) (Albers et al., 1992).

Enthusiasm for the clinical use of NMDA antagonists was fostered by early experimental work indicating that the selective blockade of NMDA receptors was adequate to prevent ischemic neuronal injury not only in cell culture but also in both focal and global animal models of cerebral ischemia (Albers et al., 1989). Subsequently, it has become evident that much of the early success obtained in global ischemia models—models that mimic cardiac arrest—was due to NMDA antagonist-induced hypothermia (Albers, 1990; Buchan, 1990). When these experiments were repeated under strict temperature control, NMDA antagonists continued to be highly effective in focal ischemia models (similar to human stroke), yet were ineffective in models of severe global ischemia.

Surprisingly, non-NMDA receptor antagonists have recently been shown to significantly attenuate neuronal necrosis in some global ischemia models, even when administered up to 12 hours after cerebral reperfusion (Pulsinelli, 1993). In contrast to previous beliefs that non-NMDA receptors did not conduct calcium ions, recent evidence indicates that AMPA receptors may have a high calcium conductance in some situations. The AMPA receptor is composed of several different subunits. Subunits GLuR1 and GLuR3 seem to allow calcium permeability; however, calcium conductance is suppressed in the presence of subunit GLuR2. Recent evidence in a rat global ischemia model suggests that down regulation of the GLuR2 subunit occurs after global ischemia. This finding may help explain the effectiveness of non-NMDA antagonists in global ischemia models.

The lack of effectiveness of NMDA antagonists in global ischemia models has not diminished the enthusiasm for their use in focal ischemia. In fact, experimental studies are nearly uniform in their agreement that NMDA antagonists can significantly reduce ischemic damage (usually by about 50%) in animal stroke models. Although these agents are most effective if administered before the onset of ischemia, they have also been shown to be effective when administered shortly after the ischemic event (Albers, 1990; Steinberg et al., 1991). In virtually all models, maximal neuroprotection is obtained in the cortex, whereas minimal benefit is observed in the striatum. Possible explanations for this finding include the higher concentration of NMDA receptors in the cortex and the poor collateral blood flow in the striatum.

Based on the encouraging results from animal stroke models, several NMDA antagonists are currently undergoing clinical trials in human stroke patients. The competitive NMDA antagonist CGS-19755 has been studied in a double-blind, placebo-controlled pilot study designed to determine the maximum tolerated dose and to obtain preliminary efficacy information (Clark et al., 1994). In addition, the noncompetitive NMDA antagonist dextrorphan, which is the major metabolite of the commonly used antitussive dextromethorphan, has undergone preliminary evaluation in stroke patients (Albers et al., 1994). Both of these compounds attenuate hypoxic neuronal injury in cell culture and reduce infarct volume in a variety of animal stroke models.

Although initial clinical testing has focused on stroke patients, the efficacy of NMDA antagonists may be limited by the difficulty in treating stroke patients emergently. In animal stroke models, NMDA antagonists have been shown to lose their effectiveness within a few hours after the onset of ischemia. Therefore, considerable effort must be made to treat human stroke patients with these agents as rapidly as possible. Furthermore, the results of animal studies suggest that greater clinical success may be achieved if patients are treated before the onset of focal ischemia. Because focal brain ischemia (secondary to both macro- and micro-embolization) occurs frequently during cardiac bypass surgery (Fish, 1988; Nussmeier and McDermott, 1988), this setting allows patients to be treated with NMDA antagonists before the ischemic insult. In addition, this setting allows for the detailed assessment of patients' neurological status before and after the ischemic event, and it is likely that NMDA antagonist side effects may be better tolerated by patients who are under general anesthesia. Pilot studies using NMDA antagonists during high-risk neurovascular surgery, such as the repair of aneurysms, arteriovenous malformations, and carotid endarterectomy, were also prompted by these advantages.

Potential Side Effects and Safety Issues

As NMDA antagonists move from the "bench top to the bedside" (Choi, 1992), several important safety issues must be considered. Inhibition of glutamate-mediated synaptic transmission may cause CNS depression, hypoventilation, and hypotension (Foutz et al., 1988; Rothman and Olney, 1986). However, because non-NMDA-mediated neurotransmission seems to account for the majority of excitatory neurotransmission in the brain, the side effect profile of selective NMDA antagonists may be acceptable (Mayer and Westbrook, 1987).

NMDA receptor activation also plays a crucial role in neuronal plasticity. The induction phase of long-term potentiation (LTP), a fundamental process in learning and memory, is inhibited in a dose-dependent manner by NMDA antagonists (Morris et al., 1989). Although interference with learning and memory may be inconsequential in the setting of acute brain ischemia or trauma, it could be a significant impediment to the chronic administration of NMDA antagonists for the treatment of neurodegenerative disorders or epilepsy.

Normal brain maturation and development also seem to be dependent on NMDA-mediated activity. NMDA antagonists can cause abnormalities in the developing visual system of the cat (Kleinschmidt et al., 1987). In addition, glutamate receptor activation may mediate the normal programmed neuronal death that occurs during development (Choi, 1992). These concerns suggest that NMDA antagonists should be used with extreme caution in pediatric populations.

Another issue that delayed clinical testing of NMDA antagonists was the concern that some of these agents can produce a dose-dependent cytopathology, consisting of cytoplasmic vacuoles in layers III and IV of the posterior cingulate and retrosplenial neocortex of rats (Olney et al., 1989). These changes have been reported predominantly with the potent NMDA antagonist MK-801, which has been demonstrated to increase glucose utilization significantly in some brain regions and may be characterized by a different side effect profile than other NMDA antagonists. This and other concerns about MK-801 have tempered the enthusiasm for its clinical use (Ginsberg, 1993).

An additional safety concern regarding NMDA antagonists is the fear that they will cause adverse psychiatric reactions. Phencyclidine is an NMDA antagonist that is known to cause severe cognitive and behavioral abnormalities in humans, including agitation and hallucinations. In addition, the anesthetic ketamine, which is also a noncompetitive NMDA antagonist, has been noted to produce PCP-like behavioral changes in humans. Finally, the commonly used antitussive dextromethorphan—a noncompetitive NMDA antagonist—can produce hallucinations in humans when administered at very high doses. Therefore, it may be necessary to institute concurrent treatment with antipsychotic medications during NMDA antagonist therapy.

INHIBITORS OF LIPID PEROXIDATION

As discussed above, pharmacological attempts to prevent the toxic accumulation of intracellular calcium can protect neurons from several experimentally induced injuries. However, since calcium can enter neurons through a variety of routes, multiple agents may be required to significantly attenuate calcium fluxes. Because the side effects associated with this polytherapy approach may be clinically intolerable, pharmacological attempts to modify the consequences of excessive intracellular calcium have been investigated extensively. Calcium buffers, inhibitors of protease or phospholipase activation, growth factors, or alterations in gene expression have shown promise as neuroprotective agents. Currently, the inhibition of oxygen radical-mediated lipid peroxidation seems to be one of the most promising approaches (Hall, 1992).

Ultimately, the most destructive pathophysiological reaction responsible for neuronal destruction seems to be peroxidation of neuronal membranes. Lipid peroxidation is a progressive process that can rapidly spread over cell membranes and cause widespread disruption of ionic gradients, eventually resulting in membrane destruction. In addition, free radicals also attack glial cells and damage microvasculature, resulting in hypoperfusion and secondary ischemic neuronal injury.

As a consequence of neuronal injury, free radicals are formed by a variety of reactions, only a few of which are discussed here. Small quantities of free radicals are generated during normal neuronal metabolism; however, they are neutralized by such enzymes as superoxide dismutase, catalase, and glutathione peroxidase, as well as by scavengers like vitamin E. Because neurons have low levels of these protective enzymes, they are vulnerable when excess free radicals are produced (Ikeda and Long, 1990).

In the acidotic and damaged CNS environment, ferritin and transferrin release iron to catalyze oxygen radical reactions, such as the Haber-Weiss reaction, which lead to the production of hydroxyl radicals (Halliwell, 1989). Hydroxyl radicals are extremely potent oxidants that can react with lipids, DNA, and proteins. Elevated intracellular calcium levels activate the arachidonic acid cascade, which results in superoxide radical production. Under conditions of tissue acidosis, superoxides are converted to the hydroperoxyl radical. This radical is highly lipid soluble and can initiate lipid peroxidation.

Lipid peroxidation occurs when a free radical removes an allelic hydrogen from an unsaturated fatty acid. Once lipid peroxidation begins, reactive lipid hydroperoxides form and react with iron to create additional lipid radicals. In damaged neurons, cellular defenses are compromised, and lipid peroxidation may rapidly spread throughout neuronal membranes. Damage to lysosomal membranes can cause the intracellular release of destructive enzymes that damage other cell membranes causing further compromise of ionic gradients.

Because of their importance, considerable effort has been devoted to the development of pharmacological agents that interfere with free radical-induced neuronal destruction (Chan et al., 1987; Levasseur et al., 1989). It has proved difficult to enhance levels of endogenous superoxide dismutase because of poor brain penetration related to its high molecular weight. Although the endogenous antioxidant vitamin E has high CNS penetration due to its lipophilicity, CNS uptake is slow, and in general, animal studies with vitamin E have not documented dramatic neuroprotective effects. A few years ago, a new class of potent inhibitors of lipid peroxidation was developed. These agents, known as 21-aminosteroids or "lazaroids," currently represent the most promising clinical strategy for preventing lipid peroxidation.

Clinical Development of Lazaroids

The development of the lazaroids began with biochemical studies that suggested that oxygen radical generation and lipid peroxidation play a prominent role in traumatic brain and spinal cord injury. In addition, several studies documented that megadoses of glucocorticosteroids, such as methylprednisolone, could improve recovery from experimental CNS injury (Bracken et al., 1990; Hall, 1987). It was hypothesized that the effectiveness of high-dose steroids might be related to their ability to inhibit lipid peroxidation, rather than to their glucocorticoid effects. Therefore, analogs of methylprednisolone that lacked the 11-beta hydroxyl (which is essential for glucocorticoid receptor binding) were prepared. These compounds continued to show antioxidant efficacy without glucocorticoid effects. Subsequently, it was found that the lipid antioxidant activity of these compounds could be enhanced considerably by substituting a complex amine in place of the 21-hydroxyl on the non-glucocorticoid steroid nucleus. These molocules, known as 21-aminosteroids or "lazaroids," are extremely potent inhibitors of iron-catalyzed lipid peroxidation in brain tissue homogenates. One of these agents, tirilazad mesylate (U-74006F), has been selected for clinical development.

Mechanism of Action of Tirilazad. Tirilazad attenuates iron-induced damage to cultured cortical neurons (Monyer et al., 1990). It seems to inhibit lipid peroxidation by scavenging lipid peroxyl radicals and preventing lipid radical chain reactions in a similar manner to vitamin E. In fact, tissue vitamin E levels are preserved when tirilazad is administered to injured CNS tissue. Tirilazad also reacts with hydroxyl radicals and apparently forms a hydroxylated metabolite, which may be a more efficient scavenger of lipid peroxyl radicals than tirilazad itself. In addition, tirilazad seems to insert within the hydrophobic core of cell membranes, resulting in decreased membrane fluidity. This "membrane-stabilizing" effect may help inhibit the propagation of lipid peroxidation by inhibiting the movement of reactive molecules.

Efficacy of Tirilazad in Experimental Brain Injury. Experimental spinal cord trauma studies have demonstrated substantial protective effects with either pre- or post-treatment with tirilazad or other lazaroids (Hall, 1992). Subsequently, lazaroids have been tested in a variety of other experimental settings, including models of traumatic brain injury and focal cerebral ischemia. A 1 mg/kg dose of tirilazad administered 30 minutes after experimentally induced brain injury in cats was associated with significant reductions in lactic acid accumulation in both the cerebral cortex and the subcortical white matter (Dimlich et al., 1990). In addition, improved functional recovery has been shown in both rat and mouse head trauma models using this agent.

A variety of experimental focal cerebral ischemia studies have also produced encouraging results. In a gerbil model, tirilazad was demonstrated to promote survival and reduce neuronal necrosis in animals that were treated before unilateral carotid occlusion. Protection against vitamin E depletion and the recovery of cerebral cortical extracellular calcium concentrations have also been shown in this model (Hall et al., 1991). Tirilazad was also effective in reducing infarct volume in a cat ischemia model, when administered 15 minutes after the onset of a temporary (1-hour) middle cerebral artery occlusion (Silvia et al., 1987). Although tirilazad was effective with a 1-hour occlusion, no benefit was apparent after a 3-hour occlusion time in a similar study. Encouraging results have been obtained using tirilazad in temporary middle cerebral artery (MCA) occlusion models in rats; however, it has been less effective or ineffective in permanent MCA occlusion models. As with NMDA antagonists, neuronal preservation is usually restricted to the ischemic penumbra.

Clinical Trials with Tirilazad. In contrast to the substantial side effects and safety concerns that have slowed the clinical development of NMDA antagonists, lazaroids seem to be remarkably well tolerated (Hall, 1992). This may be explained by the fact that lipid peroxidation is a predominantly pathological process, whereas excitatory amino acid neurotransmission is essential for a wide variety of normal processes. As a result, tirilazad has made a rapid transition from laboratory studies to clinical trials. Studies of patients with brain and spinal cord trauma, subarachnoid hemorrhage, and stroke are currently underway (Hall, 1992).

It seems that effective therapies for a variety of previously untreatable neurological disorders are on the horizon. These new approaches for protecting the cortical neuron have emerged because of dramatic progress in the molecular and cellular neurobiology of neuronal injury. The new therapies are based on a strong scientific foundation and are supported by highly encouraging preclinical studies. Human trials using many of these new approaches are currently in progress, and it is clear that close collaboration between neuroscientists and clinicians will be required to translate advances in basic neurosciences into clinical successes.

References

Abeles M. (1982). Local Cortical Circuits: An Electrophysiological Study. Springer-Verlag: Berlin.

Abeles M. (1991). *Corticonics: Neural Circuits of the Cerebral Cortex.* New York: Cambridge University Press.

Adams MD, Dubnick M, Kerlavage A, Moreno R, Kelley JM, Utterback TR, Nagle JW, Fields C, Venter JC. (1992). Sequence identification of 2,375 human brain genes. *Nature* 355:632–634.

Agmon A, Connors BW. (1989). Repetitive burst-firing neurons in the deep layers of mouse somatosensory cortex. *Neurosci Lett* 99:137–141.

Agmon A, Connors BW. (1991). Thalamocortical responses of mouse somatosensory (barrel) cortex in vitro. *Neuroscience* 41:365–379.

Agmon A, Connors BW. (1992). Correlation between intrinsic firing patterns and thalamocortical synaptic responses of neurons in mouse barrel context. *J Neurosci* 12:319–329.

Agmon A, O'Dowd DK. (1992). NMDA receptor-mediated currents are prominent in the thalamocortical synaptic response before maturation of inhibition. *J Neurophysiol* 68:345–349.

Aicardi J. (1986). *Epilepsy in Children.* New York: Raven Press.

Ajmone-Marsan C. (1969). Acute effects of topical epileptogenic agents. In: Jasper HH, Ward AA, Jr, Pope A, eds. *Basic Mechanisms of the Epilepsies.* Boston: Little, Brown; 299–319.

Akabas MH, Stauffer DA, Xu M, Karlin A. (1992). Acetylcholine receptor channel structure probed in cysteine-substitution mutants. *Science* 258:307–310.

Albers GW. (1990). Potential therapeutic uses of N-methyl-D-aspartate antagonists in cerebral ischemia. *Clin Neuropharmacol* 13:177–197.

Albers GW, Goldberg MP, Choi DW. (1989). N-methyl-D-aspartate antagonists: Ready for clinical trial in brain ischemia? *Ann Neurol* 25:398–403.

Albers GW, Saenz RE, Moses JA. (1992). Tolerability of oral dextromethorphan in patients with a history of brain ischemia. *Clin Neuropharmacol* 15: 509–514.

Albers GW, Atkinson R, Kelley R, Rosenbaum DM. (1994). Safety, tolerability and pharmacokinetics of the N-methyl-D-aspartate antagonist dextrorphan in patients with acute stroke. *Neurology* 44 (suppl 2):A270.

Alexander KA, Cimler BM, Meier KE, Storm DR. (1987). Regulation of calmodulin binding to P-57. A neurospecific calmodulin binding protein. *J Biol Chem* 262:6108–6113.

Alger BE. (1984). Characteristics of a slow hyperpolarizing synaptic potential in rat hippocampal pyramidal cells in vitro. *J Neurophysiol* 52:892–910.

Alger BE. (1991). Gating of GABAergic inhibition in hippocampal pyramidal cells. In: Wolpaw JR, Schmidt W, eds. *Activity Driven CNS Changes in Learning and Development.* New York: New York Academy of Sciences: 249–263.

Alger BE, Nicoll RA. (1979). GABA-mediated biphasic inhibitory responses in hippocampus. *Nature* 281:315–317.

Alger BE, Nicoll RA. (1982a). Feed-forward dendritic inhibition in rat hippocampal pyramidal cells studied in vitro. *J Physiol* (Lond) 328:105–123.

Alger BE., Nicoll RA. (1982b). Pharmacological evidence for two kinds of GABA receptors on rat hippocampal pyramidal cells studied in vitro. *J Physiol* (Lond) 328:125–141.

Alger BE, Teyler TJ. (1979). Long-term and short-term plasticity in CA1, CA3 and dentate regions of the rat hippocampal slice. *Brain Res* 110:463–480.

Allard T, Clark SA, Jenkins WM, Merzenich MM. (1991). Reorganization of somatosensory area 3b representations in adult owl monkeys after digital syndactyly. *J Neurophysiol* 66:1048–1058.

Allbritton NL, Meyer T, Stryer L. (1992). Range of messenger action of calcium ion and inositol 1,4,5-triphosphate. *Science* 258:1812–1815.

Alonso A, Llinás RR. (1989). Subthreshold Na^+-dependent theta-like rhythmicity in stellate cells of entorhinal cortex layer II. *Nature* 342:175–177.

Alzheimer C, Schwindt PC, Crill WE. (1993a). Postnatal development of a persistent Na^+ current in pyramidal neurons from rat sensorimotor cortex. *J Neurophysiol* 69:290–292.

Alzheimer C, Schwindt PC, Crill WE. (1993b). Modal gating of Na^+ channels as a mechanism of persistent Na+ current in pyramidal neurons from rat and cat sensorimotor cortex. *J Neurosci* 13:660–673.

Amara SG. (1992). A tale of two families. *Nature* 360:420–421.

Ambrosetto G, Gobbi G. (1975). Benign epilepsy of childhood with rolandic spikes, or a lesion? EEG during a seizure. *Epilepsia* 16:793–796.

Ambrosetto G, Viovanardi RP. (1975). Benign epilepsy of infancy with rolandic and/or middle temporal EEG paroxysms: a polygraphic study during night time sleep. *Gaslini* 7:35–37.

Amitai Y, Connors BW. (1995). Intrinsic physiology and morphology of single neurons in neocortex. In: Jones EG, Diamond IT, eds. *The Cerebral Cortex, Vol. 11, The Barrel Cortex of Rodents.* New York: Plenum Press: 299–331.

Amitai Y, Friedman A, Connors BW, Gutnick MJ. (1993). Regenerative activity in apical dendrites of pyramidal cells in neocortex. *Cereb Cortex* 3:26–38.

Andersen P, Andersson SA. (1968). *Physiological Basis of the Alpha Rhythm.* New York: Appleton-Century-Crofts.

Andersen P, Eccles JC, Loyning Y. (1964a). Location of postsynaptic inhibitory synapses on hippocampal pyramids. *J Neurophysiol* 27:592–607.

Andersen P, Eccles JC, Loyning Y. (1964b). Pathway of postsynaptic inhibition in the hippocampus. *J Neurophysiol* 27:608–619.

Andersen P, Bliss TVL, Skrede KK. (1971). Lamellar organization of hippocampal excitatory pathways. *Exp Brain Res* 13:222–238.

Andersen P, Sundberg SH, Sveen O, Wigström H. (1977). Specific long-lasting potentiation of synaptic transmission in hippocampal slices. *Nature* 266: 736–737.

Andersen P, Dingledine R, Gjerstad L, Langmoen IA, Laursen AM. (1980). Two different responses of hippocampal pyramidal cells to application of gamma-amino butyric acid. *J Physiol* (Lond) 305:279–296.

Anderson DJ. (1989). The neural crest cell lineage problem: neuropoiesis? *Neuron* 3:1–12.

Anderson KJ, Bridges RJ, Cotman CW. (1991). Increased density of excitatory amino acid transport sites in the hippocampal formation following an entorhinal lesion. *Brain Res* 562:285–290.

Andrade R. (1993). Enhancement of beta-adrenergic responses by G_i-linked receptors in rat hippocampus. *Neuron* 10:83–88.

Andrade R, Malenka RC, Nicoll RA. (1986). A G protein couples serotonin and $GABA_B$ receptors to the same channels in hippocampus. *Science* 234: 1261–1265.

Angelotti TP, MacDonald RL. (1993). Assembly of $GABA_A$ receptor subunits: $\alpha 1\beta 1$ and $\alpha 1\beta 1\gamma 2S$ subunits produce unique ion channels with dissimilar single-channel properties. *J Neurosci* 13:1429–1440.

Angelotti TP, Uhler MD, MacDonald RL. (1993). Assembly of $GABA_A$ receptor subunits: analysis of transient single-cell expression utilizing a fluorescent substrate/marker gene technique. *J Neurosci* 13:1418–1428.

Aniksztejn L, Ben-Ari Y. (1991). Novel form of long-term potentiation produced by a K^+ channel blocker in the hippocampus. *Nature* 349:67–69.

Antonini A, Stryker MP. (1993). Rapid remodeling of axonal arbors in the visual cortex. *Science* 260:1819–1821.

Araki M, McGeer PL, McGeer EG. (1984). Presumptive γ-amino-butyric acid pathways from the midbrain to the superior colliculus studied by a combined horseradish peroxidase-γ-aminobutyric acid transaminase pharmacohistochemical method. *Neuroscience* 13:433–439.

Araki T, Kiyama H, Tohyama M. (1992). The $GABA_A$ receptor gamma 1 subunit is expressed by distinct neuronal populations. *Mol Brain Res* 15:121–132.

Aram JA, Michelson HB, Wong RKS. (1991). Synchronized GABAergic IPSPs recorded in the neocortex after blockade of synaptic transmission mediated by excitatory amino acids. *J Neurophysiol* 65:1034–1041.

Armstrong-James M, Johnson R. (1970). Quantitative studies of postnatal changes in synapses in rat superficial motor cerebral cortex. *Zeitschr Zellforsch* 110: 559–568.

Armstrong-James M, Fox K, Das-Gupta A. (1992). The flow of excitation within rat S1 barrel cortex on striking a single vibrissa. *J Neurophysiol* 68:1345–1358.

Armstrong-James M, Welker E, Callahan CA. (1993). The contribution of NMDA and non-NMDA receptors to fast and slow transmission of sensory information in the rat S1 barrel cortex. *J Neurosci* 13:2149–2160.

Aroniadou VA, Teyler TJ. (1991). The role of NMDA receptors in long-term potentiation (LTP) and depression (LTD) in rat visual cortex. *Brain Res* 562: 136–143.

Aroniadou VA, Teyler TJ. (1992). Induction of NMDA receptor-independent long-term potentiation (LTP) in visual cortex of adult rats. *Brain Res* 584:169–173.

Artola A, Singer W. (1987). Long-term potentiation and NMDA receptors in rat visual cortex. *Nature* 330:649–652.

Artola A, Singer W. (1993). Long-term depression of excitatory synaptic transmission and its relationship to long-term potentiation. *Trends Neurosci* 16: 480–487.

Artola A, Brocher S, Singer W. (1990). Different voltage-dependent thresholds for inducing long-term depression and long-term potentiation in slices of rat visual cortex. *Nature* 347:69–72.

Ascher P, Choi DW, Christen Y. (1991). *Glutamate, Cell Death and Memory.* Heidelberg: Springer-Verlag.

Ashwood TJ, Lancaster B, Wheal HV. (1986). Intracellular electrophysiology of CA1 pyramidal neurones in slices of the kainic and lesioned hippocampus of the rat. *Exp Brain Res* 62:189–198.

Aston-Jones G, Bloom FE. (1981). Activity of norepinephrine-containing locus coeruleus neurons in behaving rats anticipates fluctuations in the sleep-walking cycle. *J Neurosci* 1:876–886.

Austin CP, Cepko CL. (1990). Cellular migration patterns in the developing mouse cerebral cortex. *Development* 110:713–732.

Avanzini G, deCurtis M, Panzica F, Spreafico R. (1989). Intrinsic properties of nucleus reticularis thalami neurones of the rat studied in vitro. *J Physiol* (Lond) 416:111–122.

Avoli M. (1986). Inhibitory potentials in neurons of the deep layers of the in vitro neocortical slice. *Brain Res* 370:165–170.

Avoli M, Gloor P. (1981). The effects of transient functional depression of the thalamus on spindles and on bilateral synchronous epileptic discharges of feline generalized penicillin epilepsy. *Epilepsia* 22:443–452.

Avoli M, Gloor P. (1993). Interaction of cortex and thalamus in spike and wave discharges of feline generalized penicillin epilepsy. *Exp Neurol* 76:196–217.

Avoli M, Olivier A. (1987). Bursting in human epileptogenic neocortex is depressed by an N-methyl-D-aspartate antagonist. *Neurosci Lett* 76:249–254.

Avoli M, Olivier A. (1989). Electrophysiological properties and synaptic responses in the deep layers of the human epileptogenic neocortex in vitro. *J Neurophysiol* 61:589–606.

Avoli M, Gloor P, Kostopoulos G, Gotman J. (1983). An analysis of penicillin-induced generalized spike and wave discharges using simultaneous recordings of cortical and thalamic single neurons. *J Neurophysiol* 50:819–837.

Avoli M, Louvel J, Pumain R, Olivier A. (1987). Seizure-like discharges induced by lowering [Mg^{2+}] in the human epileptogenic neocortex maintained in vitro. *Brain Res* 417:199–203.

Avoli M, Perreault P, Olivier A, Villemure JG. (1988). 4-Aminopyridine induces a long-lasting depolarizing GABAergic potential in human neocortical and hippocampal neurons maintained in vitro. *Neurosci Lett* 94:327–332.

Avoli M, Drapeau C, Perreault P, Louvel J, Pumain R. (1990a). Epileptiform activity induced by low chloride medium in the CA1 subfield of the hippocampal slice. *J Neurophysiol* 61:1747–1757.

Avoli M, Gloor P, Kostopoulos G, Naquet R. (1990b). Generalized epilepsy. In: Avoli M, Gloor P, eds. *Neurobiological Approaches*. Boston: Birkhauser.

Babb TL, Carr E, Crandall PH. (1973). Analysis of extracellular firing patterns of deep temporal lobe structures in man. *Electroencephalogr Clin Neurophysiol* 34:247–257.

Babb TL, Brown WJ, Pretorius J, Davenport C, Lieb JP, Crandall PH. (1984a). Temporal lobe volumetric cell densities in temporal lobe epilepsy. *Epilepsia* 25:729–740.

Babb TL, Lieb JP, Brown WL, Pretorius J, Crandall PH. (1984b). Distribution of pyramidal cell density and hyperexcitability in the epileptic human hippocampal formation. *Epilepsia* 25:721–728.

Babb TL, Pretorius JK, Kupfer WR, Crandall PH. (1989). Glutamate decarboxylase-immunoreactive neurons are preserved in human epileptic hippocampus. *J Neurosci* 9:2562–2574.

Babb TL, Kupfer WR, Pretorius JK, Crandall PH, Levesque MF. (1991). Synaptic reorganization by mossy fibers in human epileptic fascia dentata. *Neuroscience* 42:351–363.

Backus KH, Arigoni M, Drescher U, Scheurer L, Malherbe P, Möhler H, Benson JA. (1993). Stoichiometry of a recombinant $GABA_A$ receptor deduced from mutation-induced rectification. *Neuroreport* 5:285–288.

Bading H, Ginty DD, Greenbert ME. (1993). Regulation of gene expression in hippocampal neurons by distinct calcium signaling pathways. *Science* 260: 181–186.

Baier H, Bonhoeffer F. (1992). Axon guidance by gradients of a target-derived component. *Science* 255:472–475.

Bakay RAE, Harris, AB. (1981). Neurotransmitter, receptor and biochemical changes in monkey cortical epileptic foci. *Brain Res* 206:387–404.

Banker G, Goslin K. (1991). *Culturing Nerve Cells*. Cambridge: MIT Press.

Baraban JM, Snyder SH, Alger BE. (1985). Protein kinase C regulates ionic conductance in hippocampal pyramidal neurons: electrophysiological effects of phorbol esters. *Proc Natl Acad Sci USA* 82:2538–2542.

Baranyi A, Szente MB. (1987). Long-lasting potentiation of synaptic transmission requires postsynaptic modifications in the neocortex. *Brain Res* 423:378–384.

Baranyi A, Szente MB, Woody CD. (1991). Properties of associative long-lasting potentiation induced by cellular conditioning in the motor cortex of conscious cats. *Neuroscience* 42:321–334.

Baranyi A, Szente MB, Woody CD. (1993). Electrophysiological characterization of different types of neurons recorded in vivo in the motor cortex of the cat: I. Patterns of firing activity and synaptic responses. *J Neurophysiol* 69: 1850–1864.

Barbe MF, Levitt P. (1992). Attraction of specific thalamic input by cerebral grafts depends on molecular identity of the implant. *Proc Natl Acad Sci USA* 89: 3706–3710.

Barbour B, Szatkowski M, Ingledew N, Attwell D. (1989). Arachidonic acid induces a prolonged inhibition of glutamate into glial cells. *Nature* 342:918–920.

Barlow HB, Blakemore C, Pettigrew JD. (1967). The neural mechanism of binocular depth discrimination. *J Physiol* 193:327–342.

Barrett EF, Stevens CF. (1972a). The kinetics of transmitter release at the frog neuromuscular junction. *J Physiol* (Lond) 227:691–708.

Barrett EF, Stevens CF. (1972b). Quantal independence and uniformity of presynaptic release kinetics at the frog neuromuscular junction. *J Physiol* (Lond) 227:665–689.

Barrett JN, Crill WE. (1974). Influence of dendritic location and membrane properties on the effectiveness of synapses on cat motoneurones. *J Physiol* (Lond) 293:325–345.

Bashir ZI, Bortolotto ZA, Davies CH, Berretta N, Irving AJ, Seal AJ, Henly JM, Jane DE, Watkins JC, Collingridge GL. (1993). The synaptic activation of glutamate metabotropic receptors is necessary for the induction of LTP in the hippocampus. *Nature* 363:347–350.

Baskys A, Bernstein NK, Barolet AW, Carlen PL. (1990a). NMDA and quisqualate reduce a Ca-dependent K^+ current by a protein kinase-mediated mechanism. *Neurosci Lett* 112:76–81.

Baskys A, Reynolds JN, Carlen PL. (1990b). NMDA depolarizations and long-term potentiation are reduced in the aged rat neocortex. *Brain Res* 530: 142–146.

Baumbach HD, Chow KL. (1981). Visuocortical epileptiform discharges in rabbits: differential effects on neuronal development in the lateral geniculate nucleus and superior colliculus. *Brain Res* 209:61–76.

Beal MF, Swartz KJ, Isacson O. (1992). Developmental changes in brain kynurenic acid concentrations. *Dev Brain Res* 68:136–139.

Bear MF, Kirkwood A. (1993). Long-term potentiation in the neocortex. *Curr Opin Neurobiol* 3:197–202.

Bear MF, Press WA, Connors BW. (1992). Long-term potentiation in slices of kitten visual cortex and the effects of NMDA receptor blockade. *J Neurophysiol* 67:1–11.

Beattie CE, Siegel RE. (1993). Developmental cues modulate $GABA_A$ receptor subunit mRNA expression in cultured cerebellar granule neurons. *J Neurosci* 13:1784–1792.

Beaussart M. (1972). Benign epilepsy of children with rolandic (centro-temporal) paroxysmal foci. A clinical entity. Study of 221 cases. *Epilepsia* 13:795–811.

Becker C, Hoch W, Betz H. (1988). Glycine receptor heterogeneity in rat spinal cord during postnatal development. *EMBO J* 7:3717–3726.

Becker C, Schmieden V, Tarroni P, Strasser U, Betz H. (1992). Isoform-selective deficit of glycine receptors in the mouse mutant spastic. *Neuron* 8:283–289.

Bekenstein JW, Lothman EW. (1993). Dormancy of inhibitory interneurons in a model of temporal lobe epilepsy. *Science* 259:97–100.

Bekkers JM, Stevens CF. (1989). NMDA and non-NMDA receptors are co-localized at individual excitatory synapses in cultured rat hippocampus. *Nature* 341:230–233.

Bekkers JM, Stevens CF. (1990). Presynaptic mechanism for long-term potentiation in the hippocampus. *Nature* 346:724–729.

Bekkers JM, Richardson GB, Stevens CF. (1990). Origin of variability in quantal size in cultured hippocampal neurons and hippocampal slices. *Proc Natl Acad Sci USA* 87:5359–5362.

Bellman R, Widmann R, Olenik C, Meyer DK, Mass D, Marksyteiner J, Sperk G. (1991). Enhanced rate of expression and synthesis of neuropeptide Y after kainic acid-induced seizures. *J Neurochem* 56:525–530.
Ben-Ari Y. (1985). Limbic seizure and brain damage produced by kainic acid: mechanisms and relevance to human temporal lope epilepsy. *Neuroscience* 14:375–403.
Ben-Ari Y, Cherubini E, Krnjevic K. (1988). Changes in voltage dependence of NMDA currents during development. *Neurosci Lett* 94:88–92.
Ben-Ari Y, Cherubini E, Corradetti R, Gaiarsa J. (1989). Giant synaptic potentials in immature rat CA3 hippocampal neurones. *J Physiol* (Lond) 416:303–325.
Benardo LS. (1993a). Separate activation of fast and slow inhibitory postsynaptic potentials in rat neocortex in vitro. *J Physiol.*
Benardo LS. (1993b). $GABA_A$ receptor-mediated mechanisms contribute to frequency-dependent depression of IPSPs in the hippocampus. *Brain Res* 607: 81–88.
Benardo LS. (1994a). Recruitment of inhibition by enhanced activation of synaptic NMDA responses in the rat cerebral cortex. *Brain Res* 627:314–324.
Benardo LS. (1994b). Separate activation of fast and slow inhibitory postsynaptic potentials in rat neocortex in vitro. *J Physiol* (Lond) 259:97–100.
Benardo LS, Foster RE. (1986). Oscillatory behavior in inferior olive neurons: mechanism, modulation, cell aggregates. *Brain Res Bull* 17:773–784.
Benardo LS, Prince DA. (1982). Cholinergic excitation of mammalian hippocampal pyramidal cells. *Brain Res* 249:315–331.
Benardo LS, Masukawa LM, Prince DA. (1982). Electrophysiology of isolated hippocampal pyramidal dendrites. *J Neurosci* 2:1614–1622.
Benavides J, Claustre Y, Scatton B. (1988). L-glutamate increases internal free calcium levels in synaptoneurosomes from immature rat brain via quisqualate receptors. *J Neurosci* 8:3607–3615.
Benfenati F, Valtorta F, Rubenstein JL, Gorelick FS, Greengard P, Czernick AJ. (1992a). Synaptic vesicle-associated Ca^{2+}/calmodulin-dependent protein kinase II is a binding protein for synapsin I. *Nature* 359:417–420.
Benfenati F, Valtorta F, Chieregatti E, Greengard P. (1992b). Interaction of free and synaptic vesicle-bound synapsin I with F-actin. *Neuron* 8:377–386.
Bennett JA, Dingledine R. (1995). Topology profile for a glutamate receptor: three transmembrane domains and a channel-lining re-entrant membrane loop. *Neuron* 14:373–384.
Bennett MK, Scheller RH. (1993). The molecular machinery for secretion is conserved from yeast to neurons. *Proc Natl Acad Sci USA* 90:2559–2563.
Benninger C, Kadis J, Prince DA. (1980). Extracellular calcium and potassium changes in hippocampal slices. *Brain Res* 187:165–182.
Benson DL, Isackson PJ, Gall CM, Jones EG. (1991). Differential effects of monocular deprivation on glutamic acid decarboxylase and type II calcium-cadmodulin-dependent protein kinase gene expression in the adult monkey visual cortex. *J Neurosci* 11:31–47.
Benson DL, Huntsman MM, Jones EG. (1993). Activity dependent changes in GAD and preprotachykinin mRNAs in visual cortex of adult monkeys. *Cereb Cortex* 4:40–51.
Benveniste H, Drejer J, Schousboe A, Diemer NH. (1984). Elevation of the extracellular concentrations of glutamate and aspartate in rat hippocampus dur-

ing transient cerebral ischemia monitored by intracerebral microdialysis. *J Neurochem* 43:1369–1374.

Bergold PJ, Casaccia-Bonnefil P, Federoff HJ, Stelzer A. (1993). Transduction of CA3 neurons by a herpes virus vector containing a GluR6 subunit of kainate receptor induces epileptiform discharge. *Soc Neurosci Abstr* 19:21.

Berman NJ, Douglas RJ, Martin KAC. (1989). The conductances associated with inhibitory postsynaptic potentials are larger in visual cortical neurons in vitro than in similar neurons in intact, anesthetized rats. *J Physiol* (Lond) 418:107P.

Berman NJ, Douglas RJ, Martin KAC, Whitteridge D. (1991). Mechanisms of inhibition in cat visual cortex. *J Physiol* (Lond) 440:697–722.

Bernander O, Douglas RJ, Martin KAC, Koch C. (1991). Synaptic background activity influences spatiotemporal integration in single pyramidal cells. *Proc Natl Acad Sci USA* 88:11569–11573.

Bernier GP, Richer F, Giard N, Bouvier G, Mercier M, Turnmel A, Saint-Hilaire JM. (1990). Electrical stimulation of the human brain in epilepsy. *Epilepsia* 31:513–520.

Bernstein A, Breitman M. (1989). Genetic ablation in transgenic mice. *Mol Biol Med* 6:523–539.

Bettler B, Boulter J, Hermans-Borgmeyer I, O'Shea-Greenfield A, Deneris ES, Moll C, Borgmeyer U, Hollmann M, Heinemann S. (1990). Cloning of a novel glutamate receptor subunit, GluR5: expression in the nervous system during development. *Neuron* 5:583–595.

Bettler B, Egebjerg J, Sharma G, Pecht G, Hermans-Borgmeyer I, Moll C, Stevens CF, Heinemann S. (1992). Cloning of a putative glutamate receptor: a low affinity kainate-binding subunit. *Neuron* 8:1–20.

Bielfeldt K, Rotter JL, Jackson MB. (1992). Three potassium channels in rat posterior pituitary nerve terminals. *J Physiol* (Lond) 458:41–67.

Bienenstock EL, Cooper LN, Munro PW. (1982). Theory for the development of neuron selectivity: orientation specificity and binocular interaction in visual cortex. *J Neurosci* 2:32–48.

Biggio G, Concas A, Costa E. (1992). *GABAergic Synaptic Transmission. Molecular, Pharmacological, and Clinical Aspects.* New York: Raven Press.

Bigot D, Matus A, Hunt SP. (1991). Reorganization of the cytoskeleton in rat neurons following stimulation with excitatory amino acids in vitro. *Eur J Neurosci* 3:551–558.

Bindman L, Murphy KPS, Pockett S. (1988). Postsynaptic control of the induction of long-term changes in efficacy of transmission at neocortical synapses in slices of rat brain. *J Neurophysiol* 60:1053–1065.

Bindman L, Christofi G, Murphy K, Nowicki A. (1991). Long-term potentiation and long-term depression in the neocortex and hippocampus: an overview. In: Stone TW, ed. *Aspects of Synaptic Transmission.* London: Taylor Francis: 3–25.

Blanton MG, Kriegstein AR. (1991). Appearance of putative amino acid neurotransmitters during differentiation of neurons in embryonic turtle cerebral cortex. *J Comp Neurol* 310:571–592.

Blanton MG, Turco JJL, Kriegstein AR. (1989). Whole cell recording from neurons in slices of reptilian and mammalian cerebral cortex. *J Neurosci Meth* 30: 203–210.

Blaschke M, Keller BU, Rivosecchi R, Hollmann M, Heinemann S, Konnerth A. (1993). A single amino acid determines the subunit-specific spider toxin block of α-amino-3-hydroxy-5-methylisoxazole-4-propionate/kainate receptor channels. *Proc Natl Acad Sci USA* 90:6528–6532.

Blatz LA, Magleby KL. (1986). Single apamin-blocked Ca-activated K^+ channels of small conductance in cultured rat skeletal muscle. *Nature* 323:718–720.

Bliss TVP, Collingridge GL. (1993). A synaptic model of memory: long-term potentiation in the hippocampus. *Nature* 361:31–39.

Bliss TVP, Lomo T. (1973). Long-lasting potentiation of synaptic transmission in the dentate area of the anesthetized rabbit following stimulation of the perforant path. *J Physiol* (Lond) 232:331–356.

Bliss TVP, Burns BD, Uttley AM. (1968). Factors affecting the conductivity of pathways in the cerebral cortex. *J Physiol* (Lond) 195:339–367.

Bliss TVP, Douglas RM, Errington ML, Lynch MA. (1986). Correlation between long-term potentiation and release of endogenous amino acids from dentate gyrus of anesthetized rats. *J Physiol* (Lond) 377:391–408.

Blom S, Heijbel J, Bergfors PG. (1972). Benign epilepsy of children with centrotemporal EEG foci: Prevalence and follow-up study of 40 patients. *Epilepsia* 16:133–140.

Blue ME, Parnavelas JG. (1983). The formation and maturation of synapses in the visual cortex of the rat. II. Quantitative analysis. *J Neurocytol* 12:697–712.

Blume WT. (1982). *Atlas of Pediatric Electroencephalography.* New York: Raven Press.

Bochet P, Audinat E, Lambolez B, Crepel F, Rossier J, Iino M, Tsuzuki K, Ozawa S. (1994). Subunit composition at the single-cell level explains functional properties of a glutamate-gated channel. *Neuron* 12:383–388.

Boekhoff I, Breer H. (1992). Termination of second messenger signaling in olfaction. *Proc Natl Acad Sci USA* 89:471–474.

Bolton JS. (1910). A contribution to the localization of cerebral function, based on the clinicopathological study of mental disease. *Brain* 33:26–147.

Bolz J, Gilbert CD. (1986). Generation of end-inhibition in the visual cortex via interlaminar connections. *Nature* 320:362–365.

Bolz J, Gilbert CD. (1989). The role of horizontal connections in generating long receptive fields in the cat visual cortex. *Eur J Neurosci* 1:263–268.

Bonhaus DW, McNamara JO. (1988). N-methyl-D-aspartate receptor regulation of uncompetitive antagonist binding in rat brain membranes: kinetic analysis. *Mol Pharmacol* 34:250–255.

Bonhaus DW, McNamara JO. (1989). TCP binding: a tool for studying NMDA receptor-mediated neurotransmission in kindling. *Neurosci Biobehav Rev* 13:261–267.

Bonnerot C, Grimber G, Briand P, Nicolas JF. (1990). Patterns of expression of position-dependent integrated transgenes in mouse embryo. *Proc Natl Acad Sci USA* 87:6331–6335.

Bormann J. (1988). Electrophysiology of $GABA_A$ and $GABA_B$ receptor subtypes. *Trends Neurosci* 11:112–116.

Bormann J, Kettenmann H. (1988). Patch-clamp study of gamma-aminobutyric acid receptor Cl-channels in cultured astrocytes. *Proc Natl Acad Sci USA* 85:9336–9340.

Bortolotto ZA, Collingridge GL. (1993). Characterization of LTP induced by the activation of glutamate metabotropic receptors in area CA1 of the hippocampus. *Neuropharmacology* 32:1–9.

Bourne HR, Nicoll R. (1993). Molecular machines integrate coincident synaptic signals. *Cell* 72:65–75.

Bowery NG. (1993). $GABA_B$ receptor pharmacology. *Annu Rev Pharmacol Toxicol* 33:109–147.

Bracken MB, Shepard MJ, Collins WF. (1990). A randomized, controlled trial of methylprednisolone or naloxone in the treatment of acute spinal-cord injury. *N Engl J Med* 322:1405–1411.

Bray PF, Wiser WC. (1965). Hereditary characteristics of familial temporal-central focal epilepsy. *Pediatrics* 36:207–211.

Brocher S, Artola A, Singer W. (1992). Intracellular injection of Ca^{2+} chelators blocks induction of long-term depression in rat visual cortex. *Proc Natl Acad Sci USA* 89:123–127.

Brook JD. (1992). Molecular basis of myotonic dystrophy: expansion of a trinucleotide (CTG) repeat at the 3′ end of a transcript encoding a protein kinase family member. *Cell* 68:799–808.

Brorson JR, Manzolillo PA, Miller RJ. (1994). Ca^{2+} entry via AMPA/KA receptors and excitotoxicity in cultured cerebellar Purkinje cells. *J Neurosci* 14:187–197.

Brown AM, Sayer RJ, Schwindt P, Crill WE. (1994). P-type calcium channels in rat neocortical neurones. *J Physiol* (Lond) 475:197–205.

Brown TA, Dingledine R. (1992). NMDA and AMPA receptor components of EPSPs from rat dentate hilar interneurons. *Neurosci Abstr* 18:1356.

Brown TH, Kairiss EW, Keenan CL. (1990). Hebbian synapses: biophysical mechanisms and algorithms. *Annu Rev Neurosci* 13:475–511.

Buchan AM. (1990). Do NMDA antagonists protect against cerebral ischemia: are clinical trials warranted? *Cerebrovasc Brain Metab Rev* 2:1–26.

Buckmaster PS, Strowbridge BW, Kunkel DD, Schmiege DL, Schwartzkroin PA. (1992). Mossy cell axonal projections to the dentate gyrus molecular layer in the rat hippocampal slice. *Hippocampus* 2:349–362.

Buckmaster PS, Strowbridge BW, Schwartzkroin PA. (1993). A comparison of rat hippocampal mossy cells and CA3c pyramidal cells. *J Neurophysiol* 70: 1281–1299.

Buhl EH, Halasy K, Somogyi P. (1994). Diverse sources of hippocampal unitary inhibitory postsynaptic potentials and the number of synaptic release sites. *Nature* 368:823–828.

Bullock TH. (1993). How are more complex brains different? One view and an agenda for comparative neurobiology. *Brain Behav Evol* 41:88–96.

Burgard EC, Hablitz JJ. (1993). Developmental changes in NMDA and non-NMDA receptor-mediated synaptic potentials in rat neocortex. *J Neurophysiol* 69:230–240.

Burkhalter A. (1989). Intrinsic connections of rat primary visual cortex: laminar organization of axonal projections. *J Comp Neurol* 279:171–186.

Burnashev N, Monyer H, Seeburg PH, Sakmann B. (1992a). Divalent ion permeability of AMPA receptor channels is dominated by the edited form of a single subunit. *Neuron* 8:189–198.

Burnashev N, Schoepfer R, Monyer H, Ruppersberg JP, Gunther W, Seeburg PH, Sakmann B. (1992b). Control by asparagine residues of calcium permeability and magnesium blockade in the NMDA receptor. *Science* 257:1415–1419.

Burns DB. (1958). *The Mammalian Cerebral Cortex*. London: Edward Arnold.

Busch C, Sakmann B. (1990). Synaptic transmission in hippocampal neurons: numerical reconstruction of quantal IPSCs. *Cold Spring Harb Symp Quant Biol* 55:69–80.

Bush PC, Douglas RJ. (1991). Synchronization of bursting action potential discharge in a model network of neocortical neurons. *Neural Comp* 3:19–30.

Bush PC, Sejnowski TJ. (1993). Reduced compartmental models of neocortical pyramidal cells. *J Neurosci Meth* 46:159–166.

Bush PC, Sejnowski TJ. (1994). Effects of inhibition and dendritic saturation in simulated neocortical pyramidal cells. *J Neurophysiol* 71:2183–2193.

Byrne MC, Gottlieb R, McNamara JO. (1980). Amygdala kindling induces muscarinic cholinergic receptor decline in a highly specific distribution within the limbic system. *Exp Neurol* 69:85–98.

Cajal SR. (1899). Estudios sobre la corteza cerebral humana II: Estructura de la corteza motriz del hombre y mamíferos superiores. *Rev Trim Micrográf* 4: 117–200.

Cajal SR. (1909). *Histologie du Système Nerveux de L'homme et des Vertébrés*, Vol. II. Paris: Maloine.

Callaway EM, Katz LC. (1991). Effects of binocular deprivation on the development of clustered horizontal connections in cat striate cortex. *Proc Natl Acad Sci USA* 88:745–749.

Calvin WH, Sypert GW. (1976). Fast and slow pyramidal tract neurons: an intracellular analysis of their contrasting repetitive firing properties in the cat. *J Neurophysiol* 39:420–434.

Calvin WH, Sypert GW, Ward AA Jr. (1968). Structured timing patterns within bursts from epileptic neurons in undrugged monkey cortex. *Exp Neurol* 21:535–549.

Calvin WH, Ojemann GA, Ward AA Jr. (1973). Human cortical neurons in epileptogenic foci: comparison of interictal firing patterns to those of "epileptic" neurons in animals. *Electroencephalogr Clin Neurophysiol* 34:337–351.

Campbell AW. (1905). *Histological Studies on the Localization of Cerebral Function*. London: Cambridge University Press.

Cannon SC, Strittmater SM. (1993). Functional expression of sodium channel mutations identified in families with periodic paralysis. *Neuron* 10:317–326.

Capogna M, Gähwiler BH, Thompson SM. (1993). Mechanism of m-opioid receptor-mediated presynaptic inhibition in the rat hippocampus in vitro. *J Physiol* (Lond) 470:539–558.

Carey LJ, Fisken RA, Powell TPS. (1973). Observations on the growth of cells in the lateral geniculate nucleus of the cat. *Brain Res* 52:359–362.

Carlin RK, Grab DJ, Cohen RS, Siekevitz P. (1980). Isolation and characterization of postsynaptic densities from various brain regions: enrichment of different types of postsynaptic densities. *J Cell Biol* 86:831–843.

Carmignoto G, Vicini S. (1992). Activity-dependent decrease in NMDA receptor responses during development of the visual cortex. *Science* 258:1007–1011.

Casaccia-Bonnefil P, Benedikz E, Shen H, Stelzer A, Edelstein D, Geschwind M, Brownlee M, Federoff HJ, Bergold PJ. (1993). Localized gene transfer into organotypic hippocampal slice cultures and acute hippocampal slices. *J Neurosci Meth* 50:341–351.

Caskey CT, Pizzutti A, Fu YH, Fenwick RG, Nelson DL. (1992). Triplet repeat mutations in human disease. *Science* 256:784–788.

Catarsi S, Drapeau P. (1993). Tyrosine kinase-dependent selection of transmitter responses induced by neuronal contact. *Nature* 363:353–356.

Cauller LJ, Connors BW. (1992). Functions of very distal dendrites: experimental and computational studies of layer I synapses on neocortical pyramidal cells. In: McKenna T, Davis J, Zornetzer SF, eds. *Single Neuron Computation.* San Francisco: Academic Press: 199–230.

Cauller LJ, Connors BW. (1994). Synaptic physiology of horizontal afferents to layer I of primary somatosensory cortex in rats. *J Neurosci* 14:751–762.

Cavazos JE, Sutula TP. (1990). Progressive neuronal loss induced by kindling: a possible mechanism for mossy fiber synaptic reorganization and hippocampal sclerosis. *Brain Res* 527:1–6.

Cavazos JE, Golarai G, Sutula TP. (1991). Mossy fiber synaptic reorganization induced by kindling: time course of development, progression, and permanence. *J Neurosci* 11:2795–2803.

Cavazzuti GB, Cappella L, Nalin A. (1980). Longitudinal study of epileptiform EEG patterns in normal children. *Epilepsia* 21:43–55.

Caviness VS. (1976). Patterns of cell and fiber distribution in the neocortex of the reeler mouse. *J Comp Neurol* 170:435–448.

Caviness VS. (1982). Neocortical histiogenesis in normal and reler mice: a developmental study based on [^{3}H] thymidine autoradiography. *Dev Brain Res* 4:293–302.

Cepko C. (1988). Retrovirus vectors and their applications in neurobiology. *Neuron* 1:345–353.

Chagnac-Amitai Y, Connors BW. (1989a). Synchronized excitation and inhibition driven by intrinsically bursting neurons in neocortex. *J Neurophysiol* 62: 1149–1162.

Chagnac-Amitai Y, Connors BW. (1989b). Horizontal spread of synchronized activity in neocortex and its control by GABA-mediated inhibition. *J Neurophysiol* 61:747–758.

Chagnac-Amitai Y, Luhmann HJ, Prince DA. (1990). Burst generating and regular spiking layer 5 pyramidal neurons of rat neocortex have different morphological features. *J Comp Neurol* 296:598–613.

Chan PH, Longar S, Fishman RA. (1987). Protective effects of liposome-entrapped superoxide dismutase on posttraumatic brain edema. *Ann Neurol* 21:540–547.

Charnet P, Labarca C, Leonard RJ, Vogelaar NJ, Czyzyk L, Gouin A, Davidson N, Lester HA. (1990). An open-channel blocker interacts with adjacent turns of a-helices in the nicotinic acetylcholine receptor. *Neuron* 2:87–95.

Charpak S, Gähwiler BH. (1991). Glutamate mediates a slow synaptic response in hippocampal slice culture. *Proc Roy Soc Lond* 243:221–226.

Charpak S, Gähwiler BH, Do KQ, Knöpfel T. (1990). Potassium conductances in hippocampal neurons blocked by excitatory amino acid transmitters. *Nature* 347:765–767.

Chavez RA, Hall ZW. (1991). The transmembrane topology of the amino terminus of the a subunit of the nicotinic acetylcholine receptor. *J Biol Chem* 266: 15532–15538.

Chavez RA, Hall ZW. (1992). Expression of fusion proteins of the nicotinic acetylcholine receptor from mammalian muscle identifies the membrane-spanning regions in the alpha and delta subunits. *J Cell Biol* 116:385–393.

Chavez-Noriega LE, Stevens CF. (1994). Increased transmitter release at excitatory synapses produced by direct activation of adenylate cyclase in rat hippocampal slices. *J Neurosci* 14:310–317.

Chen L, Huang L. (1992). Protein kinase C reduces Mg^{2+} block of NMDA-receptor channels as a mechanism of modulation. *Nature* 356:521–523.

Chen LS, Weingart JB, McNamara JO. (1990). Biochemical and radiohistochemical analyses of alpha-2 adrenergic receptors in the kindling model of epilepsy. *J Pharmacol Exp Ther* 253:1272–1277.

Chen QX, Wong RKS. (1991). Intracellular Ca^{2+} suppressed a transient potassium current in hippocampal neurons. *J Neurosci* 11:337–343.

Chen QX, Stelzer A, Kay AR, Wong RKS. (1990). $GABA_A$ receptor function is regulated by phosphorylation in acutely dissociated guinea-pig hippocampal neurons. *J Physiol* (Lond) 420:207–221.

Cherlow DG, Dymond AM, Crandall PH, Walter RD, Serafetinides EA. (1977). Evoked response and after-discharge thresholds to electrical stimulation in temporal lobe epileptics. *Arch Neurol* 34:527–531.

Cherubini E, Herrling PL, Lanfumey L, Stanzione P. (1988). Excitatory amino acids in synaptic excitation of rat striatal neurones in vitro. *J Physiol* (Lond) 400:677–690.

Cherubini E, Gaiarsa JL, Ben-Ari Y. (1991). GABA: an excitatory transmitter in early postnatal life. *Trends Neurosci* 14:515–519.

Chervin RD, Pierce PA, Connors BW. (1988). Periodicity and directionality in the propagation of epileptiform discharges across neocortex. *J Neurophysiol* 60:1695–1713.

Choi DW. (1988). Glutamate neurotoxicity and diseases of the nervous system. *Neuron* 1:623–634.

Choi D. (1992a). Excitotoxic cell death. *J Neurobiol* 23:1261–1276.

Choi DW. (1992b). Bench to bedside: the glutamate connection. *Science* 258:241–243.

Choi DW, Rothman SM. (1990). The role of glutamate neurotoxicity in hypoxic-ischemic neuronal death. *Annu Rev Neurosci* 13:17–182.

Choi E, Xia Z, Villacres EC, Storm DR. (1993). The regulatory diversity of the mammalian adenyly cyclases. *Curr Opin Cell Biol* 5:269–273.

Chow KL, Baumbach HD, Glanzman DL. (1978). Abnormal development of lateral geniculate neurons in rabbit subjected to either eyelid closure or corticofugal paroxysmal discharges. *Brain Res* 146:151–158.

Christofi G, Nowicky AV, Bolsover SR, Bindman LJ. (1993). The postsynaptic induction of nonassociative long-term depression of excitatory synaptic transmission in rat hippocampal slices. *J Neurophysiol* 69:219–229.

Clark WM, Coull BM. (1994). Randomized trail of CGS19755, a glutamate antagonist, in acute ischemic stroke treatment. *Neurology* 44 (suppl 2):A270.

Clements JD, Lester RAJ, Tong G, Jahr CE, Westbrook GL. (1992). The time course of glutamate in the synaptic cleft. *Science* 258:1498–1501.

Code RA, Winer JD. (1985). Commisural neurons in layer III of cat primary auditory cortex (AI): pyramidal and non-pyramidal cell input. *J Comp Neurol* 242:485–510.

Coenen AML, Vendrik AJH. (1972). Determination of the transfer ratio of cat's geniculate neurons through quasi-intracellular recordings and the relation with the level of alertness. *Exp Brain Res* 14:227–242.

Cohen GA, Doze VA, Madison DV. (1992). Opioid inhibition of GABA release from presynaptic terminals of rat hippocampal interneurons. *Neuron* 9: 325–335.

Collard KJ, Edwards R, Liu Y. (1993). Changes in synaptosomal glutamate release during postnatal development in the rat hippocampus and cortex. *Dev Brain Res* 71:37–43.

Collingridge G. (1987). Synaptic plasticity. The role of NMDA receptors in learning and memory. *Nature* 330:604–605.

Collingridge GL, Bliss T. (1987). NMDA receptors, their role in long term potentiation. *Trends Neurosci* 10:288–293.

Collingridge GL, Kehl SJ, McLennan H. (1983). Excitatory amino acids in synaptic transmission in the Schaffer collateral-commissural pathway of the rat hippocampus. *J Physiol* (Lond) 334:33–46.

Collingridge GL, Herron CE, Lester RAJ. (1988a). Frequency-dependent N-methyl-D-aspartate receptor-mediated synaptic transmission in rat hippocampus. *J Physiol* (Lond) 399:301–312.

Collingridge GL, Herron CE, Lester RAJ. (1988b). Synaptic activation of N-methyl-D-aspartate receptors in the Schaffer collateral-commisural pathway of rat hippocampus. *J Physiol* (Lond) 399:283–300.

Colmers WF, Lukowiak K, Pittman QJ. (1987). Presynaptic action of neuropeptide Y in area CA1 of the rat hippocampal slice. *J Physiol* (Lond) 383:285–299.

Colmers WF, Lukowiak K, Pittman QJ. (1988). Neuropeptide Y action in the rat hippocampal slice: site and mechanism of presynaptic inhibition. *J Neurosci* 8:3827–3837.

Conel JL. (1939). *The Postnatal Development of the Human Cerebral Cortex.* Vol. I. Cambridge: Harvard University Press.

Conel JL. (1941). *The Postnatal Development of the Human Cerebral Cortex.* Vol. II. Cambridge: Harvard University Press.

Conel JL. (1947). *The Postnatal Development of the Human Cerebral Cortex.* Vol. III. Cambridge: Harvard University Press.

Conn PJ, Patel J. (1994). *The Metabotropic Glutamate Receptors.* Totowa, NJ: Humana Press, Inc.

Connor JA, Stevens CF. (1971). Prediction of repetitive firing behavior from voltage clamp data on an isolated neurone soma. *J Physiol* 213:31–53.

Connors BW. (1984). Initiation of synchronized neuronal bursting in neocortex. *Nature* 310:685–687.

Connors BW. (1992). $GABA_A$- and $GABA_B$-mediated processes in visual cortex. In: Mize RR, Marc R, Sillito A, eds. *Mechanisms of GABA in the Visual System. Prog Brain Res* 90:335–348.

Connors BW, Amitai Y. (1993). Generation of epileptiform discharge by local circuits of neocortex. In: Schwartzkroin PA, ed. *Epilepsy: Models, Mechanisms and Concepts.* New York: Cambridge University Press: 388–423.

Connors BW, Gutnick MJ. (1990). Intrinsic firing patterns of diverse neocortical neurons. *Trends Neurosci* 13:99–104.

Connors BW, Malenka RC, Silva LR. (1988). Two inhibitory postsynaptic potentials, and $GABA_A$ and $GABA_B$ receptor-mediated responses in neocortex of rat and cat. *J Physiol* (Lond) 406:443–468.

Connors BW, Gutnick MJ, Prince DA. (1982). Electrophysiological properties of neocortical neurons in vitro. *J Neurophysiol* 48:1302–1320.

Connors BW, Benardo LS, Prince DA. (1983). Coupling between neurons of the developing rat neocortex. *J Neurosci* 3:773–782.

Constantine-Paton M, Cline HT, Debski E. (1990). Patterned activity synaptic convergence and the NMDA receptor in developing visual pathways. *Annu Rev Neurosci* 13:129–154.

Conti FA, Rustioni A, Petrusz P, Towle AC. (1987). Glutamate-positive neurons in the somatic sensory cortex of rats and monkeys. *J Neurosci* 7:1887–1901.

Cornish SM, Wheal HV. (1989). Long-term loss of paired pulse inhibition in the kainic acid-lesioned hippocampus of the rat. *Neuroscience* 28:563–571.

Coulter DA. (1992a). Pentylenetetrazol effects on mouse thalamocortical rhythms and synaptic potentials in vitro. *Epilepsia* 33 (suppl 3):21(abstr).

Coulter DA. (1992b). Physiological studies of thalamocortical rhythms, recorded in vitro in a brain slice preparation. *Soc Neurosci Abstr* 18:1391.

Coulter DA. (1994). Neurophysiological studies in animal models of absence. In: Duncan JS, Panaylotopoulos CP, eds. *Typical Absences and Related Syndromes.* New York: Churchill-Livingstone Press.

Coulter DA, Lee C. (1993). Thalamocortical rhythm generation *in vitro*: extra- and intracellular recordings in mouse thalamocortical slices perfused with low Mg^{2+} medium. *Brain Res* 631:137–142.

Coulter DA, Zhang Y. (1993). Differential dependence of two distinct types of generalized epileptiform discharges on intact thalamocortical connections in mouse thalamocortical slices. *Epilepsia* (abstr).

Coulter DA, Zhang Y. (1994). Thalamocortical rhythm generation in vitro: physiological mechanisms, pharmacological control and relevance to generalized absence epilepsy. In: Malafosso A, Genton P, Hirsch E, Marescaux D, Broglin D, Bernasconi R, eds. *Idiopathic Generalized Epilepsies: Clinical, Experimental, and Genetic Aspects.* London: John Libbey Press.

Coulter DA, Huguenard JR, Prince DA. (1989a). Characterization of ethosuximide reduction of low-threshold calcium current in thalamic neurons. *Ann Neurol* 25:582–593.

Coulter DA, Huguenard JR, Prince DA. (1989b). Calcium currents in rat thalamocortical relay neurones: kinetic properties of the transient, low-threshold current. *J Physiol* (Lond) 414:587–604.

Coulter DA, Huguenard JR, Prince DA. (1989c). Specific petit mal anticonvulsants reduce calcium currents in thalamic neurons. *Neurosci Lett* 98:74–78.

Coulter DA, Huguenard JR, Prince DA. (1990). Differential effects of petit mal anticonvulsants and convulsants on thalamic neurones. *Br J Pharmacol* 100:800–806.

Coyle JT, Schwarcz R. (1976). Lesion of striatal neurones with kainic acid provides a model for Huntington's chorea. *Nature* 263:244–246.

Crabtree JW, Chow KL, Ostrach LH, Baumbach HD. (1981). Development of receptive field properties in the visual cortex of rabbits subjected to early epileptiform cortical discharges. *Brain Res* 227:269–281.

Crabtree JW, Ostrach LH, Campbell BG, Chow KL. (1983). Long-term effects of early cortical epileptiform activity on development of visuocortical receptor fields in the rabbit. *Dev Brain Res* 9:1–9.

Cragg BG. (1975). The development of synapses in the visual system of the cat. *J Comp Neurol* 160:147–166.

Crain BJ, Chang K, McNamara JO. (1987). An *in vitro* autoradiographic analysis of mu and delta opioid binding in the hippocampal formation of kindled rats. *Brain Res* 412:343–351.

Crandall PH, Walter RD, Rand RW. (1963). Clinical applications of studies on sterotactically implanted electrodes in temporal-lobe epilepsy. *J Neurosurg* 20:827–840.

Creutzfeldt O. (1956). Reaktionen einzelner Neuronen des morotischen Cortex der Katze auf elektrische Reize. III. Bahnung der Primarentladung und konvulsive Nachentladungen nach Schienreizen. *Arch Psychiat Nerrenkrank* 194:205–217.

Crunelli V, Leresche N. (1991). A role for $GABA_B$ receptors in excitation and inhibition of thalamocortical cells. *Trends Neurosci* 14:16–21.

Crunelli V, Lightowler S, Pollard CE. (1989). A T-type Ca^{2+} current underlies low-threshold Ca^{2+} potentials in cells of the cat and rat lateral geniculate nucleus. *J Physiol* (Lond) 413:543–561.

Cummins TR, Zhou J, Sigworth FJ, Ukomadu U, Stephen M, Ptacek LJ, Leppert MF, Agnew WS. (1993). Functional consequences of a sodium channel mutation causing hyperkalemic periodic paralysis. *Neuron* 10:667–678.

Curran T, Morgan JI. (1991). Stimulus transcription coupling in the nervous system: involvement of the inducible protooncogenes fos and jun. *Annu Rev Neurosci* 14:421–451.

Cusick CG, Wall JT, Whiting JH, Wiley RG. (1990). Temporal progression of cortical reorganization following nerve injury. *Brain Res* 537:355–358.

Dash PK, Karl KA, Colicos MA, Prywes R, Kandel ER. (1991). cAMP response element-binding protein is activated by Ca^{2+}/calmodulin- as well as cAMP-dependent protein kinase. *Proc Natl Acad Sci USA* 88:5061–5065.

Dasheiff RM, McNamara JO. (1982). Intradentate colchicine retards the development of amygdala kindling. *Ann Neurol* 11:347–352.

Davenport CJ, Brown WJ, Babb TL. (1990). Sprouting of GABAergic and mossy fiber axons in dentat gyrus following intrahippocampal kainate in the rat. *Exp Neurol* 109:180–190.

Davies CH, Starkey SJ, Possa MF, Collingridge GL. (1991). $GABA_B$ autoreceptors regulate the induction of LTP. *Nature* 349:609–611.

Davies SN, Lester RAJ, Reymann KG, Collingridge GL. (1989). Temporally distinct pre- and postsynaptic mechanisms maintain long-term potentiation. *Nature* 338:500–503.

Daw NW, Stein PSG, Fox K. (1993). The role of NMDA receptors in information processing. *Annu Rev Neurosci* 16:207–222.

Debski EA, Cline HT, Constantine-Paton M. (1990). Activity-dependent turning and the NMDA receptor. *J Neurobiol* 21:18–32.

DeCamilli P. (1993). Exocytosis goes with a SNAP. *Nature* 364:387–388.

DeCarlos JA, Lopez-Mascaraque L, Valverde F. (1985). Development, morphology and topography of chandelier cells in the auditory cortex of the cat. *Dev Brain Res* 22:293–300.

DeFelipe J. (1993). Neocortical neuronal diversity: chemical heterogeneity revealed by colocalization studies of classic neurotransmitters, neuropeptides, calcium-binding proteins, and cell surface molecules. *Cereb Cortex* 3:273–289.

DeFelipe J, Fairen A. (1982). A type of basket cells in superficial layers of the cat visual cortex. A Golgi-electron microscope study. *Brain Res* 244:9–16.

DeFelipe J, Farinas I. (1992). The pyramidal neuron of the cerebral cortex: morphological and chemical characteristics of the synaptic inputs. *Progr Neurobiol* 39:563–607.

DeFelipe J, Jones EG. (1988). *Cajal on the Cerebral Cortex*. Oxford: Oxford University Press.

DeFelipe J, Jones EG. (1992). High-resolution light and electron microscopic immunocytochemistry of colocalized GABA and calbindin D-28k in somata and double bouquet cell axons of monkey somatosensory cortex. *Eur J Neurosci* 4:46–60.

DeFelipe J, Hendry SHC, Jones EG, Schmechel D. (1985). Variability in the terminations of GABAergic chandelier cell axons on initial segments of pyramidal cell axons in the monkey sensory-motor cortex. *J Comp Neurol* 231:364–384.

DeFelipe J, Conley M, Jones EG. (1986). Long-range focal collateralization of axons arising from corticocortical cells in monkey sensory-motor cortex. *J Neurosci* 6:3749–3766.

DeFelipe J, Hendry SHC, Hashikawa T, Molinari M, Jones EG. (1990). A microcolumnar structure of monkey cerebral cortex revealed by immunocytochemical studies of double bouquet cell axons. *Neuroscience* 37:655–673.

Degen R, Degen HE, Kuhlmann HP. (1988). Die Rolandi-Epilepsie-ein häufiges Anfallsleiden des Kindesalters. *Nervenarzt* 59:19–25.

Deisz RA, Prince DA. (1989). Frequency-dependent depression of inhibition in guinea pig neocortex in vitro by $GABA_B$ receptor feed-back on GABA release. *J Physiol* (Lond) 412:513–542.

Deisz RA, Fortin G, Zieglgänsberger W. (1991). Voltage dependence of excitatory postsynaptic potentials of rat neocortical neurons. *J Neurophysiol* 65: 371–382.

Deitch JS, Smith KC, Lee CL, Swann JW, Turner JN. (1990). Confocal scanning laser microscope images of hippocampal neurons intracellularly labeled with biocytin. *J Neurosci Meth* 33:61–76.

De Koninck Y, Mody I. (1994). Noise analysis of miniature IPSCs in adult rat brain slices; properties and modulation of synaptic $GABA_A$ receptor channels. *J Neurophysiol* 71:1318–1335.

de Lanerolle NC, Kim JH, Robbins RJ, Spencer DD. (1989). Hippocampal interneuron loss and plasticity in human temporal lobe epilepsy. *Brain Res* 495: 387–395.

de Lanerolle NC, Brines M, Williamson A, Kim JH, Spencer DD. (1992). Neurotransmitters and their receptors in human temporal-lobe epilepsy. *Epilepsy Res* 7:235–250.

Deschenes M. (1981). Dendritic spikes induced in fast pyramidal tract neurons by thalamic stimulation. *Exp Brain Res* 43:304–308.

Deschenes M, Labelle A, Landry P. (1979). Morphological characterization of slow and fast pyramidal tract cells in the cat. *Brain Res* 178:251–274.

Deschenes M, Roy JP, Steriade M. (1982). Thalamic bursting mechanism: an inward slow current revealed by membrane hyperpolarization. *Brain Res* 239:289–293.

Diamond IT. (1979). The subdivisions of neocortex: a proposal to revise the traditional view of sensory, motor and association areas. *Progr Psychobiol Psychol* 8:1–42.

Dichter MA, Ayala GF. (1987). Cellular mechanisms of epilepsy: a status report. *Science* 237:157–164.

Dichter M, Spencer WA. (1969a). Penicillin-induced interictal discharges from the cat hippocampus. I. Characteristics and topographical features. *J Neurophysiol* 32:649–662.

Dichter M, Spencer WA. (1969b). Penicillin-induced interictal discharges from the cat hippocampus. II. Mechanisms underlying origin and restriction. *J Neurophysiol* 32:663–687.

Dickinson E. (1960). *The Complete Poems of Emily Dickinson.* Boston: Little, Brown.

DiFrancesco D. (1993). Pacemaker mechanisms in cardiac tissue. *Annu Rev Physiol* 55:455–472.

Dimlich RVW, Tornheim PA, Kindel RM, Hall ED, Braughler JM, McCall JM. (1990). Effects of a 21-aminosteroid (U-74006F) on cerebral metabolities and edema after severe experimental head trauma. *Adv Neurol* 52:365–375.

Dingledine R. (1984). *Brain Slices.* New York: Plenum Press.

Dingledine R, Gjerstad L. (1980). Reduced inhibition during epileptiform activity in the vitro hippocampal slice. *J Physiol* (Lond) 305:297–313.

Dingledine R, Hynes MA, King GL. (1986). Involvement of N-methyl-D-aspartate receptors in epileptiform burst firing in the rat hippocampal slice. *J Physiol* (Lond) 380:175–189.

Dingledine R, Hume RI, Heinemann SF. (1992). Structural determinants of barium permeation and rectification in non-NMDA glutamate receptor channels. *J Neurosci* 12:4080–4087.

DiPaola M, Kao PN, Karlin A. (1990). Mapping the alpha-subunit site photolabeled by the noncompetitive inhibitor [3H]quinacrine azide in the active state of the nicotinic acetylcholine receptor. *J Biol Chem* 265:11017–11029.

Djorup A, Jahnsen H, Laursen AM. (1981). The dendritic response to GABA in CA1 of the hippocampal slice. *Brain Res* 219:196–201.

Dodt H, Pawelzik H, Zieglgänsberger W. (1991). Actions of noradrenaline on neocortical neurons in vitro. *Brain Res* 545:307–311.

Dolphin AC, Errington ML, Bliss TVP. (1982). Long-term potentiation of the perforant path *in vivo* is associated with increased glutamate release. *Nature* 297:496–498.

Donoghue JP, Kitai ST. (1981). A collateral pathway to the neostriatum from corticofugal neurons of the rat sensory-motor cortex: an intracellular HRP study. *J Comp Neurol* 201:1–13.

Douglas RJ, Martin KAC. (1990a). Control of neuronal output by inhibition at the axon initial segment. *Neural Comp* 2:283–292.

Douglas RJ, Martin KAC. (1990b). Neocortex. In: Shepherd G, ed. *Synaptic Organization of the Brain.* New York: Oxford University Press: 220–248.

Douglas RJ, Martin KAC. (1991). A functional microcircuit for cat visual cortex. *J Physiol* (Lond) 440:735–769.

Douglas RJ, Martin KAC. (1992). Exploring cortical microcircuits: a combined anatomical, physiological, and computational approach. In: McKenna T, Davis J, Zornetzer SF, eds. *Single Neuron Computation.* San Diego: Academic Press: 381–412.

Douglas RK, Martin KAC, Whitteridge D. (1988). Selective responses of visual cortical cells do not depend on shunting inhibition. *Nature* 332:642–644.

Douglas RJ, Martin KA, Whitteridge D. (1991). An intracellular analysis of the visual responses of neurones in cat visual cortex. *J Physiol* (Lond) 440: 659–696.

Doze VA, Cohen GA, Madison DV. (1991). Synaptic localization of adrenergic disinhibition in the rat hippocampus. *Neuron* 6:889–900.

Dragunow M, Goulding M, Faull RL, Ralph R, Mee E, Frith R. (1990). Induction of c-fos mRNA and protein in neurons and glia after traumatic brain injury: pharmacological characterization. *Exp Neurol* 107:236–248.

Dudek FE, Snow RW, Taylor CP. (1986). Role of electrical interactions in synchronization of epileptiform bursts. In: Delgado-Escueta AV, Ward AA, Woodbury DM, Porter RJ, eds. *Advances in Neurology.* New York: Raven Press: 593–617.

Dudek SM, Bear MF. (1992). Homosynaptic long-term depression in area CA1 of the hippocampus and effects of N-methyl-D-aspartate receptor blockade. *Proc Natl Acad Sci USA* 89:4363–4367.

Dugich-Djordjevic MM, Tocco G, Willoughby DA, Najm I, Pasinetti G, Thompson RF, Baudry M, Lapchak PA, Hefti F. (1992). BDNF mRNA expression in the developing rat brain following kainic acid-induced seizure activity. *Neuron* 8:1127–1138.

Dun FT, Feng TP. (1940). Studies on the neuromuscular junction. XIX. Retrograde discharges from motor nerve endings in veratrinized muscle. *Clin J Physiol* 15:405–432.

Durand GM, Bennett MVL, Zukin RS. (1993). Splice variants of the N-methyl-D-aspartate receptor NR1 identify domains involved in regulation by polyamines and protein kinase C. *Proc Natl Acad Sci USA* 90:6731–6735.

Durand GM, Gregor P, Zheng X, Bennett MV, Uhl GR, Zukin RS. (1992). Cloning of an apparent splice variant of the rat N-methyl-D-aspartate receptor NMDAR1 with altered sensitivity to polyamines and activators of protein kinase C. *Proc Natl Acad Sci USA* 89:9359–9363.

Dutar P, Nicoll RA. (1988). A physiological role for $GABA_B$ receptors in the central nervous system. *Nature* 332:156–158.

Dykes RW, Lamour Y, Diadori P, Landry P, Dutar P. (1988). Somatosensory cortical neurons with identifiable electrophysiological signature. *Brain Res* 441:48–59.

Ebersole JS, Chatt AB. (1986). Spread and arrest of seizures: The importance of layer 4 in laminar interactions during neocortical epileptigenesis. In: Delgado-Escueta AV, Ward AA, Woodbury DM, Porter RJ, eds. *Advances in Neurology*, Vol 44. New York: Raven Press: 515–558.

Eccles JC. (1957). *The Physiology of Nerve Cells.* Baltimore: Johns Hopkins Press.

Echlin FA. (1959). The supersensitivity of chronically "isolated" cerebral cortex as a mechanism in focal epilepsy. *Electroencephalogr Clin Neurophysiol* 11:697–722.

Eckhorn R, Bauer R, Jordan W, Brosch M, Kruse W, Munk M, Reitboeck HJ. (1988). Coherent oscillations: a mechanism for feature linking in the visual cortex? *Biol Cyber* 60:121–130.

Edwards FA. (1991). LTP is a long term problem. *Nature* 350:217–272.

Edwards FA, Konnerth A, Sakmann B, Takahashi T. (1989). A thin slice preparation for patch clamp recordings from neurones of the mammalian central nervous system. *Pflügers Arch* 414:600–612.

Edwards FA, Konnerth A, Sakmann B. (1990). Quantal analysis of inhibitory synaptic transmission in the dentate gyrus of rat hippocampal slices: a patch-clamp study. *J Physiol* (Lond) 430:213–249.

Egebjerg J, Heinemann SF. (1993). Ca^{2+} permeability of unedited and edited versions of the kainate selective glutamate receptor GluR6. *Proc Natl Acad Sci USA* 90:755–759.

Egebjerg J, Bettler B, Hermans-Borgmeyer I, Heinemann S. (1991). Cloning of a cDNA for a glutamate receptor subunit activated by kainate but not AMPA. *Nature* 351:745–748.

Elkabes S, Cherry JA, Schoups AA, Black IB. (1993). Regulation of protein kinase C activity by sensory deprivation in the olfactory and visual symptoms. *J Neurochem* 60:1835–1842.

Engel AK, Konig P, Kreiter AK, Shillen TB, Singer W. (1992). Temporal coding in the visual cortex: new vistas on integration in the nervous system. *Trends Neurosci* 15:218–226.

Engel J Jr. (1983). Functional localization of epileptogenic lesions. *Trends Neurosci* 6:60–65.

Erdö SL, Wolff JR. (1990a). Postnatal development of the excitatory amino acid system in visual cortex of the rat. Changes in uptake and levels of asparate and glutamate. *Int J Dev Neurosci* 8:205–208.

Erdö SL, Wolff JR. (1990b). Postnatal development of the excitatory amino acid system in visual cortex of the rat. Changes in ligand binding to NMDA, quisqualate and kainate receptors. *Int J Dev Neurosci* 8:199–204.

Ernfors P, Bengzon J, Kokaia Z, Persson H, Lindvall O. (1991). Increased levels of messenger RNAs for neurotrophic factors in the brain during kindling epileptogenesis. *Neuron* 7:165–176.

Evans GA. (1989). Dissecting mouse development with toxigenics. *Genes Dev* 3: 259–263.

Eysel UT, Muche T, Worgotter F. (1988). Lateral interactions at directing selective striate neurones in the cat demonstrated by local cortical inactivation. *J Physiol* (Lond) 399:657–675.

Faber DS, Young WS, Legendre P, Korn H. (1992). Intrinsic quantal variability due to stochastic properties of receptor-transmitter interactions. *Science* 258:1494–1498.

Fairen A, Valverde F. (1980). A specialized type of neuron in the visual cortex of cat: a Golgi and electron microscope study of chandelier cells. *J Comp Neurol* 194:761–779.

Fairen A, DeFelipe J, Martinez-Ruiz R. (1981). The Golgi-EM procedure: a tool to study neocortical interneurons. In: *Eleventh International Congress of Anatomy: Glial and Neuronal Cell Biology.* New York: Alan R Liss: 291–301.

Fairen A, DeFelipe J, Regidor J. (1984). Nonpyramidal neurons: general account. In: Peters A, Jones EG, eds. *The Cerebral Cortex. Vol I, Cellular Components of the Cerebral Cortex.* New York: Plenum Press: 201–253.

Farinas I, DeFelipe J. (1991a). Patterns of synaptic input on corticocortical and corticothalamic cells in the cat visual cortex. I. The cell body. *J Comp Neurol* 304:53–69.

Farinas I, DeFelipe J. (1991b). Patterns of synaptic input on corticocortical and corticothalamic cells in the cat visual cortex. II. The axon initial segment. *J Comp Neurol* 304:70–77.

Farrant M, Feldmeyer D, Takahashi T, Cull-Candy SG. (1994). NMDA receptor channel diversity in the developing cerebellum. *Nature* 368:335–339.

Fazeli MS, Corbet J, Dunn MJ, Dolphin AC, Bliss TVP. (1993). Changes in protein synthesis accompanying long term potentiation in the dentate gyrus in vivo. *J Neurosci* 13:1346–1353.

Feldman M. (1984). Morphology of the neocortical pyramidal neuron. In: Peters A, Jones EG, eds. *Cerebral Cortex, Vol. 1. Cellular Components of the Cerebral Cortex*. New York: Plenum: 123–200.

Feldman ML, Peters A. (1978). The forms of non-pyramidal neurons in the visual cortex of the rat. *J Comp Neurol* 179:761–794.

Ferster D. (1986). Orientation selectivity of synaptic potentials in neurons of cat primary visual cortex. *J Neurosci* 6:1284–1301.

Ferster D, Jagadeesh B. (1992). EPSP-IPSP interactions in cat visual cortex studied with in vivo whole-cell patch recording. *J Neurosci* 12:1262–1274.

Ferster D, Lindstrom S. (1983). An intracellular analysis of geniculocortical connectivity in area 17 of the cat. *J Physiol* (Lond) 342:181–215.

Finch DM, Jackson MB. (1990). Presynaptic enhancement of synaptic transmission in hippocampal cell cultures by phorbol esters. *Brain Res* 518:269–273.

Fink DJ, Sternberg LR, Weber PC, Matam M, Goins WF, Glorioso JC. (1992). In vivo expression of b-galactosidase in hippocampal neurons by HSV-mediated gene transfer. *Hum Gene Ther* 3:11–19.

Finley JC, Grossman GH, Dimeo P, Petrusz P. (1978). Somatostatin-containing neurons in the rat brain: widespread distribution revealed by immunocytochemistry after pretreatment with pronase. *Am J Anat* 153:483–488.

Finley JC, Maderdrut JL, Roger LJ, Petrusz P. (1981). The immunocytochemical localization of somatostatin-containing neurons in the rat central nervous system. *Neuroscience* 6:2173–2192.

Fish KJ. (1988). Microembolization: etiology and prevention. In: Hilberman M, ed. *Brain Injury and Protection During Heart Surgery.* Boston: Martinus Nijhoff Publishing: 67–83.

Fleshman JW, Segev I, Burke RE. (1988). Electrotonic architecture of type-identified alpha-motoneurons in the cat spinal cord. *J Neurophysiol* 60:60–85.

Flint AC, Connors BW. (1993). Two distinct forms of synchronous rhythms mediated by different glutamate receptors and separate populations of neurons in neocortex. *Soc Neurosci Abstr* 19:20.

Foehring RC, Waters RS. (1991). Contributions of low-threshold calcium current and anomalous rectifier (I_h) to slow depolarizations underlying burst firing in human neocortical neurons in vitro. *Neurosci Lett* 124:17–21.

Foehring RC, Wyler AR. (1990). Two patterns of firing in human neocortical neurons. *Neurosci Lett* 110:279–285.

Foehring RC, Schwindt PC, Crill WE. (1989). Norepinephrine selectively reduces slow Ca^{2+}- and Na^{+}-mediated K^{+} currents in cat neocortical neurons. *J Neurophysiol* 61:245–256.

Foehring RC, Lorenzon NM, Herron P, Wilson CJ. (1991). Correlation of physiologically and morphologically identified neuronal types in human association cortex in vitro. *J Neurophysiol* 66:1825–1837.

Forsythe ID, Clements JD. (1990). Presynaptic glutamate receptors depress excitatory monosynaptic transmission between mouse hippocampal neurones. *J Physiol* (Lond) 429:1–16.

Fosse VM, Heggelund P, Fonnum F. (1989). Postnatal development of glutamatergic, GABAergic and cholinergic neurotransmitter phenotypes in the visual cortex, lateral geniculate nucleus, pulvinar and superior colliculus in cats. *J Neurosci* 9:426–435.

Foutz AS, Champagnat J, Denavit-Saubie M. (1988). N-methyl-D-aspartate (NMDA) receptors control respiratory off-switch in cat. *Neurosci Lett* 87: 221–226.

Fox K, Sato H, Daw N. (1989). The location and function of NMDA receptors in cat and kitten visual cortex. *J Neurosci* 9:2433–2454.

Fox K, Sato H, Daw N. (1990). The effect of varying stimulus intensity on NMDA-receptor activity in cat visual cortex. *J Neurophysiol* 64:1413–1429.

Fox K, Daw N, Sato H, Czepita D. (1992). The effect of visual experience on development of NMDA receptor synaptic transmission in kitten visual cortex. *J Neurosci* 12:2672–2684.

Franceschetti S, Buzio S, Sancini G, Panzica F, Avanzini G. (1993). Expression of intrinsic bursting properties in neurons of maturing sensorimotor cortex. *Neurosci Lett* 162:25–28.

Franck JE. (1993). Cell death plasticity, and epilepsy: insights provided by experimental models of hippocampal sclerosis. In: Schwartzkroin PA, ed. *Epilepsy: Models, Mechanisms and Concepts.* Cambridge: Cambridge University Press: 281–303.

Franck JE, Schwartzkroin PA. (1985). Do kainate-lesioned hippocampi become epileptogenic? *Brain Res* 329:309–313.

Franck JE, Kunkel DD, Basking DG, Schwartzkroin PA. (1988). Inhibition in kainate-lesioned hyperexcitable hippocampi: physiologic, autoradiographic and immunocytochemical evidence. *J Neurosci* 8:1991–2002.

Fregnac Y, Imbert M. (1984). Development of neuronal selectivity in primary visual cortex of cat. *Physiol Rev* 64:325–434.

French CR, Sah P, Buckett KJ, Gage PW. (1990). A voltage-dependent persistent sodium current in mammalian hippocampal neurons. *J Gen Physiol* 95: 1139–1145.

Freund TF, Antal M. (1988). GABA-containing neurons in the septum control inhibitory interneurons in the hippocampus. *Nature* 336:170–173.

Freund TF, Martin KAC, Smith AD, Somogyi P. (1983). Glutamate decarboxylase-immunoreactive terminals of Golgi-impregnated axoaxonic cells and of presumed basket cells in synaptic contact with pyramidal neurons of the cat's visual cortex. *J Comp Neurol* 221:263–278.

Friauf E, Shatz CJ. (1991). Changing patterns of synaptic input to subplate and cortical plate during development of visual cortex. *J Neurophysiol* 66: 2059–2071.

Fried I, Kim J, Spencer DD. (1992). Hippocampal pathology in patients with intractable seizures and temporal lobe masses. *J Neurosurg* 76:735–740.

Friedman A. (1991). *Active and Passive Properties of Neocortical Neurons and Their Role in Determining Neuronal Firing Pattern.* Beersheva, Israel: Ben-Gurion University. Thesis.

Friedman A, Gutnick MJ. (1987). Low-threshold calcium electrogenesis in neocortical neurons. *Neurosci Lett* 81:117–122.

Friedman A, Gutnick MJ. (1989). Intracellular calcium and control of burst generation in neurons of guinea-pig neocortex in vitro. *Eur J Neurosci* 1:374–381.

Friedman A, Arens J, Heinemann U, Gutnick MJ. (1992). Slow depolarizing afterpotentials in neocortical neurons are sodium and calcium dependent. *Neurosci Lett* 135:13–17.

Frost JD Jr, Hrachovy RA, Glaze DG. (1992). Spike morphology in childhood focal epilepsy: relationship to syndromic classification. *Epilepsia* 33:5531–5536.

Frotscher M. (1985). Mossy fibres form synapses with identified pyramidal basket cells in the CA3 region of the guinea-pig hippocampus: a combined Golgi electron microscope study. *J Neurocytol* 14:245–259.

Frush DP, Giacchino JL, McNamara JO. (1986). Evidence implicating dentate granule cells in development of entorhinal kindling. *Exp Neurol* 92:92–101.

Fugita Y. (1979). Evidence for the existence of inhibitory postsynaptic potentials in dendrites and their functional significance in hippocampal pyramidal cells of adult rabbits. *Brain Res* 175:59–69.

Fukuda A, Prince DA. (1992). Postnatal development of electrogenic sodium pump activity in rat hippocampal neurons. *Dev Brain Res* 65:101–114.

Fukunaga K, Soderling TR, Miyamoto E. (1992). Activation of calcium/calmodulin-dependent protein kinase II and protein kinase C by glutamate in cultured rat hippocampal neurons. *J Biol Chem* 267:22527–22533.

Fukunaga K, Stoppini L, Miyamoto E, Muller D. (1993). Long-term potentiation is associated with an increased activity of Ca^{2+}/calmodulin-dependent protein kinase. *J Biol Chem* 268:7863–7867.

Gabbott PLA, Martin KAC, Whitteridge D. (1987). Connections between pyramidal neurons in layer 5 of cat visual cortex (area 17). *J Comp Neurol* 259: 364–381.

Gähwiler BH, Brown DA. (1985). $GABA_B$-receptor-activated K^+ current in voltage-clamped CA3 pyramidal cells in hippocampus. *Proc Natl Acad Sci USA* 82:1558–1562.

Gaiarsa JL, Corradetti R, Cherubini E, Ben-Ari Y. (1990). The allosteric glycine site of the N-methyl-D-aspartate receptor modulates GABAergic-mediated synaptic events in neonatal rat CA3 hippocampal neurons. *Proc Natl Acad Sci USA* 87:343–346.

Gall CM, Lauterborn J. (1992). The dentate gyrus: a model system for studies of neurotrophin regulation. *Epilepsy Res* 7(suppl):171–185.

Gall CM, Sumikawa K, Lynch G. (1990). Levels of mRNA for a putative kainate receptor are affected by seizures. *Proc Natl Acad Sci USA* 87:7643–7647.

Gall CM, Lauterborn J, Bundman M, Murray K, Isackson P. (1991). Seizures and the regulation of neurotrophic factor and neuropeptide gene expression in brain. In: Anderson E, Leppik, eds. *Genetic Strategies in Epilepsy Research.* Amsterdam: Elsevier.

Gamrani H, Harandi M, Dubois MP, Calas A. (1984). Direct electron microscopic evidence for the coexistence of GABA uptake and endogenous serotonin in the same rat central neurons by coupled radioautographic and immunocytochemical procedures. *Neurosci Lett* 48:25–30.

Gao W, Liu XL, Hatten ME. (1992). The weaver gene encodes a non-autonomous signal for CNS neuronal differentiation. *Cell* 68:841–854.

Geddes JW, Chang-Chui H, Cooper SM, Lott IT, Cotman CW. (1986). Density and distribution of NMDA receptors in the human hippocampus in Alzheimer's disease. *Brain Res* 399:156–161.

Geddes JW, Cahan LD, Cooper SM, Choi BH, Kim RC, Cotman CW. (1990). Altered distribution of excitatory amino acid receptors in temporal epilepsy. *Exp Neurol* 108:214–220.

Ghose GM, Freeman RD. (1992). Oscillatory discharge in the visual system: does it have a functional role? *J Neurophysiol* 68:1558–1574.

Gibbs EL, Gillen HW, Gibbs FA. (1954). Disappearance and migration of epileptic foci in childhood. *Am J Dis Child* 88:596–603.

Gilbert CD. (1992). Horizontal integration and cortical dynamics. *Neuron* 9:1–13.

Gilbert CD. (1993). Circuitry, architecture, and functional dynamics of visual cortex. *Cereb Cortex* 3:373–386.

Gilbert CD, Wiesel TN. (1983). Clustered intrinsic connections in cat visual cortex. *J Neurosci* 3:1116–1133.

Gilbert CD, Wiesel TN. (1992). Receptive field dynamics in adult primary visual cortex. *Nature* 356:150–152.

Gilbert CS. (1983). Microcircuitry of the visual cortex. *Annu Rev Neurosci* 6: 217–247.

Ginsberg MD. (1993). Emerging strategies for the treatment of ischemic brain injury. In: Waxman SG, ed. *Molecular and Cellular Approaches to the Treatment of Neurological Disease.* New York: Raven Press.

Giraudat J, Dennis M, Heidman T, Haumont P, Lederer F, Changeux J. (1987). Structure of the high-affinity binding site for noncompetitive blockers of the acetylcholine receptor: [^{3}H]chlorpromazine labels homologous residues of the beta and delta chains. *Biochemistry* 26:2410–2418.

Giraudat J, Galzi J, Revah F, Changeux J, Haumont P, Lederer F. (1989). The noncompetitive block [^{3}H]cholopromazine labels segment M2 but not segment M1 of the nicotinic acetylcholine receptor alpha-subunit. *FEBS Lett* 253:190–198.

Gisselmann G, Sewing S, Madsen BW, Mallart A, Angaut-Pettit D, Mulle-Holtkamp F, Ferrus A, Pongs O. (1989). The interference of truncated with normal potassium channel subunits leads to abnormal behavior in transgenic drosophila. *EMBO J* 8:2359.

Gloor P, Fariello RG. (1988). Generalized epilepsy: some of its cellular mechanisms differ from those of focal epilepsy. *Trends Neurosci* 11:63–68.

Goddard GV. (1967). The development of epileptic seizures through brain stimulation at low intensity. *Nature* 214:1020–1021.

Goddard GV, McIntyre EC, Leech CK. (1969). A permanent change in brain function resulting from daily electrical stimulation. *Exp Neurol* 25:295–330.

Goldensohn ES, Purpura DP. (1963). Intracellular potentials of cortical neurons during focal epileptogenic discharges. *Science* 139:840–842.

Goldowitz D, Koch J. (1986). Performance of normal and neurological mutant mice on radial arm maze and active avoidance tasks. *Behav Neural Biol* 46:216–226.

Goodman CS, Shatz CJ. (1993). Developmental mechanisms that generate precise patterns of neuronal connectivity. *Cell* 72:77–98.

Grafstein B, Sastry PB. (1957). Some preliminary electrophysiological studies on chronic neuronally isolated cerebral cortex. *Electroencephalogr Clin Neurophysiol* 9:723–725.

Grant SGN, O'Dell TJ, Karl KA, Stein PL, Soriano P, Kandel ER. (1993). Impaired long term potentiation, spatial learning, and hippocampal development in fyn mutant mice. *Science* 258:1903–1910.
Gray CM, König P, Engel AK, Singer W. (1989). Oscillatory responses in cat visual cortex exhibit inter-columnar synchronization which reflects global stimulus properties. *Nature* 338:334–337.
Gray CM, Engel AK, König P, Singer W. (1992). Synchronization of oscillatory neuronal responses in cat striate cortex: temporal properties. *Vis Neurosci* 8:337–347.
Gregory DL, Wong PK. (1984). Topographical analysis of the centrotemporal discharges in benign rolandic epilepsy of childhood. *Epilepsia* 25:705–711.
Green WN, Ross AF, Claudio T. (1991). Acetylcholine receptor assembly is stimulated by phosphorylation of its gamma subunit. *Neuron* 7:659–666.
Greengard P, Jen J, Nairn AC, Stevens CF. (1991). Enhancement of the glutamate response by cAMP-dependent protein kinase in hippocampal neurons. *Science* 253:1135–1137.
Greengard P, Valtorta F, Czernik AJ, Benfenati F. (1993). Synaptic vesicle phosphoproteins and regulation of synaptic function. *Science* 259:780–785.
Greenwood RS, Godar S, Reaves TA, Hayward JN. (1981). Cholecystokinin in hippocampal pathways. *J Comp Neurol* 203:335–350.
Grigonis AM, Murphy EH. (1994). The effects of epileptic cortical activity on the development of callosal projections. *Dev Brain Res* 77:251–255.
Grisar T, Guillaume D, Delgado-Escueta AV. (1992). Contribution of Na^+, K^+-ATPase to focal epilepsy: a brief review. *Epilepsy Res* 12:141–149.
Grossman Y, Parnas I, Spira ME. (1979a). Differential conduction block in branches of a bifurcating axon. *J Physiol* (Lond) 295:283–305.
Grossman Y, Parnas I, Spira ME. (1979b). Mechanism involved in differential conduction of potentials at high frequency in a branching axon. *J Physiol* (Lond) 295:307–322.
Grover LM, Teyler TJ. (1990). Two components of long-term potentiation induced by different patterns of afferent activation. *Nature* 347:477–479.
Grover LM, Lambert NA, Schwartzkroin PA, Teyler TJ. (1993). Role of HCO_3-ions in depolarizing $GABA_A$ receptor-mediated responses in pyramidal cells of rat hippocampus. *J Neurophysiol* 69:1541–1555.
Groves AK, Barnett SC, Franklin RJM, Crang AJ, Mayer M, Blakemore WF, Noble M. (1993). Repair of demyelinated lesions by transplantation of purified O-2A progenitor cells. *Nature* 362:453–455.
Grundfest H. (1966). Heterogeneity of excitable membrane: electrophysiological and pharmacological evidence and some consequences. *Ann NY Acad Sci* 137:901–949.
Gruner JE, Hirsch JC, Sotelo C. (1974). Ultrastructural features of the isolated suprasylvian gyrus. *J Comp Neurol* 154:1–27.
Guberman A, Gloor P, Sherwin AL. (1975). Response of generalized penicillin epilepsy in the cat to ethosuximide and diphenylhydantoin. *Neurology* 25: 758–764.
Gunel M, Thomas PG, Cornell-Bell AH, Brines ML, Spencer DD, DeLanerolle N. (1991). Human glia cultured from epileptogenic foci demonstrate increased basal and glutamate-induced calcium fluxes. *Epilepsia* 32 (suppl 3):66.
Gustafsson B. (1979). Changes in motoneurone electrical properties following axotomy. *J Physiol* (Lond) 293:197–215.

Gustafsson B, Wigström H, Abraham WC, Huang Y. (1987). Long-term potentiation in the hippocampus using depolarizing current pulses as the conditioning stimulus to single volley synaptic potentials. *J Neurosci* 7:774–780.

Guthrie PB, Segal M, Kater SB. (1991). Independent regulation of calcium revealed by imaging dendritic spines. *Nature* 354:76–80.

Gutierrez AR, Brick JF, Bodensteiner J. (1990). Dipole reversale: an ictal feature of benign partial epilepsy with centrotemporal spikes. *Epilepsia* 31:544–548.

Gutnick MJ, Friedman A. (1986). Synaptic and intrinsic mechanisms of synchronization and epileptogenesis in the neocortex. *Exp Brain Res* 114(suppl): 327–335.

Gutnick MJ, Prince DA. (1972). Thalamocortical relay neurons: antidromic invasion of spikes from cortical epileptogenic focus. *Science* 176:424–426.

Gutnick MJ, Prince DA. (1974). Effects of projected cortical epileptiform discharges on neuronal activities in cat VPLI. Interictal discharge. *J Neurophysiol* 37:1310–1327.

Gutnick MJ, Prince DA. (1981). Dye-coupling and possible electrotonic coupling in the guinea pig neocortical slice. *Science* 211:67–70.

Gutnick MJ, Connors BW, Prince DA. (1982). Mechanisms of neocortical epileptogenesis in vitro. *J Neurophysiol* 48:1321–1335.

Haas HL. (1982). Cholinergic disinhibition in hippocampal slices of rat. *Brain Res* 233:200–204.

Hablitz JJ. (1988). Spontaneous ictal-like discharges and sustained potentials shifts in the developing rat neocortex. *J Neurophysiol* 58:1052–1065.

Hablitz JJ, Johnston D. (1981). Endogenous nature of spontaneous bursting in hippocampal pyramidal neurons. *Cell Mol Neurobiol* 1:325–334.

Hablitz JJ, Langmoen IA. (1986). N-methyl-D-aspartate receptor antagonists reduce synaptic excitation in the hippocampus. *J Neurosci* 7:102–106.

Haglund MM, Schwartzkroin PA. (1990). Role of Na-K pump potassium regulation and IPSPs in seizures and spreading depression in immature rabbit hippocampal slices. *J Neurophysiol* 63:225–239.

Haglund MM, Stahl WL, Kunkel DD, Schwartzkroin PA. (1985). Developmental and regional differences in the localization of Na, K-ATPase activity in the rabbit hippocampus. *Brain Res* 343:198–203.

Haglund MM, Berger MS, Kunkel DD, Franck JE, Ghatan S, Ojemann GA. (1992). Changes in gamma-aminobutyric acid and somatostatin in epileptic cortex associated with low-grade gliomas. *J Neurosurg* 77:209–216.

Halasy K, Somogyi P. (1993). Distribution of GABAergic synapses and their targets in the dentate gyrus of rat: a quantitative immunoelectron microscopic analysis. *J Hirnforsch* 34:299–308.

Hall ED. (1987). Beneficial effects of the 21-aminosteroid U74006F in acute CNS trauma and hypovolemic shock. *Acta Anaesth Belg* 38:421–425.

Hall ED. (1992). Novel inhibitors of iron-dependent lipid peroxidation for neurodegenerative disorders. *Ann Neurol* 32:137–142.

Hall ED, Pazara KE, Braughler JM. (1991). Effect of tirilazad mesylate on postischemic lipid peroxidation and recovery of extracellular calcium in gerbils. *Stroke* 22:361–366.

Halliwell B. (1989). Oxidants and the central nervous system: some fundamental questions. *Acta Neurol Scand* 126:23–33.

Halpern LM. (1972). Chronically isolated aggregates of mammalian cerebral cortical neurons studied in situ. In: Purpura DP, Penry JK, Tower D, Woodbury DM, Walter R, eds. *Experimental Models of Epilepsy—A Manual for the Laboratory Worker.* New York: Raven Press: 197–221.

Hamill OP, Marty A, Neher E, Sakmann B, Sigworth FJ. (1981). Improved patch-clamp techniques for high resolution current recording from cells and cell-free membrane patches. *Pfluegers Arch* 391:85–100.

Hamill OP, Huguenard JR, Prince DA. (1991). Patch-clamp studies of voltage-gated currents in identified neurons of the rat cerebral cortex. *Cereb Cortex* 1:48–61.

Hamlyn LH. (1963). An electron microscope study of pyramidal neurons in the Ammon's horn of the rabbit. *J Anat* 97:189–210.

Hamos JE, Davis TL, Sterling P. (1983). Four types of neurons in layer IVab of cat cortical area 17 accumulate 3H-GABA. *J Comp Neurol* 217:449–457.

Hanson PI, Schulman H. (1992). Neuronal Ca^{2+}/calmodulin-dependent protein kinases. *Annu Rev Biochem* 61:559–601.

Hanson PI, Meyer T, Stryer L, Schulman H. (1994). Dual role of calmodulin in autophosphorylation of multifunctional CaM kinase may underlie decoding of calcium signals. *Neuron* 12:943–956.

Hara M, Inoue M, Yasakura T, Ohnishi S, Mikami Y, Inagaki C. (1992). Uneven distribution of intracellular Cl^- in rat hippocampal neurons. *Neurosci Lett* 143:135–138.

Harada K, Yoshimura T, Nakjima K, Ito H, Ebina Y, Shingai R. (1992). N-methyl-D-aspartate increases cytosolic Ca^{2+} via G proteins in cultured hippocampal neurons. *Am J Physiol* 262:C870–C875.

Hardiman O, Burke T, Phillips J, Murphy S, et al. (1988). Microdysgenesis in resected temporal neocortex: incidence and clinical significance in focal epilepsy. *Neurology* 38:1041–1047.

Harris EW, Cotman CW. (1986). Long-term potentiation of guinea pig mossy fiber responses is not blocked by N-methyl-D-aspartate antagonists. *Neurosci Lett* 70:132–137.

Hauser WA, Annegers JF, Kurland LT. (1991). Prevalence of epilepsy in Rochester, Minnesota: 1940–1980. *Epilepsia* 32:429–445.

Hazelton B, Mitchell B, Tupper J. (1979). Calcium, magnesium, and growth control in the W1-38 human fibroblast cell. *J Cell Biol* 83:487–498.

Hebb DO. (1949). *The Organization of Behavior.* New York: John Wiley & Sons.

Heijbel J. (1976). Benign epilepsy of children with centro-temporal EEG foci. Clinical, genetic and neurophysiological studies. *Umeå University Medical Dissertations*, New Series No. 17, 31 pp.

Heijbel J, Blom S, Rasmuson M. (1975). Benign epilepsy of childhood with centro-temporal EEG foci: a genetic study. *Epilepsia* 16:285–293.

Heinemann U, Lux HD, Gutnick MJ. (1977). Extracellular free calcium and potassium during paroxysmal activity in the cerebral cortex of the cat. *Exp Brain Res* 27:237–243.

Heinemann U, Konnerth A, Pumain R, Waldman WJ. (1986). Extracellular calcium and potassium concentration changes in chronic epileptic brain tissue. *Adv Neurol* 44:41–61.

Helekar S, Noebels JL. (1991). Synchronous hippocampal bursting unmasks latent network excitability alterations in a epileptic gene mutation. *Proc Natl Acad Sci USA* 88:4736–4740.

Helekar SA, Noebels JL. (1992). A burst dependent excitability defect elicited by potassium at the developmental onset of spike-wave seizures in the tottering mutant. *Dev Brain Res* 65:205–210.

Helekar S, Noebels JL. (1993). Analysis of voltage-gated and synaptic conductances contributing to a gene-linked prolongation of depolarizing shifts in the epileptic mutant mouse tottering. *J Neurophysiol* 71:1–10.

Hendry SHC, Jones EG. (1981). Size and distribution of neurons incorporating tritiated GABA in monkey sensory-motor cortex. *J Neurosci* 1:390–408.

Hendry SHC, Jones EG. (1986). Reduction in number of immunostained GABAergic neurones in deprived-eye dominance columns of monkey area 17. *Nature* 320:750–753.

Hendry SHC, Jones EG. (1988). Activity-dependent regulation of GABA expression in the visual cortex of adult monkeys. *Neuron* 1:701–712.

Hendry SHC, Kennedy MB. (1986). Immunoreactivity for a calcodulin-dependent protein kinase is selectively increased in macaque striate cortex after monocular deprivation. *Proc Natl Acad Sci USA* 83:1536–1540.

Hendry SHC, Jones EG, DeFelipe J, Schmechel D, Brandon C, Emson PC. (1984a). Neuropeptide-containing neurons of the cerebral cortex are also GABAergic. *Proc Natl Acad Sci USA* 81:6526–6530.

Hendry SHC, Jones EG, Emson PC. (1984b). Morphology distribution and synaptic relations of somatostatin and neuropeptide y-immunoreactive neurons in rat and monkey neocortex. *J Neurosci* 4:2497–2517.

Hendry SHC, Schwark HD, Jones EG, Yan J. (1987). Numbers and proportions of GABA-immunoreactive neurons in different areas of monkey cerebral cortex. *J Neurosci* 7:1503–1519.

Hendry SHC, Jones EG, Burstein N. (1988a). Activity-dependent regulation of tachykinin-like immunoreactivity in neurons of monkey visual cortex. *J Neurosci* 8:1225–1238.

Hendry SHC, Jones EG, Hockfield S, McKay RDG. (1988b). Neuronal populations stained with the monoclonal antibody Cat-301 in the mammalian cerebral cortex and thalamus. *J Neurosci* 8:518–542.

Hendry SHC, Jones EG, Emson PC, Lawson DEM, Heizmann CW, Streit P. (1989). Two classes of cortical GABA neurons defined by differential calcium binding protein immunoreactivities. *Exp Brain Res* 76:467–472.

Hendry SHC, Fuchs J, deBlas AL, Jones EG. (1990). Distribution and plasticity of immunocytochemically localized $GABA_A$ receptors in adult monkey visual cortex. *J Neurosci* 10:2438–2450.

Hepler JR, Gilman AG. (1992). G proteins. *Trends Biochem Sci* 17:383–387.

Herdegen T, Sandkuhler J, Gass P, Kiessling M, Bravo R, Zimmerman M. (1993). JUN, FOS, KROX, CREB transcription factor proteins in the rat cortex: basal expression and induction by spreading depression and epileptic seizures. *J Comp Neurol* 333:271–288.

Herlitze S, Raditsch M, Ruppersberg JP, Jahn W, Monyer H, Schoepfer R, Witzemann V. (1993). Argiotoxin detects molecular differences in AMPA receptor channels. *Neuron* 10:1131–1140.

Hernandez-Cruz A, Pape H-C. (1989). Identification of two calcium currents in acutely dissociated neurons from the rat lateral geniculate nucleus. *J Neurophysiol* 61:1270–1283.

Hersch SM, White EL. (1982). A quantitative study of the thalamocortical and other synapses in layer 4 of pyramidal cells projecting from mouse SmI cortex to the caudate-putamen nucleus. *J Comp Neurol* 211:217–255.
Hessler NA, Shirker AM, Malinow R. (1993). The probability of transmitter release at a mammalian central synapse. *Nature* 366:569–572.
Hestrin S, Sah P, Nicoll RA. (1990). Mechanisms generating the time course of dual component excitatory synaptic currents recorded in hippocampal slices. *Neuron* 5:247–253.
Hestrin S. (1992a). Developmental regulation of NMDA receptor-mediated synaptic currents at a central synapse. *Nature* 357:686–689.
Hestrin S. (1992b). Activation and desensitization of glutamate-activated channels mediating fast excitatory synaptic currents in the visual cortex. *Neuron* 9: 991–999.
Hestrin S. (1993). Different glutamate receptor channels mediate fast excitatory synaptic currents in inhibitory and excitatory cortical neurons. *Neuron* 11: 1083–1091.
Heyer C, Llinás R. (1977). Control of rhythmic firing in normal and axotomized cat spinal motoneurons. *J Neurophysiol* 40:480.
Higashi H, Tanaka E, Nishi S. (1991). Synaptic responses of guinea-pig cortical neurons in vitro. *J Neurophysiol* 65:822–833.
Higuchi M, Single FN, Kohler M, Sommer B, Sprengel R, Seeburg PH. (1993). RNA editing of AMPA receptor subunit GluR-B: a base-paired intron-exon structure determines position and efficiency. *Cell* 75:1361–1370.
Hille B. (1992). G protein-coupled mechanisms and nervous signaling. *Neuron* 9: 187–195.
Hillman D, Chen S, Aung TT, Cherksey B, Sugimori M, Llinás R. (1991). Localization of P-type calcium channels in the central nervous system. *Proc Natl Acad Sci USA* 88:7076–7080.
Hinds HL, Ashley CT, Sutcliffe JS, Nelson DL, Warren ST, Houseman DE, Schalling M. (1993). Tissue-specific expression of FMR-1 provides evidence for a functional role in fragile X syndrome. *Nature Gen* 36:43.
Hirsch JA, Gilbert CD. (1993). Long-term changes in synaptic strength along specific intrinsic pathways in the cat visual cortex. *J Physiol* (Lond) 461:247–262.
Hirsch JC, Crepel F. (1990). Use-dependent changes in synaptic efficacy in rat prefrontal neurons in vitro. *J Physiol* (Lond) 427:31–49.
Hirsch JC, Crepel F. (1991). Blockade of NMDA receptors unmasks a long-term depression in synaptic efficacy in rat prefrontal neurons *in vitro*. *Exp Brain Res* 85:621–624.
Hirsch JC, Crepel F. (1992). Postsynaptic calcium is necessary for the induction of LTP and LTD of monosynaptic EPSPs in prefrontal neurons: an *in vitro* study in the rat. *Synapse* 10:173–175.
Hirsch JC, Fourment A, Marc ME. (1983). Sleep-related variations of membrane potential in the lateral geniculate body relay neurons of the cat. *Brain Res* 259:308–312.
Hockfield S, McKay R. (1985). Identification of major cell classes in the developing mammalian nervous system. *J Neurosci* 5:3310–3329.
Hodgkin AL, Huxley AF. (1952). A quantitative description of membrane current and its application to conduction and excitation in nerve. *J Physiol* (Lond) 117:500–544.

Hoffman SN, Prince DA. (1995). Epileptogenesis in immature neocortical brain slices induced by 4-aminopyridine. *Dev Brain Res*, in press.

Hoffman SN, Salin PA, Prince DA. (1994). Chronic neocortical epileptogenesis in vitro. *J Neurophysiol* 71:1762–1773.

Hoffman WH, Haberly LB. (1989). Bursting induces persistent all-or-none EPSPs by an NMDA-dependent process in piriform cortex. *J Neurosci* 91:206–215.

Hollman M, Heinemann S. (1994). Cloned glutamate receptors. *Annu Rev Neurosci* 17:31–108.

Hollmann M, Boulter J, Maron C, Beasley L, Sullivan J, Pecht G, Heinemann S. (1993). Zinc potentiates agonist-induced currents at certain splice variants of the NMDA receptor. *Neuron* 10:943–954.

Holmes WR, Woody CD. (1989). Effects of uniform and non-uniform synaptic "activation distributions" on the cable properties of modeled cortical pyramidal neurons. *Brain Res* 505:12–22.

Hosford DA, Crain BJ, Cai Z, Bonhaus DW, Friedman AH, Okazaki MM, Nadler JV, McNamara JO. (1991). Increased AMPA-sensitive quisqualate receptor binding and reduced NMDA receptor binding in epileptic human hippocampus. *J Neurosci* 11:428–434.

Hosford DA, Clark S, Cao Z, Wilson W Jr, Lin F-H, Morrisett RA, Huin A. (1992). The role of $GABA_B$ receptor activity in absence seizures of lethargic (lh/lh) mice. *Science* 257:398–401.

Hotson JR, Prince DA. (1981). Penicillin- and barium-induced epileptiform bursting in hippocampal neurons: actions on Ca^{++} and K^+ potentials. *Ann Neurol* 10:11–17.

Hotson JR, Prince DA, Schwartzkroin PA. (1979). Anomalous inward rectification in hippocampal neurons. *J. Neurophysiol* 42:889–895.

Hounsgaard J, Nicholson C. (1983). Potassium accumulation around individual Purkinje cells in cerebellar slices from the guinea-pig. *J Physiol* (Lond) 340: 359–388.

Houser CR. (1990). Granule cell dispersion in the dentate gyrus of humans with temporal lobe epilepsy. *Brain Res* 535:195–204.

Houser CR, Hendry SHC, Jones EG, Vaughn JE. (1983). Morphological diversity of immunocytochemically identified GABA neurons in the monkey sensory-motor cortex. *J Neurocytol* 12:617–638.

Houser CR, Miyashiro JE, Schwartz BE, Walsh GO, Rich JR, Delgado-Escueta AV. (1990). Altered patterns of dynorphin immunoreactivity suggest mossy fiber reorganization in human hippocampal epilepsy. *J Neurosci* 10:267–282.

Howe JR, Sutor B, Zieglgänsberger W. (1987a). Characteristics of long-duration inhibitory postsynaptic potentials in rat neocortical neurons *in vitro*. *Cell Mol Neurobiol* 7:1–18.

Howe JR, Sutor B, Zieglgänsberger W. (1987b). Baclofen reduces post-synaptic potentials of rat cortical neurones by an action other than its hyperpolarizing action. *J Physiol* (Lond) 384:539–569.

Huang Y, Malenka RC. (1993). Examination of TEA-induced synaptic enhancement in area CA1 of the hippocampus: the role of voltage-dependent Ca^{2+} channels in the induction of LTP. *J Neurosci* 13:568–576.

Hubel DH, Wiesel TN. (1965). Binocular interaction in striate cortex of kittens reared with artificial squint. *J Neurophysiol* 28:1041–1059.

Hubel DH, Wiesel TN. (1970). The period of susceptibility to the physiological effects of unilateral eye closure in kittens. *J Physiol* 206:419–436.

Hucho F, Oberthur W, Lottspeich F. (1986). The ion channel of the nicotinic acetylcholine receptor is formed by the homologous helices M II of the receptor subunits. *FEBS Lett* 205:137–142.

Huguenard JR, Prince DA. (1992a). A novel T-type current underlies prolonged Ca^{2+}-dependent burst firing in GABAergic neurons of rat thalamic reticular nucleus. *J Neurosci* 12:3804–3817.

Huguenard JR, Prince DA. (1992b). Multiple sites of action of succinimides and related anticonvulsants. *Epilepsia* 3(suppl):30.

Huguenard JR, Prince DA. (1994a). Intrathalamic rhythmicity studied *in vitro*: nominal T current modulation causes robust anti-oscillatory effects. *J Neurosci* 14:5485–5502.

Huguenard JR, Prince DA. (1994b). Clonazepam suppresses $GABA_B$ mediated inhibition in thalamic relay neurons through effects in nucleus reticularis. *J Neurophysiol* 71:2576–2581.

Huguenard JR, Hamill OP, Prince DA. (1988). Developmental changes in Na^+ conductances in rat neocortical neurons: appearance of a slowly inactivating component. *J Neurophysiol* 59:778–795.

Huguenard JR, Hamill OP, Prince DA. (1989). Sodium channels in dendrites of rat cortical pyramidal neurons. *Proc Natl Acad Sci USA* 86:2473–2477.

Hume RI, Dingledine R, Heinemann SF. (1991). Identification of a site in glutamate receptor subunits that controls calcium permeability. *Science* 253: 1028–1031.

Humphrey DR, Corrie WS. (1978). Properties of pyramidal tract neuron system within a functionally defined subregion of primate motor cortex. *J Neurophysiol* 41:216–243.

Huntley GW, Jones EG. (1991). Relationship of intrinsic connections to forelimb movement representations in monkey motor cortex: a correlative anatomic and physiological study. *J Neurophysiol* 66:390–413.

Huntsman MM, Isackson PJ, Jones EG. (1994). Lamina-specific expression and activity-dependent regulation of seven $GABA_A$ receptor subunit mRNAs in monkey visual cortex. *J Neurosci* 14:2236–2259.

Hwa GGC, Avoli M. (1992a). Excitatory postsynaptic potentials recorded from regular-spiking cells in layers II/III or rat sensorimotor cortex. *J Neurophysiol* 67:728–737.

Hwa GGC, Avoli M. (1992b). Excitatory synaptic transmission mediated by NMDA and non-NMDA receptors in the superficial/middle layers of the epileptogenic human neocortex maintained in vitro. *Neurosci Lett* 143:83–86.

Iino M, Ozawa S, Tsuzuki K. (1990). Permeation of calcium through excitatory amino acid receptor channels in cultured rat hippocampal neurones. *J Physiol* (Lond) 424:151–165.

Ideka Y, Long DM. (1990). The molecular basis of brain injury and brain edema: the role of oxygen free radicals. *Neurosurgery* 25:1–11.

Ikura M, Clore GM, Fronenborn AM, Zhu G, Klee CB, Bax A. (1992). Solution structure of a calmodulin-target peptide complex by multidimensional NMR. *Science* 256:632–638.

Imoto K, Busch C, Sakmann B, Mishina M, Konno T, Nakai J, Bujo H, Mori Y, Fukuda K, Numa S. (1988). Rings of negatively charged amino acids determine the acetylcholine receptor channel conductance. *Nature* 335:645–648.

Innocenti GM, Fiore L. (1976). Morphological correlates of visual field transformation in the corpus callosum. *Neurosci Lett* 2:245–252.

Inoue M, Hara M, Zeng X, Hirose T, Ohnishi S, Yasukura T, Uriu T, Omori K, Minato A, Inagaki C. (1991). An ATP-driven Cl^- pump regulates Cl^- concentrations in rat hippocampal neurons. *Neurosci Lett* 134:75–78.

Insel TR, Miller LP, Gelhard RE. (1990). The ontogeny of excitatory amino acid receptors in the rat forebrain. I. N-methyl-D-aspartate and quisqualate receptors. *Neuroscience* 35:31–43.

Isaacson JS, Solis JM, Nicoll RA. (1993). Local and diffuse synaptic actions of GABA in the hippocampus. *Neuron* 10:165–175.

Isackson PJ, Huntsman MM, Murray KD, Gall CM. (1991). BDNF mRNA expression is increased in adult rat forebrain after limbic seizures: temporal patterns of induction distinct from NGF. *Neuron* 6:937–948.

Ishii T, Moriyoshi K, Sugihara H, Sakarada K, Kadotani H, Yokoi M, Akazawa C, Shigomoto R, Mizuno N, Masu M. (1993). Molecular characterization of the family of the N-methyl-D-aspartate receptors. *J Biol Chem* 268:2836–2843.

Ishijima B, Hori T, Yoshimasu N, Fukushima T, Hirakawa K, Sekino H. (1975). Neuronal activities in human epileptic foci and surrounding areas. *Electroencephalogr Clin Neurophysiol* 39:643–650.

Isokawa M, Levesque MF. (1991). Increased NMDA responses and dendritic degeneration in human epileptic hippocampal neurons in slices. *Neurosci Lett* 132:212–216.

Isokawa M, Avanzini G, Finch DM, Babb TL, Levesque MF. (1991). Physiologic properties of human dentate granule cells in slices prepared from epileptic patients. *Epilepsy Rev* 9:242–250.

Isokawa M, Levesque MF, Babb TL, Engel J. (1993). Single mossy fiber axonal systems of human dentate granule cells studied in hippocampal slices from patients with temporal lobe epilepsy. *J Neurosci* 13:1511–1522.

Isokawa-Akesson M, Wilson CL, Babb TL. (1987). Diversity in periodic pattern of firing in human hippocampal neurons. *Exp Neurol* 98:137–151.

Ito S, Cherubini E. (1991). Strychnine-sensitive glycine responses of neonatal rat hippocampal neurones. *J Physiol* (Lond) 440:67–83.

Izant JG. (1983). The role of calcium ions during mitosis. *Chromosoma* 88:1–10.

Jack JJ, Redman SJ, Wong K. (1981). The components of synaptic potentials evoked in cat spinal motoneurones by impulses in single group Ia afferents. *J Physiol* (Lond) 321:65–96.

Jackson JH. (1931). Selected writings of John Hughlings Jackson, Vol. 1. In: *On Epilepsy and Epileptiform Convulsions*. London: Hodder and Stoughton.

Jacobs KM, Donoghue JP. (1991). Reshaping the cortical map by unmasking latent intracortical connections. *Science* 251:944–947.

Jacobs KM, Hess G, Connors BW, Donoghue JP. (1995). Intracortical vertical and horizontal response patterns produced by layer V stimulation in rat motor cortex in vitro. *J Neurophysiol* (in press).

Jahnsen H, Llinás R. (1984a). Electrophysiological properties of guinea-pig thalamic neurones: an in vitro study. *J Physiol* (Lond) 349:205–226.

Jahnsen H, Llinás R. (1984b). Ionic basis for the electroresponsiveness and oscillatory properties of guinea-pig thalamic neurones in vitro. *J Physiol* (Lond) 349:227–247.

Jasper HH. (1970). Physiopathological mechanisms of post-traumatic epilepsy. *Epilepsia* 11:73–80.

Jenkins WM, Merzenich M. M, Ochs MT, Allard T, Guic-Robles E. (1990). Functional reorganization of primary somatosensory cortex in adult owl monkeys after behaviorally controlled tactile stimulation. *J Neurophysiol* 63: 82–104.

Jia W, Beaulieu C, Huang FL, Cynader MS. (1990). Protein kinase C immunoreactivity in kitten visual cortex is developmentally regulated and input-dependent. *Brain Res* 57:209–221.

Johnston D. (1976). Voltage clamp reveals basis for calcium regulation of bursting pacemaker potentials in Aplysia neurons. *Brain Res* 107:418–423.

Johnston D, Brown TH. (1981). Giant synaptic potential hypothesis for epileptiform activity. *Science* 211:294–297.

Johnston D, Williams S, Jaffe D, Gray R. (1992). NMDA-receptor-independent long-term potentiation. *Annu Rev Physiol* 54:489–505.

Johnston, MV. (1988). Biochemistry of neurotransmitters in cortical development. In: Peters A, Jones EG, eds. *Cerebral Cortex*. New York: Plenum: 211–236.

Jonas P, Major G, Sakmann B. (1993). Quantal components of unitary EPSCs at the mossy fibre synapse on CA3 pyramidal cells of rat hippocampus. *J Physiol* (Lond) 472:615–663.

Jonas P, Racca C, Sakmann B, Seeburg PH, Monyer H. (1994). Differences in Ca^{2+} permeability of AMPA-type glutamate receptor channels in neocortical neurons caused by differential GluR-B subunit expression. *Neuron* 12:1281–1289.

Jones EG. (1975). Varieties and distribution of non-pyramidal cells in the somatic sensory cortex of the squirrel monkey. *J Comp Neurol* 160:205–268.

Jones EG. (1984). Laminar distribution of cortical efferent cells. In: Peters A, Jones EG, eds. *Cerebral Cortex, Vol 1, Cellular Components of the Cerebral Cortex*. New York: Plenum: 521–556.

Jones EG. (1985). *The Thalamus*. New York: Plenum.

Jones EG. (1993). GABAergic neurons and their role in cortical plasticity in primates. *Cereb Cortex* 3:361–372.

Jones EG, Hendry SHC. (1984). Basket cells. In: Peters A, Jones EG, eds. *The Cerebral Cortex, Vol I, Cellular Components of the Cerebral Cortex*. New York: Plenum: 309–336.

Jones EG, Huntley GW, Benson DL. (1993). Alpha calcium/calmodulin dependent protein kinase II selectively expressed in a subpopulation of excitatory neurons in monkey sensory-motor cortex: comparison with GAB-67 expression. *J Neurosci*.

Jones KA, Baughman RW. (1988). NMDA- and non-NMDA receptor components of excitatory synaptic potentials recorded from cells in layer V of rat visual cortex. *J Neurosci* 8:3522–3534.

Jones KA, Baughman RW. (1991). Both NMDA and non-NMDA subtypes of glutamate receptors are concentrated at synapses on cerebral cortical neurons in culture. *Neuron* 7:593–603.

Jordan SJ, Jeffreys JGR. (1992). Sustained and selective block of IPSPs in brain slices from rats made epileptic by intrahippocampal tetanus toxin. *Epilepsy Res* 11:119–129.

Kaas JH, Merzenich MM, Killackey HP. (1983). The reorganization of somatosensory cortex following peripheral nerve damage in adult and developing mammals. *Annu Rev Neurosci* 6:325–356.

Kaila K, Voipio J. (1987). Postsynaptic fall in intracellular pH induced by GABA-activated bicarbonate conductance. *Nature* 330:163–165.

Kaila K, Pasternack M, Saarikoski J, Voipio J. (1989). Influence of GABA-gated bicarbonate conductance on potential, current and intracellular chloride in crayfish muscle fibres. *J Physiol* (Lond) 416:161–181.

Kajitani T, Ueoka K, Nakamura M, Kumanomidou Y. (1981). Febrile convulsions and rolandic discharges. *Brain Dev* 3:351–359.

Kallen RG, Sheng ZH, Yang J, Chen L, Rogart RB, Barchi RL. (1990). Primary structure and expression of a sodium channel characteristic of denervated and immature rat skeletal muscle. *Neuron* 4:233–242.

Kanaseki T, Ikeuchi Y, Sugiura H, Yamuchi T. (1991). Structural features of Ca^{2+}/calmodulin-dependent protein kinase II revealed by electron microscopy. *J Cell Biol* 115:1049–1060.

Kandel ER, Spencer WA. (1961a). Excitation and inhibition of single pyramidal cells during hippocampal seizure. *Exp Neurol* 4:162–179.

Kandel ER, Spencer WA. (1961b). Electrophysiology of hippocampal neurons. I. Afterpotentials and repetitive firing. *J Neurophysiol* 24:243–259.

Kandel ER, Spencer WA, Brimley FJ. (1961). Electrophysiology of hippocampal neurons. I. Sequential invasion and synaptic organization. *J Neurophysiol* 24:225–242.

Kasper EM, Larkman AU, Lübke J, Blakemore C. (1994a). Pyramidal neurons in layer 5 of the rat visual cortex. I. Correlation among cell morphology, intrinsic electrophysiological properties, and axon targets. *J Comp Neurol* 339:459–474.

Kasper EM, Larkman AU, Lübke J, Blakemore C. (1994b). Pyramidal neurons in layer 5 of the rat visual cortex. II. Development of electrophysiological properties. *J Comp Neurol* 339:459–474.

Kato H, Ito Z, Matusoka S, Sakurai Y. (1973). Electrical activities of neurons in the sliced human cortex in vitro. *Electroencephalogr Clin Neurophysiol* 35: 457–462.

Kato N. (1993). Dependence of long-term depression on postsynaptic metabotropic glutamate receptors in visual cortex. *Proc Natl Acad Sci USA* 90:3650–3654.

Kato N, Artola A, Singer W. (1991). Developmental changes in the susceptibility to long-term potentiation of neurones in rat visual cortex slice. *Dev Brain Res* 60:43–50.

Kato Y. (1977). Clinical electroencephalographic studies on patients with abnormal EEG patterns occurring in the rolandic region. *J Nagoya Med Assoc* 100:31–42.

Katz B. (1969). *The Release of Neural Transmitter Substances.* Liverpool: Liverpool University Press.

Katz B, Miledi R. (1965). The measurements of synaptic delay, and the time course of acetylcholine release as the neuromuscular junction. *Proc Roy Soc Lond* 161:483–695.

Kauer JA, Malenka RC, Nicoll RA. (1988). NMDA application potentiates synaptic transmission in the hippocampus. *Nature* 334:250–252.

Kawaguchi Y. (1992). Receptor subtypes involved in callosally-induced postsynaptic potentials in rat frontal agranular cortex in vitro. *Exp Brain Res* 88: 33–40.

Kawaguchi Y. (1993). Groupings of nonpyramidal and pyramidal cells with specific physiological and morphological characteristics in rat frontal cortex. *J Neurophysiol* 69:416–431.

Kawaguchi Y, Hama K. (1988). Physiological heterogeneity of nonpyramidal cells in rat hippocampal CA1 region. *Exp Brain Res* 72:494–502.

Kawaguchi Y, Kubota Y. (1993). Correlation of physiological subgroupings of nonpyramidal cells with parvalbumin- and calbindin D28k-immunoreactive neurons in layer V of rat frontal cortex. *J Neurophysiol* 70:387–396.

Kawajiri S, Dingledine R. (1993). Multiple structural determinants of voltage-dependent magnesium block in recombinant NMDA receptors. *Neuropharmacology* 32:1203–1211.

Keegan K, Noebels JL. (1993). In vitro electrophysiology of spontaneous and induced epileptiform discharges reveals increased cortical excitability in the mutant mouse, stargazer. *Neurosci Abstr* 19:1031.

Keinanen K, Wisen W, Sommer B, Werner P, Herb A, Verdoorn TA, Sakmann B, Seeburg PH. (1990). A family of AMPA-selective glutamate receptors. *Science* 249:556–560.

Kellaway P. (1975). Afferent input: a critical factor in the ontogenesis of brain electrical activity. In: Burch N, Altshuler HL, eds. *Behavior and Brain Electrical Activity.* New York: Plenum Press: 391–420.

Kellaway P. (1981). The incidence, significance and natural history of spike foci in children. In: Henry CE, ed. *Current Clinical Neurophysiology: Update on EEG and Evoked Potentials.* New York: Elsevier: 151–175.

Kellaway P. (1982). Maturational and biorhythmic changes in the electroencephalogram. In: Anderson VE, Hauser WA, Penry JK, Sing CF, eds. *Genetic Basis of the Epilepsies.* New York: Raven Press: 21–33.

Kellaway P. (1985). Sleep and epilepsy. *Epilepsia* 26:S15–S30.

Kellaway P. (1991). Midline central spikes in 3 Hz spike-and-wave and in benign rolandic epilepsy. *Electroencephalogr Clin Neurophysiol* 79:27P.

Kellaway P, Frost JD Jr. (1983). Biorhythmic modulation of epileptic events. In: Pedley TA, Meldrum BS, eds. *Recent Advances in Epilepsy.* Edinburgh: Churchill Livingstone: 139–154.

Kellaway P, Hrachovy RA. (1983). Status epilepticus in newborns: a perspective on neonatal seizures. In: Delgado-Escueta AV, Wasterlain CG, Treiman DM, et al, eds. *Status Epilepticus: Mechanisms of Brain Damage and Treatment.* New York: Raven Press: 93–99.

Kellaway P, Bloxsom A, MacGregor M. (1955). Occipital foci associated with retrolental fibroplasia and other forms of retinal loss in children. *Electroencephalogr Clin Neurophysiol* 7:469–470.

Kellenberger S, Malherbe P, Sigel E. (1992). Function of the alpha 1 beta 2 gamma 2S gamma-aminobutyric acid type A receptor is modulated by protein kinase C via multiple phosphorylation sites. *J Biol Chem* 267:25660–25663.

Keller A, White EL. (1986). Distribution of glutamic acid decarboxylase-immunoreactive structures in the barrel region of mouse somatosensory cortex. *Neurosci Lett* 66:245–250.

Keller A, White EL. (1987). Synaptic organization of GABAergic neurons in the mouse SmI cortex. *J Comp Neurol* 262:1–12.

Keller BU, Hollmann M, Heinemann S, Konnerth A. (1992). Calcium influx through subunits GluR1/GluR3 of kainate/AMPA channels is regulated by cAMP dependent protein kinase. *EMBO J* 11:891–896.

Kelly KM, Gross RA, MacDonald RL. (1990). Valproic acid selectively reduces the low-threshold (T) calcium current in rat nodose neurons. *Neurosci Lett* 116:233–238.

Kelso SR, Ganong AH, Brown TH. (1986). Hebbian synapses in hippocampus. *Proc Natl Acad Sci USA* 83:5326–5330.

Kennedy MD, Bennett MK, Erondu NE. (1983). Biochemical and immunochemical evidence that the "major postsynaptic density protein" is a subunit of a calmodulin-dependent protein kinase. *Proc Natl Acad Sci USA* 80:7357–7361.

Kim GY, Shatz CJ, McConnell SK. (1991). Morphology of pioneer and follower growth cones in the developing cerebral cortex. *J Neurobiol* 22:629–642.

Kim HG, Connors BW. (1993). Apical dendrites of the neocortex: correlation between sodium- and calcium-dependent spiking and pyramidal cell morphology. *J Neurosci* 13:5301–5311.

Kim JH, Guimaraes PO, Shen MY, Masukawa LM, Spencer DD. (1990). Hippocampal neuronal density in temporal lobe epilepsy with and without gliomas. *Acta Neuropathol* 80:41–45.

Kimura F, Nishigori A, Shirokawa T, Tsumoto T. (1989). Long-term potentiation and N-methyl-D-aspartate receptors in the visual cortex of young rats. *J Physiol* (Lond) 414:125–144.

Kimura F, Tsumoto T, Nishigori A, Yoshimura Y. (1990). Long-term depression but not potentiation is induced in Ca^{2+}-chelated visual cortex neurons. *Neurosci Report* 1:65–68.

King GL, Dingledine R, Giacchino JL, McNamara JO. (1985). Abnormal neuronal excitability in hippocampal slices from kindled rats. *J Neurophysiol* 54: 1295–1304.

Kingsley DM, Rinchik EM, Russell LB, Ottiger H, Sutcliffe JG, Copeland NG, Jenkins NA. (1990). Genetic ablation of a mouse gene expressed specifically in brain. *EMBO J* 9:395–399.

Kirkwood A, Dudek SM, Gold JT, Aisenman CD, Bear MF. (1993). Common forms of synaptic plasticity in hippocampus and neocortex in vitro. *Science* 260:1518–1521.

Kirsch J, Wolters I, Triller A, Betz H. (1993). Gephyrin antisense oligonucleotides prevent glycine receptor clustering in spinal neurons. *Nature* 366:745–748.

Kisvarday SF, Martin KAC, Freund TF, Magloczky ZF, Whitteridge D, Somogyi P. (1986). Synaptic targets of HRP-filled layer III pyramidal cells in the cat striate cortex. *Exp Brain Res* 64:541–552.

Kisvarday ZF, Eysel UT. (1992). Cellular organization of reciprocal patchy networks in layer III of cat visual cortex (area 17). *Neuroscience* 46:275–286.

Kisvarday ZF, Martin KA, Whitteridge D, Somogyi P. (1985). Synaptic connections of intracellularly filled clutch cells: a type of small basket cell in the visual cortex of the cat. *J Comp Neurol* 241:111–137.

Kisvarday ZF, Adams CBT, Smith AD. (1986). Synaptic connections of axo-axonic (chandelier) cells in human epileptic temporal cortex. *Neuroscience* 19: 1179–1186.

Kisvarday ZF, Beaulieu C, Eysel UT. (1993). Network of GABAergic large basket cells in cat visual cortex (area 18): implication for lateral disinhibition. *J Comp Neurol* 327:398–415.

Klar A, Baldassara M, Jessell TM. (1992). F-spondin: a gene expressed at high levels in the floor plate encodes a secreted protein that promotes neuronal cell adhesion and neurite extension. *Cell* 69:95–110.

Klee CB. (1991). Concerted regulation of protein phosphorylation and dephosphorylation by calmodulin. *Neurochem Res* 16:1059–1065.

Kleinschmidt A, Bear MF, Singer W. (1987). Blockade of "NMDA" receptors disrupts experience-dependent plasticity of kitten striate cortex. *Science* 238: 355–358.

Knowles WD, Schwartzkroin PA. (1981). Local circuit synaptic interactions in hippocampal brain slices. *J Neurosci* 1:318–322.

Knowles WD, Awad IA, Nayel MH. (1992). Differences of *in vitro* electrophysiology of hippocampal neurons from epileptic patients with mesiotemporal sclerosis versus structural lesions. *Epilepsia* 33:601–609.

Koch C, Douglas R, Wehmeier U. (1990). Visibility of synaptically induced conductance changes: theory and simulations of anatomically characterized cortical pyramidal cells. *J Neurosci* 10:1728–1744.

Koch MC, Steinmayer K, Lorenz C, Ricker K, Wolf F, Otto M, Zoll B, Lehmann-Horn F, Grzeschik K, Jentsch TJ. (1992). The skeletal muscle chloride channel in dominant and recessive human myotonia. *Science* 257:797–800.

Kohler M, Burnashev N, Sakmann B, Seeburg PH. (1993). Determinants of Ca^{2+} permeability in both TM1 and TM2 of high affinity kainate receptor channels: diversity by RNA editing. *Neuron* 10:491–500.

Köhr G, De Koninck Y, Mody I. (1993). Properties of NMDA receptor channels in neurons acutely isolated from epileptic (kindled) rats. *J Neurosci* 13: 3612–3627.

Koike H, Okada Y, Oshima T. (1968a). Accommodative properties of fast and slow pyramidal tract cells and their modification by different levels of their membrane potential. *Exp Brain Res* 5:189–201.

Koike H, Okada Y, Oshima T, Takahashi K. (1968b). Accommodative behavior of cat pyramidal tract cells investigated with intracellular injection of currents. *Exp Brain Res* 5:173–188.

Komatsu Y, Iwakiri M. (1991). Postnatal development of neuronal connections in cat visual cortex studied by intracellular recording in slice preparation. *Brain Res* 540:14–24.

Komatsu Y, Iwakiri M. (1992). Low-threshold Ca^{2+} channels mediate induction of long-term potentiation in kitten visual cortex. *J Neurophysiol* 67:401–410.

Komatsu Y, Nakajima S, Toyama K. (1991). Induction of long-term potentiation without participation of N-methyl-D-aspartate receptors in kitten visual cortex. *J Neurophysiol* 65:20–32.

Komuro H, Rakic P. (1992). Selective role of N-type calcium channels in neuronal migration. *Science* 257:806–809.

Komuro H, Rakic P. (1993). Modulation of neuronal migration by NMDA receptors. *Science* 260:95–97.

König P, Schillen TB. (1991). Stimulus-dependent assembly formation of oscillatory responses: I. Synchronization. *Neural Comp* 3:155–166.

Konnerth A, Heinemann U, Yaari Y. (1986a). Nonsynaptic epileptogenesis in the mammalian hippocampus in vitro. I. Development of seizure-like activity in low extracellular calcium. *J Neurophysiol* 56:409–423.

Konnerth A, Lux HD, Heinemann U. (1986b). Ionic properties of burst generation in hippocampal pyramidal cell somata in vitro. *Exp Brain Res* 14:368–374.

Korn H, Faber DS. (1991). Quantal analysis and synaptic efficacy in the CNS. *Trends Neurosci* 14:439–445.

Korn H, Bausela F, Charpier S, Faber DS. (1993). Synaptic noise and multiquantal release at dendritic synapses. *J Neurophysiol* 70:1249–1254.

Kornhuber J, Retz W, Riederer P, Heinsen H, Fritze J. (1988). Effect of antemortem and postmortem factors on [^{3}H] glutamate binding in the human brain. *Neurosci Lett* 93:312–317.

Korpi ER, Seeburg PH. (1993). Natural mutation of $GABA_A$ receptor $\alpha 6$ subunit alters benzodiazepine binding affinity but not allosteric GABA effect. *Eur J Pharmacol* 247:23–27.

Kostyuk PG, Molokanova EA, Pronchuk NF, Savchenko AN, Verkhratsky AN. (1992). Different actions of ethosuximide on low- and high-threshold calcium currents in rat sensory neurons. *Neuroscience* 51:755–758.

Kraus JE, Yeh G, Bonhaus DW, Nadler JV, McNamara JO. (1994). Kindling induces the long-lasting expression of a novel population of NMDA receptors in hippocampal region CA3. *J Neurosci* 14:4196–4205.

Kriebel ME, Gross CE. (1974). Multimodal distribution of frog miniature endplate potentials in adult, denervated and tadpole leg muscle. *J Gen Physiol* 64: 85–103.

Kriegstein AR, Suppes T, Prince DA. (1987). Cellular and synaptic physiology and epileptogenesis of developing rat neocortical neurons *in vitro*. *Dev Brain Res* 34:161–171.

Kritzer MF, Cowey A, Somogyi P. (1992). Patterns of inter- and intralaminar GABAergic connections distinguish striate (V1) and extrastriate (V2,V4) visual cortices and their functionally specialized subdivisions in the rhesus monkey. *J Neurosci* 12:4545–4564.

Krnjevic K. (1974). Chemical nature of synaptic transmission in vertebrates. *Physiol Rev* 54:418–540.

Krnjevic K, Phillis JW. (1963). Iontophoretic studies of neurones in the mammalian cerebral cortex. *J Physiol* (Lond) 165:274–304.

Krnjevic K, Randic M, Straughan DW. (1966). Nature of a cortical inhibitory process. *J Physiol* (Lond) 184:49–77.

Kubalek E, Ralston S, Lindstrom J, Unwin N. (1987). Location of subunits within the acetylcholine receptor by electron image analysis of tubular crystals from Torpedo mammorata. *J Cell Biol* 105:9–18.

Kullman DM, Nicoll RA. (1992). Long-term potentiation is associated with increases in quantal content and quantal amplitude. *Nature* 357:240–244.

Kullman DM, Perkel DJ, Manabe T, Nicoll RA. (1992). Ca^{2+} entry via postsynaptic voltage-sensitive Ca^{2+} channels can transiently potentiate excitatory synaptic transmission in the hippocampus. *Neuron* 9:1175–1183.

Kumar A, Schliebs R. (1992). Postnatal laminar development of cholinergic receptors, protein kinase C and dihydrophyridine-sensitive calcium antagonist binding in rat visual cortex. Effect of visual deprivation. *Int J Dev Neurosci* 10:491–504.

Kupfermann I. (1991). Functional studies of cotransmission. *Physiol Rev* 71:683–732.

Lacaille JC. (1991). Postsynaptic potentials mediated by excitatory and inhibitory amino acids in interneurons of stratum pyramidale of the CA1 region of rat hippocampal slices in vitro. *J Neurophysiol* 56:1441–1454.

Lacaille JC, Schwartzkroin PA. (1988a). Stratum lacunosum-moleculare interneurons of hippocampal CA1 region. I. Intracellular response characteristics, synaptic responses, and morphology. *J Neurosci* 8:1400–1410.

Lacaille JC, Schwartzkroin PA. (1988b). Stratum lacunosum-moleculare interneurons of hippocampal CA1 region. II. Intrasomatic and intradendritic recordings of local circuit synaptic interactions. *J Neurosci* 8:1411–1424.

Lacaille JC, Williams S. (1990). Membrane properties of interneurons in stratum oriensalveus of the CA1 region of rat hippocampus *in vitro*. *Neuroscience* 36:349–359.

Lacaille JC, Mueller AL, Kunkel DD, Schwartzkroin PA. (1987). Local circuit interactions between oriens/alveus interneurons and CA1 pyramidal cells in hippocampal slices: electrophysiology and morphology. *J Neurosci* 7:1979–1993.

Laiwand R, Werman R, Yarom Y. (1988). Electrophysiology of degenerating neurones in the vagal motor nucleus of the guinea-pig following axotomy. *J Physiol* (Lond) 404:749–766.

LaMantia AS, Rakic P. (1990). Axon overproduction and elimination in the corpus callosum of the developing rhesus monkey. *J Neurosci* 10:2156–2175.

Lambert NA, Wilson WA. (1993). Heterogeneity in presynaptic regulation of GABA release from hippocampal inhibitory neurons. *Neuron* 11:1057–1067.

Lanaud P, Maggio R, Gale K, Grayson DR. (1993). Temporal and spatial patterns of expression of c-fos, zif/268, c-jun and jun-B mRNAs in rat brain following seizures evoked focally from the deep prepiriform cortex. *Exp Neurol* 119:20–31.

Lancaster B, Adams PR. (1986). Calcium-dependent current generating the afterhyperpolarization of hippocampal neurons. *J Neurophysiol* 55:1268–1272.

Lancaster B, Wheal HV. (1984). Chronic failure of inhibition of the CA1 area of the hippocampus following kainic acid lesions of the CA3/4 area. *Brain Res* 295:317–324.

Lannes B, Micheletti G, Vergnes M, Marescaux C, Depaulis A, Warter JM. (1988). Relationship between spike-wave discharges and vigilance levels in rats with spontaneous petit mal-like epilepsy. *Neurosci Lett* 34:187–191.

Larkman AU. (1991a). Dendritic morphology of pyramidal neurons in the visual cortex of the rat. I. Branching patterns. *J Comp Neurol* 306:306–319.

Larkman AU. (1991b). Dendritic morphology of pyramidal neurones of the visual cortex of the rat. III. Spine distributions. *J Comp Neurol* 306:332–343.

Larkman AU, Mason A. (1990). Correlations between morphology and electrophysiology of pyramidal neurons in slices of rat visual cortex. I. Establishment of cell classes. *J Neurosci* 10:1407–1414.

Larkman AU, Stratford K, Jack J. (1991). Quantal analysis of excitatory synaptic action and depression in hippocampal slices. *Nature* 350:344–347.

Larkman AU, Hannay T, Stratford K, Jack J. (1992). Presynaptic release probability influences the locus of long-term potentiation. *Nature* 360:70–73.

Larson J, Wong D, Lynch G. (1986). Patterned stimulation at the theta frequency is optimal for the induction of hippocampal long-term potentiation. *Brain Res* 368:347–350.

Lauder JM, Han YKM, Henderson P, Verdoorn T, Towle AC. (1986). Prenatal ontogeny of the GABAergic system in the rat brain: an immunocytochemical study. *Neuroscience* 19:465–493.

Laurie DJ, Wisden W, Seeburg PH. (1992). The distribution of thirteen subunits III. Embryonic and postnatal development. *J Neurosci* 12:4151–4172.

Le Gal La Salle G, Robert JJ, Berrard S, Ridoux V, Stratford-Perricaudet LD, Perricaudet M, Mallet J. (1993). An adenovirus vector for gene transfer into neurons and glia in the brain. *Science* 259:988–990.

Lechleiter J, Girard S, Clapham D, Peralta E. (1991). Subcellular patterns of calcium determined by G protein residues of muscarinic receptors. *Nature* 350:505–508.

Lee SM, Weisskopf MG, Ebner FF. (1991). Horizontal long-term potentiation of responses in rat somatosensory cortex. *Brain Res* 544:303–310.

Lehky SR, Sejnowski TJ. (1990). Neural network model of visual cortex for determining surface curvature from images of shaded surfaces. *Proc Roy Soc Lond* B 240:251–278.

Lehky SR, Sejnowski TJ, Desimone R. (1992). Predicting responses of nonlinear neurons in monkey striate cortex to complex patterns. *J Neurosci* 12: 3568–3581.

Leonard RJ, Labarca CG, Charnet P, Davidson N, Lester HA. (1988). Evidence that the M2 membrane-spanning region lines the ion channel pore of the nicotinic receptor. *Science* 242:1578–1581.

Lerea LS, McNamara JO. (1993). Inontropic glutamate receptor subtypes activate c-fos transcription by distinct calcium-requiring intracellular signaling pathways. *Neuron* 10:31–41.

Lerman P, Apter N. (1981). Familiality of benign focal epilepsy of childhood. *Harefuah* 100:160–161.

Lerman P, Kivity-Ephraim S. (1981). Focal epileptic EEG discharges in children not suffering from clinical epilepsy: etiology, clinical significance and management. *Epilepsia* 22:551–558.

Lerman P, Kivity S. (1975). Benign focal epilepsy of childhood. A follow-up study of 100 recovered patients. *Arch Neurol* 32:261–264.

Lester RAJ, Jahr CE. (1992). NMDA channel behavior depends on agonist affinity. *J Neurosci* 12:635–643.

Lester RAJ, Clements JD, Westbrook GL, Jahr CE. (1990). Channel kinetics determine the time course of NMDA receptor-mediated synaptic currents. *Nature* 346:565–567.

Levasseur JE, Patterson J, Ghatak NR, Kontos HA, Choi SC. (1989). Combined effect of respirator-induced ventilation and superoxide dismutase in experimental brain injury. *J Neurosurg* 71:573–577.

LeVay S. (1973). Synaptic patterns in the visual cortex of the cat and monkey. Electron microscopy of Golgi preparations. *J Comp Neurol* 150:53–86.

Levitt P, Cooper ML, Rakic P. (1981). Coexistence of neuronal and glial precursor cells in the cerebral ventricular zone of the fetal monkey: an ultrastructural immunoperoxidase analysis. *J Neurosci* 1:27–39.

Li C. (1955). Functional properties of cortical neurons with particular reference to strychninization. *Electroencephalogr Clin Neurophysiol* 7:475–483.

Li C. (1959). Cortical intracellular potentials and their responses to strychnine. *J Neurophysiol* 22:436–450.

Li C, Jasper H. (1953). Microelectrode studies of the electrical activity of the cerebral cortex in the cat. *J Physiol* (Lond) 121:117–140.

Li S, McInnis MG, Margolis RL, Antonarakis SE, Ross CA. (1993). Novel triplet repeat containing genes in human brain: cloning, expression, and length polymorphisms. *Genomics* 16:572–579.

Li Y, Ersurumlu RS, Chen C, Jhaveri S, Tonegawa S. (1994). Whisker-related neuronal patterns fail to develop in the trigeminal brainstem nuclei of NMDAR1 knockout mice. *Cell* 76:427–437.

Liao D, Jones A, Malinow R. (1992). Direct measurements of quantal changes underlying long-term potentiation in CA1 hippocampus. *Neuron* 9:1089–1097.

Lin C, Lu SM, Schmechel DE. (1986). Glutamic acid decarboxylase and somatostatin immunoreactivities in rat visual cortex. *J Comp Neurol* 244:369–393.

Lisman JE. (1989). A mechanism for the Hebb and the anti-Hebb processes underlying learning and memory. *Proc Natl Acad Sci USA* 86:9574–9578.

Lisman JE, Harris KM. (1993). Quantal analysis and synaptic anatomy—integrating two views of hippocampal plasticity. *Trends Neurosci* 16:141–147.

Liu Z, Vergnes M, Depaulis A, Marescaux C. (1992). Involvement of intrathalamic $GABA_B$ neurotransmission in the control of absence seizures in the rat. *Neuroscience* 48:87–93.

Livingstone MS, Hubel DH. (1981). Effects of sleep and arousal on the processing of visual information in the cat. *Nature* 291:554–561.

Livsey C. T, Costa E, Vicini S. (1993). Glutamate-activated currents in outside-out patches from spiny versus aspiny hilar neurons of rat hippocampal slices. *J Neurosci* 13:5324–5333.

Llinás R. (1981). Mechanisms of neuronal excitability: the basis of epileptogenesis. *Arch Neurol* 38:732.

Llinás R. (1988). The intrinsic electrophysiological properties of mammalian neurons: insights into central nervous system function. *Science* 242:1654–1664.

Llinás R, Jahnsen H. (1982). Electrophysiology of mammalian thalamic neurones in vitro. *Nature* 297:406–408.

Llinás R, Sugimori M. (1980a). Electrophysiological properties of in vitro Purkinje cell dendrites in mammalian cerebellar slices. *J Physiol* (Lond) 305:197–213.

Llinás R, Sugimori M. (1980b). Electrophysiological properties of in vitro Purkinje cell somata in mammalian cerebellar slices. *J Physiol* (Lond) 305:171–195.

Llinás R, Yarom Y. (1981). Electrophysiology of mammalian inferior olivary neurones in vitro. Different types of voltage-dependent ionic conductances. *J Physiol* (Lond) 315:549–567.

Llinás R, Baker R, Precht W. (1974). Blockage of inhibition by ammonium acetate action on chloride pump in cat trochlear motoneurons. *J Neurophysiol* 37:522–532.

Llinás R, Sugimori M, Lin J-W, Cherksey B. (1989). Blocking and isolation of a calcium channel from neurons in mammals and cephalopods utilizing a toxin fraction (FTX) from funnel-web spider poison. *Proc Natl Acad Sci USA* 86:1689–1693.

Lloyd KG, Munari C, Bossi L, Stoeffels C, Talairach J, Morselli PL. (1981). Biochemical evidence for the alterations of GABA-mediated synaptic transmission in pathological brain tissue (stereo EEG or morphological definition) from epileptic patients. In: Morselli PL, Lloyd KG, Loscher W, Meldrum B, Reynolds EH, eds. *Neurotransmitters, Seizures, and Epilepsy.* New York: Raven Press: 325–338.

Lockton JW, Holmes O. (1983). Penicillin epilepsy in the rat: the responses of different layers of the cortex cerebri. *Brain Res* 259:79–89.

Lorente de Nó R. (1949). Cerebral cortex: architecture, intracortical connections, motor projections. In: Fulton JF, ed. *Physiology of the Nervous System.* Oxford: Oxford University Press: 288–313.

Lorenzon NM, Foehring RC. (1990). Relationship between repetitive firing and afterhyperpolarizations in human neocortical neurons. *J Neurophysiol* 350: 363.

Lothman EW, Hatlelid JM, Zorumski CF, Conry JA, Moon PF, Perlin JB. (1985). Kindling with rapidly recurring hippocampal seizures. *Brain Res* 360:83–91.

LoTurco JJ, Kriegstein AR. (1991). Clusters of coupled neuroblasts revealed in embryonic neocortex. *Science* 252:563–566.

LoTurco JJ, Blanton MG, Kriegstein AR. (1991). Initial expression and endogenous activation of NMDA channels in early neocortical development. *J Neurosci* 11:792–799.

LoTurco JJ, Mody I, Kriegstein AR. (1990). Differential activation of glutamate receptors by spontaneously released transmitter in slices of neocortex. *Neurosci Lett* 114:265–271.

Lowenstein DH, Thomas MJ, Smith DH, McIntosh TK. (1992). Selective vulnerability of dentate hilar neurons following traumatic brain injury: a potential mechanistic link between head trauma and disorders of the hippocampus. *J Neurosci* 12:4846–4853.

Lüders H, Lesser RP, Dinner DS, Morris HH III. (1987). Benign focal epilepsy of childhood. In: Luders H, Lesser RP, eds. Epilepsy: Electroclinical Syndromes. London: Springer-Verlag: 303–346.

Luhmann HJ, Heinemann U. (1992). Hypoxia-induced functional alterations in adult rat neocortex. *J Neurophysiol* 67:798–811.

Luhmann HJ, Prince DA. (1990a). Control of NMDA receptor-mediated activity by GABAergic mechanisms in mature and developing rat neocortex. *Dev Brain Res* 54:287–290.

Luhmann HJ, Prince DA. (1990b). Transient expression of polysynaptic NMDA receptor-mediated activity during neonatal development. *Neurosci Lett* 111:109–115.

Luhmann HJ, Prince DA. (1991). Postnatal maturation of the GABAergic system in rat neocortex. *J Neurophysiol* 65:247–263.

Lund JS. (1987). Local circuit neurons of macaque monkey striate cortex: I. Neurons of laminae 4C and 5A *J Comp Neurol* 257:60–92.

Luskin MB, Pearlman AL, Sanes JR. (1988). Cell lineage in the cerebral cortex of the mouse studied in vivo and in vitro with a recombinant retrovirus. *Neuron* 1:635–647.

Luskin MB, Parnavelas JG, Barfield JA. (1993). Neurons, astrocytes, and oligodendrocytes of the rat cerebral cortex originate from separate progenitor cells: an ultrastructural analysis of clonally related cells. *J Neurosci* 13(4); 1730–1750.

Lux HD. (1971). Ammonium and chloride extrusion: hyperpolarizing synaptic inhibition in spinal motoneurons. *Science* 173:555–557.

Lynch G, Cotman C. W. (1975). The hippocampus as a model for studying anatomical plasticity in the adult brain. In: Isaacson RLA, ed. The Hippocampus, Vol I. New York: Plenum: 123–154.

Lynch G, Dunwiddie T, Gribkoff V. (1977). Heterosynaptic depression: a postsynaptic correlate of long-term potentiation. *Nature* 266:737–739.

Lynch G, Larson J, Kelso S, Barrioneuvo G, Schottler F. (1983). Intracellular injections of EGTA block induction of hippocampal long-term potentiation. *Nature* 305:719–721.
Lytton WW, Sejnowski TJ. (1991). Simulations of cortical pyramidal neurons synchronized by inhibitory interneurons. *J Neurophysiol* 66:1059–1079.
MacDermott AB, Mayer ML, Westbrook GL, Smith SJ, Barker JL. (1986). NMDA-receptor activation increases cytoplasmic calcium concentration in cultured spinal cord neurones. *Nature* 321:519–522.
MacDonald RL, Meldrum BS. (1989). Principles of antiepileptic drug action. In: Levy RH, Dreifuss FE, eds. Antiepileptic Drugs. New York: Raven Press.
MacLennan DH, Phillips MS. (1992). Malignant hyperthermia. *Science* 256: 1789–1794.
MacNicol M, Schulman H. (1992a). Cross-talk between protein kinase C and multifunctional Ca^{2+}/calmodulin-dependent protein kinase. *J Biol Chem* 267:12197–12201.
MacNicol M, Schulman H. (1992b). Multiple Ca^{2+} signaling pathways converge on CaM kinase in PC12 cells. *FEBS Lett* 304:237–248.
Maconochie DJ, Zempel JM, Steinbach JH. (1994). How quickly can $GABA_A$ receptors open? *Neuron* 12:61–71.
MacVicar BA, Dudek FE. (1980a). Dye-coupling between CA3 pyramidal cells in slices of rat hippocampus. *Brain Res* 196:494–497.
MacVicar BA, Dudek FE. (1980b). Local synaptic circuits in rat hippocampus: interactions between pyramidal cells. *Brain Res* 184:220–223.
Madison DV, Nicoll RA. (1988a). Enkephalin hyperpolarizes interneurones in the rat hippocampus. *J Physiol* (Lond) 398:123–130.
Madison DV, Nicoll RA. (1988b). Norepinephrine decreases synaptic inhibition in the rat hippocampus. *Brain Res* 442:131–138.
Madison DV, Malenka RC, Nicoll RA. (1991). Mechanisms underlying long-term potentiation of synaptic transmission. *Annu Rev Neurosci* 14:379–397.
Malenka RC. (1991). Postsynaptic factors control the duration of synaptic enhancement in area CA1 of the hippocampus. *Neuron* 6:53–60.
Malenka RC. (1993). Long-term depression: not so depressing after all. *Proc Natl Acad Sci USA* 90:3121–3123.
Malenka RC, Nicoll RA. (1993). NMDA-receptor-dependent synaptic plasticity: multiple forms and mechanisms. *Trends Neurosci* 16:521–527.
Malenka RC, Ayoub GS, Nicoll RA. (1987). Phorbol esters enhance transmitter release in rat hippocampal slices. *Brain Res* 403:198–203.
Malenka RC, Lancaster B, Zucker RS. (1992). Temporal limits on the rise in postsynaptic calcium required for the induction of long-term potentiation. *Neuron* 9:121–128.
Malenka RC, Madison DV, Nicoll RA. (1986). Potentiation of synaptic transmission in the hippocampus by phorbol esters. *Nature* 321:695–697.
Malenka RC, Kauer JA, Perkel DJ, Nicoll RA. (1989). The impact of postsynaptic calcium on synaptic transmission—its role in long-term potentiation. *Trends Neurosci* 12:444–450.
Malenka RC, Kauer JA, Zucker RJ, Nicoll RA. (1988). Postsynaptic calcium is sufficient for potentiation of hippocampal synaptic transmission. *Science* 242:81–84.
Malenka RC, Kocsis JD, Ransom BR, Waxman SG. (1981). Modulation of parallel fiber excitability by postsynaptically mediated changes in extracellular potassium. *Science* 214:339–341.

Malgaroli A, Tsien RW. (1992). Glutamate-induced long-term potentiation of the frequency of miniature synaptic currents in cultured hippocampal neurones. *Nature* 357:134–139.

Malinow R. (1991). Transmission between pairs of hippocampal slice neurons: quantal levels, oscillations, and LTP. *Science* 252:722–724.

Malinow R, Miller JP. (1986). Postsynaptic hyperpolarization during conditioning reversibly blocks induction of long-term potentiation. *Nature* 320:529–530.

Malinow R, Schulman H, Tsien RW. (1989). Inhibition of postsynaptic PKC or CaMKII blocks induction but not expression of LTP. *Science* 245:862–866.

Malouf AT, Robbins CA, Schwartzkroin PA. (1990). Phaclofen inhibition of the slow inhibitory postsynaptic potential in hippocampal slice cultures: a possible role for the $GABA_B$-mediated inhibitory postsynaptic potential. *Neuroscience* 35:53–61.

Manabe T, Renner P, Nicoll RA. (1992). Postsynaptic contribution to long-term potentiation revealed by the analysis of miniature synaptic currents. *Nature* 355:50–55.

Mandel G. (1992). Tissue-specific expression of the voltage-sensitive calcium channel. *J Membrane Biol* 125:193–205.

Manor Y, Koch C, Segev I. (1991). Effect of geometrical irregularities on propagation delay in axonal trees. *Biophys J* 60:1424–1437.

Marescaux C, Micheletti G, Vergnes M, Depaulis A, Rumbach L, Warter JM. (1984a). Biphasic effects of Ro 15-1788 on spontaneous petit mal-like seizures in rats. *Eur J Pharmacol* 102:355–359.

Marescaux C, Micheletti G, Vergnes M, Depaulis A, Rumbach L, Warter JM. (1984b). A model of chronic spontaneous petit mal-like seizures in the rat: comparison with pentylenetetrazol-induced seizures. *Epilepsia* 25:326–331.

Margerison JH, Corsellis JAN. (1966). A clinical, electroencephalographic and neuropathological study of the brain in epilepsy with particular reference to the temporal lobes. *Brain* 89:499–520.

Marin-Padilla M. (1967). Number and distribution of apical dendritic spines of the layer V pyramidal cells in man. *J Comp Neurol* 131:475–490.

Marin-Padilla M. (1969). Origin of the pericellular baskets of the pyramidal cells of the human motor cortex: a Golgi study. *Brain Res* 14:633–646.

Marin-Padilla M. (1970). Prenatal and early postnatal ontogenesis of the human motor cortex: a Golgi study. II. The basket-pyramidal system. *Brain Res* 23:185–191.

Marin-Padilla M. (1972). Double origin of the pericellular baskets of the pyramidal cells of the human motor cortex: a Golgi study. *Brain Res* 38:1–12.

Marin-Padilla M. (1987). The chandelier cell of the human visual cortex: a Golgi study. *J Comp Neurol* 256:61–70.

Marin-Padilla M. (1992). Ontogenesis of the pyramidal cell of the mammalian neocortex and developmental cytoarchitectonics: a unifying theory. *J Comp Neurol* 321:223–240.

Markram H, Sakmann B. (1994). Calcium transients in apical dendrites evoked by single sub-threshold excitatory post-synaptic potentials via low voltage-activated calcium channels. *Proc Natl Acad Sci USA* 91:5207–5211.

Markram H, Helm PJ, Sakmann B. (1994). Dendritic calcium transients evoked by single back-propagating action potentials in neocortical pyramidal neurones. *J Physiol* (Lond) (in press).

Marksteiner J, Ortler M, Bellmann R, Sperk G. (1990). Neuropeptide Y biosynthesis is markedly induced in mossy fibers during temporal lope epilepsy of the rat. *Neurosci Lett* 112:143–148.

Martin D, McNamara JO, Nadler JV. (1992). Kindling enhances sensitivity of CA3 hippocampal pyramidal cells to NMDA. *J Neurosci* 12:1928–1935.

Martin JB. (1993). Molecular genetics in neurology. *Ann Neurol* 34:757–773.

Martin KAC. (1988). From single cells to simple circuits in the cerebral cortex. *Q J Exp Physiol* 73:637–702.

Martin KAC, Whitteridge D. (1984). Form, function and intracortical projections of spiny neurons in the striate visual cortex of the cat. *J Physiol* (Lond) 353:463–504.

Mason A, Larkman A. (1990). Correlations between morphology and electrophysiology of pyramidal neurons in slices of rat visual cortex. II. Electrophysiology. *J Neurosci* 10:1415–1428.

Mason A, Nicoll A, Stratford K. (1991). Synaptic transmission between individual pyramidal neurons of the rat visual cortex in vitro. *J Neurosci* 11:72–84.

Masu M, Tanabe Y, Tsuchida K, Shigemoto R, Nakanishi S. (1991). Sequence and expression of a metabotropic glutamate receptor. *Nature* 349:760–765.

Masukawa LM, Prince DA. (1984). Synaptic control of excitability in isolated dendrites of hippocampal neurons. *J Neurosci* 4:217–227.

Masukawa LM, Higashima M, Kim JH, Spencer DD. (1989). Epileptiform discharges evoked in hippocampal brain slices from epileptic patients. *Brain Res* 493:168–174.

Masukawa LM, Higashima M, Hart GJ, Spencer DD, O'Connor MJ. (1991). NMDA receptor activation during epileptiform responses in the dentate gyrus of epileptic patients. *Brain Res* 562:177–180.

Masukawa LM, Uruno K, Sperling M, O'Connor MJ, Burdette LJ. (1992). The functional relationship between antidromically evoked field responses of the dentate gyrus and mossy fiber reorganization in temporal lobe epileptic patients. *Brain Res* 579:119–127.

Mathern GW, Kupfer WR, Pretorius JK, Babb TL, Levesque MF. (1992). Onset and patterns of hippocampal sprouting in the rat kainate seizure model: evidence for progressive cell loss and neo-innervation in regio inferior and superior. *Dendron* 1:69–84.

Matsumoto H, Ajmone-Marsan C. (1964). Cortical cellular phenomena in experimental epilepsy: interictal manifestations. *Exp Neurol* 9:286–304.

Matthews DA, Cotman C, Lynch G. (1976). An electron microscopic study of lesion-induced synaptogenesis in the dentate gyrus of the adult rat. II. Reappearance of morphologically normal synaptic contacts. *Brain Res* 115:23–41.

Mattson MP, Kater SB. (1989). Excitatory and inhibitory neurotransmitters in the generation and degeneration of hippocampal neuroarchitecture. *Brain Res* 478:337–348.

Mattson MP, Dou P, Kater J. (1988). Outgrowth-regulating actions of glutamate in isolated hippocampal pyramidal neurons. *J Neurosci* 8:2087–2100.

Mattson MP, Pychlik B, You JS, Sisken JE. (1991). Sensitivity of cultured human embryonic cerebral cortical neurons to excitatory amino acid-induced calcium influx and neurotoxicity. *Brain Res* 542:97–106.

Mayer ML, Westbrook GL. (1987). The physiology of excitatory amino acids in the vertebrate central nervous system. *Progr Neurobiol* 28:197–276.

Mayer ML, Westbrook GL, Guthrie PB. (1984). Voltage-dependent block by Mg^{2+} of NMDA responses in spinal cord neurones. *Nature* 309:261–263.

Mayer ML, MacDermott AB, Westbrook GL, Smith SJ, Barker JL. (1987). Agonist- and voltage-gated calcium entry in cultured mouse spinal cord neurons under voltage clamp measured using arsenazo III. *J Neurosci* 7: 3230–3244.

McBain CJ, Dingledine R. (1993). Heterogeneity of synaptic glutamate receptors on CA3 stratum radiatum interneurones of rat hippocampus. *J Physiol* (Lond) 462:373–392.

McCarren M, Alger BE. (1985). Use-dependent depression of IPSPs in rat hippocampal pyramidal cells in vitro. *J Neurophysiol* 53:557–571.

McConnell SK. (1988). Development and decision making in the mammalian cerebral cortex. *Brain Res Rev* 13:1–23.

McConnell SK. (1991). The generation of neuronal diversity in the central nervous system. *Annu Rev Neurosci* 14:269–300.

McConnell SK, Kaznowski CE. (1991). Cell cycle dependence of laminar determination in developing neocortex. *Science* 254:282–285.

McCormick DA. (1989a). Cholinergic and noradrenergic modulation of thalamocortical processing. *Trends Neurosci* 12:215–221.

McCormick DA. (1989b). GABA as an inhibitory neurotransmitter in human cerebral cortex. *J Neurophysiol* 62:1018–1027.

McCormick DA. (1991). Cellular mechanisms underlying cholinergic and noradrenergic modulation of neuronal firing mode in the cat and guinea pig dorsal lateral geniculate nucleus. *J Neurosci* 12:278–289.

McCormick DA. (1992). Neurotransmitter actions in the thalamus and cerebral cortex and their role in neuromodulation of thalamocortical activity. *Progr Neurobiol* 39:337–388.

McCormick DA, Feeser HR. (1990). Functional implications of burst firing and single spike activity lateral geniculate relay neurons. *Neuroscience* 39:103–113.

McCormick DA, Pape H. (1988). Acetylcholine inhibits identified interneurons in the cat lateral geniculate nucleus. *Nature* 334:246–248.

McCormick DA, Pape H. (1990a). Noradrenergic and serotonergic modulation of a hyperpolarization-activated cation current in thalamic relay neurons. *J Physiol* (Lond) 431:319–342.

McCormick DA, Pape H. (1990b). Properties of a hyperpolarization-activated cation current and its role in rhythmic oscillation in thalamic relay neurons. *J Physiol* (Lond) 431:291–318.

McCormick DA, Prince DA. (1986a). Mechanisms of action of acetylcholine in the guinea-pig cerebral cortex *in vitro*. *J Physiol* (Lond) 375:169–194.

McCormick DA, Prince DA. (1986b). Acetylcholine induces burst firing in thalamic reticular neurones by activating a potassium conductance. *Nature* 319:402–405.

McCormick DA, Prince DA. (1987a). Actions of acetylcholine in the guinea-pig and cat medial and lateral geniculate nuclei, *in vitro*. *J Physiol* (Lond) 392: 147–165.

McCormick DA, Prince DA. (1987b). Postnatal development of electrophysiological properties of rat cerebral cortical pyramidal neurones. *J Physiol* (Lond) 393:743–762.

McCormick DA, Prince DA. (1988). Noradrenergic modulation of firing pattern in guinea pig and cat thalamic neurons, *in vitro*. *J Neurophysiol* 59:978–996.

McCormick DA, von Krosigk M. (1992). Corticothalamic activation modulates thalamic firing through lutamate "metabotropic" receptors. *Proc Natl Acad Sci USA* 89:2774–2778.

McCormick DA, Wang Z. (1991). Serotonin and noradrenaline excite GABAergic neurones of the guinea-pig and cat nucleus reticularis thalami. *J Physiol* (Lond) 442:235–255.

McCormick DA, Williamson A. (1989). Convergence and divergence of neurotransmitter action in human cerebral cortex. *Proc Natl Acad Sci USA* 86: 8098–8102.

McCormick DA, Williamson A. (1991). Modulation of neuronal firing mode in cat and guinea pig LGNd by histamine: possible cellular mechanisms of histaminergic control of arousal. *J Neurosci* 11:3188–3199.

McCormick DA, Wang Z, Huguenard J. (1993). Neurotransmitter control of neocortical neuronal activity and excitability. *Cereb Cortex* 3:387–398.

McCormick DA, Connors BW, Lighthall JW, Prince DA. (1985). Comparative electrophysiology of pyramidal and sparsely spiny neurons of the neocortex. *J Neurophysiol* 54:782–806.

McDonald JK, Speciale SG, Parnavelas JG. (1987). The laminar distribution of glutamate decarboxylase and choline acetyltransferase in the adult and developing visual cortex of the rat. *Neuroscience* 21:825–832.

McDonald JW, Johnston MV. (1990). Physiological and pathophysiological roles of excitatory amino acids during central nervous system development. *Brain Res Rev* 15:41–70.

McDonald JW, Garofalo EA, Hood T, Sackellares JC, Gilman S, McKeever PE, Troncoso JC, Johnston MV. (1991). Altered excitatory and inhibitory amino acid receptor binding in hippocampus of patients with temporal lobe epilepsy. *Ann Neurol* 29:529–541.

McGeer EG, McGeer PL. (1976). Duplication of biochemical changes of Huntington's chorea by intrastriatal injections of glutamic and kainic acids. *Nature* 263:517–519.

McGlade-McCulloh E, Yamamoto H, Tan S, Brickey DA, Soderling TR. (1993). Phosphorylation and regulation of glutamate receptors by calcium/calmodulin-dependent protein kinase II. *Nature* 362:640–642.

McGuire BA, Hornung J, Gilbert CD, Wiesel TN. (1984). Patterns of synaptic input to layer 4 of cat striate cortex. *J Neurosci* 4:3021–3033.

McGuire BA, Gilbert CD, Rivlin PK, Wiesel TN. (1991). Targets of horizontal connections in macaque primary visual cortex. *J Comp Neurol* 305:370–392.

McLachlan RS, Avoli M, Gloor P. (1985). Transition from spindles to generalized spike and wave discharges in the cat: simultaneous single-cell recordings in cortex and thalamus. *Exp Neurol* 85:413–425.

McNamara JO. (1978). Selective alterations of regional beta-adrenergic receptor binding in the kindling model of epilepsy. *Exp Neurol* 61:582–591.

McNamara JO. (1992). The neurobiological basis of epilepsy. *Trends Neurosci* 15: 357–359.

McNamara JO. (1994). Cellular and molecular basis of epilepsy. *J Neurosci* 14: 3413–3425.

McNamara JO, Bonhaus DW, Shin C. (1993). The kindling model of epilepsy. In: Schwartzkroin PD, ed. *Epilepsy: Models, Mechanisms, and Concepts*. Cambridge: Cambridge University Press: 27–47.

McNamara JO, Morrisett R, Nadler JV. (1992). Recent advances in understanding mechanisms of the kindling model. *Adv Neurol* 57:555–560.

McNamara JO, Peper AM, Patrone V. (1980). Repeated seizures induce long-term increase in hippocampal benzodiazepine receptors. *Proc Natl Acad Sci USA* 77:3029–3032.

McNamara JO, Rigsbee LC, Galloway MT. (1983). Evidence that substantia nigra is crucial to neural network of kindled seizures. *Eur J Pharmacol* 86:485–486.

McNamara JO, Russel RD, Rigsbee L, Bonhaus DW. (1988). Anticonvulsant and antiepileptogenic actions of MK-801 in the kindling and electroshock models. *Neuropharmacology* 27:563–568.

McNamara JO, Bonhaus DW, Shin C, Crain BJ, Gellman RL, Giacchino JL. (1985). The kindling model of epilepsy: a critical review. *CRC Crit Neurobiol* 1:341–392.

McNaughton BL. (1993). The mechanism of expression of long-term enhancement of hippocampal synapses: current issues and theoretical implications. *Annu Rev Physiol* 55:375–396.

Meador WE, Means AR, Quiocho FA. (1992). Target enzyme recognition by calmodulin: 2.4. A structure of a calmodulin-peptide complex. *Science* 257: 1251–1255.

Meencke HJ. (1983). The density of dystopic neurons in the white matter of the gyrus frontalis inferior in epilepsies. *J Neurol* 230:178–181.

Meencke HJ, Janz D. (1984). Neuropathological findings in primary generalized epilepsy: a study of eight cases. *Epilepsia* 25:8–21.

Meencke HJ, Janz D. (1985). The significance of microdysgenesia in primary generalized epilepsy: an answer to the considerations of Lyon and Gastaut. *Epilepsia* 26:368–371.

Meencke H.-J, Veith G. (1992). Migration disturbances in epilepsy. *Epilepsy Res* 9(suppl):31–40.

Mehler MF, Rozental R, Dougherty M, Spray DC, Kessler JA. (1993). Cytokine regulation of neuronal differentiation of hippocampal progenitor cells. *Nature* 362:62–65.

Mehta PP, Bertram JS, Loewenstein WR. (1986). Growth inhibition of transformed cells correlates with their junctional communication with normal cells. *Cell* 44:187–196.

Mello L, Cavalheiro EA, Tan AM, Kupfer WR, Pretorius JK, Babb TL, Finch DM. (1993). Circuit mechanisms of seizures in the pilocarpine model of chronic epilepsy: cell loss and mossy fiber sprouting. *Epilepsia* 34:985–995.

Merlin LR, Wong RKS. (1993). Synaptic modifications accompanying epileptogenesis in vitro: long-term depression of GABA-mediated inhibition. *Brain Res* 627:330–340.

Merlio J, Emfors P, Kokaia Z, Middlemas DS, Bengzon J, Kodaia M, Smith M, Siesjo BK, Hunter T, Lindvall O, Persson H. (1993). Increased production of the TrkB protein tyrsine kinase receptor after brain insults. *Neuron* 10: 151–167.

Messenheimer JA, Harris EW, Steward O. (1979). Sprouting fibers gain access to circuitry transynaptically altered by kindling. *Exp Neurol* 64:469–481.

Meyer T, Hanson PI, Stryer L, Schulman H. (1992). Calmodulin trapping by calcium-calmodulin-dependent protein kinase. *Science* 256:1199–1201.

Michelson HB, Wong RKS. (1991). Excitatory synaptic responses mediated by $GABA_A$ receptors in the hippocampus. *Science* 253:1420–1423.

Miles R. (1990). Synaptic excitation of inhibitory cells by single CA3 hippocampal pyramidal cells of the guinea pig in vitro. *J Physiol* (Lond) 428:611–677.

Miles R, Wong RKS. (1984). Unitary inhibitory synaptic potentials in the guinea pig hippocampus in vitro. *J Physiol* (Lond) 356:97–113.

Miles R, Wong RKS. (1986). Excitatory synaptic interactions between CA3 neurones in the guinea pig hippocampus. *J Physiol* (Lond) 373:397–418.

Miles R, Wong RKS. (1987a). Inhibitory control of local excitatory circuits in the guinea pig hippocampus. *J Physiol* (Lond) 388:611–629.

Miles R, Wong RKS. (1987b). Latent synaptic pathways revealed after tetanic stimulation in the hippocampus. *Nature* 329:724–726.

Miller LP, Johnson AE, Gelhard RE, Insel TR. (1990). The ontogeny of excitatory amino acid receptors in the rat forebrain. II. Kainic acid receptors. *Neuroscience* 35:45–51.

Miller M, Peters A. (1981). Maturation of the rat visual cortex. II. Combined Golgi-electron microscope study of pyramidal neurons. *J Comp Neurol* 203:555–573.

Miller MW. (1986). Maturation of rat visual cortex. III. Postnatal morphogenesis and synaptogenesis of local circuit neurons. *Dev Brain Res* 25:271–285.

Miller MW. (1988). Development and maturation of cerebral cortex. In: Peters A, Jones EG, eds. *Cerebral Cortex*. New York: Plenum: 133–166.

Miller MW, Chiaia NL, Rhoades RW. (1990). Intracellular recording and injection study of corticopinal neurons in the rat somatosensory cortex: effect of prenatal exposure to alcohol. *J Comp Neurol* 297:91–105.

Miller RJ. (1991). Metabotropic excitatory amino acid receptors reveal their true colors. *TIPS* 12:365–367.

Mirski MA, Ferrendelli JA. (1986). Anterior thalamic mediation of generalized pentylenetetrazol seizures. *Brain Res* 399:212–223.

Misgeld U, Deisz RA, Dodt HU, Lux HD. (1986). The role of chloride transport in postsynaptic inhibition of hippocampal neurons. *Science* 232:1413–1415.

Misgeld U, Muller W, Brunner H. (1989). Effects of (−) baclofen on inhibitory neurons in the guinea pig hippocampal slice. *Pflugers Arch* 414:139–144.

Mishina M, Takai T, Imoto K, Noda M, Takahashi T, Numa S, Methfessel C, Sakmann B. (1986). Molecular distinction between fetal and adult forms of muscle acetylcholine receptor. *Nature* 321:406–411.

Misson JP, Edwards MA, Yamamoto M, Caviness VS. (1988). Identification of radial glial cells within the developing murine central nervous system: studies based upon a new immunohistochemical marker. *Dev Brain Res* 44: 95–108.

Mistry DK, Hablitz JJ. (1990). Nystatin-perforated patch recordings disclose NMDA-induced outward currents in cultured neocortical neurons. *Brain Res* 535:318–322.

Miyakawa H, Ross WN, Jaffe D, Callaway JC, Lasser-Ross N, Lisman JE, Johnston D. (1992). Synaptically activated increase in Ca^{2+} concentration in hippocampal CA1 pyramidal cells are primarily due to voltage-gated Ca^{2+} channels. *Neuron* 9:1163–1173.

Miyoshi R, Kito S, Doudou N, Nomoto T. (1991). Effect of age on alpha-amino-3-hydroxi-5-methylisoxazole-4-propionic acid (AMPA) binding sites in the rat brain studied by *in vitro* autoradiography. *Neurochem Res* 16:849–854.

Mody I, Heinneman U. (1987). NMDA receptors of dentate gyrus granule cells participate in synaptic transmission following kindling. *Nature* 326:701–704.

Mody I, Stanton PK, Heinemann U. (1988). Activation of N-methyl-D-aspartate receptors parallels changes in cellular and synaptic properties of dentate gyrus granule cells after kindling. *J Neurophysiol* 59:1033–1054.

Mody I, Tanelian DL, MacIver MB. (1991). Halothane enhances tonic neuronal inhibition by elevating intracellular calcium. *Brain Res* 538:319–323.

Mody I, Heinemann U, MacDonald JF, Salter MW. (1992). Recruitment of NMDA receptors into synaptic transmission after kindling-induced epilepsy and its possible mechanism. *Epilepsy Res* 8(suppl):307–310.

Mody I, Reynolds JN, Salter MW, Carlen PL, MacDonald JF. (1990). Kindling-induced epilepsy alters calcium currents in granule cells of rat hippocampal slices. *Brain Res* 531:88–94.

Molloy SS, Kennedy MB. (1991). Autophosphorylation of type II Ca^{2+}/calmodulin-dependent protein kinase in cultures of postnatal rat hippocampal slices. *Proc Natl Acad Sci USA* 88:4756–4760.

Monaghan DT, Cotman CW. (1985). Distribution of N-methyl-D-aspartate-sensitive L-[^{3}H] glutamate-binding sites in rat brain. *J Neurosci* 5:2909–2919.

Monaghan DT, Bridges RJ, Cotman CW. (1989). The excitatory amino acid receptors: their classes, pharmacology, and distinct properties in the function of the central nervous system. *Annu Rev Pharmacol Toxicol* 29:365–402.

Montoro RJ, Lopez-Barneo J, Jassik-Gerschenfeld D. (1988). Differential burst firing modes in neurons of the mammalian visual cortex in vitro. *Brain Res* 460:168–172.

Monyer H, Hartley DM, Choi DW. (1990). Aminosteroids attenuate excitotoxid neuronal injury in cortical cell cultures. *Neuron* 5:121–126.

Monyer H, Seeburg PH, Wisden W. (1991). Glutamate-operated channels: developmentally early and mature forms arise by alternative splicing. *Neuron* 6: 799–810.

Monyer H, Sprengel R, Schoepfer R, Herb A, Higuchi M, Lomeli H, Burnashev N, Sakmann B, Seeburg PH. (1992). Heteromeric NMDA receptors: molecular and functional distinction of subtypes. *Science* 256:1217–1221.

Monyer H, Burnashev N, Laurie DJ, Sakmann B, Seeburg PH. (1994). Developmental and regional expression in the rat brain and functional properties of four NMDA receptors. *Neuron* 12:529–540.

Moorman JR, Kirsch GE, VanDongen AMJ, Joho RH, Brown AM. (1990). Fast and slow gating of sodium channels encoded by a single mRNA. *Neuron* 4:243–252.

Morgan JI, Cohen DR, Hempstead JL, Curran T. (1987). Mapping patterns of c-fos expression in the central nervous system after seizure. *Science* 237: 192–197.

Mori H, Masaki H, Yamakura T, Mishina M. (1992). Identification by mutagenesis of a Mg^{2+}-block site of the NMDA receptor channel. *Nature* 358: 673–675.

Morino-Wannier P, Fujita SC, Jones EG. (1992). GABAergic neuronal populations in monkey primary auditory cortex defined by co-localized calcium binding proteins and surface antigens. *Exp Brain Res* 88:422–432.

Morris RGM, Davis S, Butcher SP. (1989). The role of NMDA receptors in learning and memory. In: Watkins JC, Collingridge GL, eds. *The NMDA Receptor*. Oxford: Oxford University Press: 137–151.

Morris RGM, Anderson E, Lynch GS, Baudry M. (1986). Selective impairment of learning and blockade of long-term potentiation by an N-methyl-D-aspartate receptor antagonist, AP5. *Nature* 319:774–76.

Morrisett RA, Chow C, Nadler JV, McNamara JO. (1989). Biochemical evidence for enhanced sensitivity to N-methyl-D-aspartate in the hippocampal formation of kindled rats. *Brain Res* 496:25–28.

Moshe SL, Albala BJ, Ackermann RF, Engel J Jr. (1983). Increased seizure susceptibility of the immature brain. *Dev Brain Res* 7:81–85.

Mott DD, Lewis DV. (1991). Facilitation of the induction of long-term potentiation by $GABA_B$ receptors. *Science* 252:1718–1720.

Mount HTJ, Dreyfus CF, Black B. (1993). Purkinje cell survival is differentially regulated by metabotropic and ionotropic excitatory amino acid receptors. *J Neurosci* 13:3173–3179.

Moure JMB, Kellaway P. (1980). The clinical correlates of multiple spike foci. *Electroencephalogr Clin Neurophysiol* 49:1P–2P.

Mouritzen-Dam A. (1980). Epilepsy and neuron loss in the hippocampus. *Epilepsia* 21:617–629.

Mueller AL, Taube JS, Schwartzkroin PA. (1984). Development of hyperpolarizing inhibitory postsynaptic potentials and hyperpolarizing response to gamma-aminobutyric acid in rabbit hippocampus studied in vitro. *J Neurosci* 4: 860–867.

Mugnaini E, Oertel WH. (1985). An atlas of the distribution of GABAergic neurons and terminals in the rat CNS as revealed by GAD immunohistochemistry. In: Bjorklund AA, ed. *Handbook of Chemical Neuroanatomy. Vol. 4, GABA and Neuropeptides in the CNS, Part I.* Amsterdam: Elsevier Science Publishers: 436–608.

Mulkey RM, Malenka RC. (1992). Mechanisms underlying homosynaptic long-term depression in area CA1 of the hippocampus. *Neuron* 9:967–975.

Mulle C, Madariaga A, Deschenes M. (1986). Morphology and electrophysiological properties of reticularis thalami neurons in cat: in vivo study of a thalamic pacemaker. *J Neurosci* 6:2134–2145.

Muller D, Joly M, Lynch G. (1988). Contributions of quisqualate and NMDA receptors to the induction and expression of LTP*Science* 242:1694–1697.

Muller W, Connor JA. (1991). Synaptic Ca^{2+} responses in dendritic spines: the spine as an individual neuronal compartment. *Nature* 354:73–76.

Muller W, Misgeld U. (1990). Inhibitory role of dentate hilus neurons in guinea pig hippocampal slice. *J Neurophysiol* 64:46–56.

Muller W, Petrozzino JJ, Griffith LC, Danho W, Connor JA. (1992). Specific involvement of Ca^{2+}-calmodulin kinase II in cholinergic modulation of neuronal responsiveness. *J Neurophysiol* 68:2264–2269.

Muller-Paschinger I, Tombol T, Petsche H. (1983). Chandelier neurons within the rabbits' cerebral cortex: a Golgi study. *Anat Embryol* 166:149–159.

Mulligan KA, van Brederode JFM, Hendrickson AE. (1989). The lectin Vicia villosa labels a distinct subset of GABAergic cells in macaque visual cortex. *Vis Neurosci* 2:63–72.

Murphy TH, Worley PF, Nakabeppu Y, Chrissty B, Gastel J, Baraban JM. (1991). Synaptic regulation of immediate early gene expression in primary cultures of cortical neurons. *J Neurochem* 57:1862–1872.

Mutani R, Futamachi KJ, Prince DA. (1974). Potassium activity in immature cortex. *Brain Res* 75:27–39.

Nadler JV, Perry BW, Cotman CW. (1978). Intraventricular kainic acid preferentially destroys hippocampal pyramidal cells. *Nature* 271:676–677.

Nadler JV, Perry BW, Gentry C, Cotman CW. (1980). Loss and reacquisition of hippocampal synapses after selective destruction of CA3-CA4 afferents with kainic acid. *Brain Res* 191:387–403.

Nadler JV, Thompson MA, McNamara JO. (1994). Kindling reduces sensitivity of CA3 hippocampal pyramidal cells to competitive NMDA receptor antagonists. *Neuropharmacology* 33:147–153.

Naegle JR, Barnstable CJ. (1989). Molecular determinants of GABAergic local-circuit neurons in the visual cortex. *Trends Neurosci* 12:28–34.

Naegle JR, Arimatsu Y, Schwartz P, Barnstable CJ. (1988). Selective staining of a subset of GABAergic neurons in cat visual cortex by monoclonal antibody VC1.1. *J Neurosci* 8:79–89.

Nagai T, McGeer PL, McGeer EG. (1983). Distribution of GABA-T-intensive neurons in the rat forebrain and midbrain. *J Comp Neurol* 218:220–238.

Nahm WK, Noebels JL. (1993). Immediate-early gene protein expression in a mutant mouse model of spike-wave epilepsy, stargazer. *Neurosci Abstr* 19: 1030.

Nakagawa F, Schulte BA, Spicer SS. (1986). Selective cytochemical demonstration of glycoconjugate-containing terminal N-acetylgalactosamine on some brain neurons. *J Comp Neurol* 243:280–290.

Nakanishi S. (1992). Molecular diversity of glutamate receptors and implications for brain function. *Science* 258:597–603.

Nayeem N, Green TP, Martin IL, Barnard EA. (1994). Quaternary structure of the native $GABA_A$ receptor determined by electron microscopic image analysis. *J Neurochem* 62:815–818.

Neafsy EJ. (1990). The complete ratunculus: Output organization of layer V on the cerebral cortex. In: Kolb B, Tees RC, eds. *The Cerebral Cortex of the Rat.* Cambridge, MA: MIT Press: 197–212.

Nedivi E, Hevroni D, Naot D, Israeli D, Citri Y. (1993). Numerous candidate plasticity-related genes revealed by differential cDNA cloning. *Nature* 363: 718–721.

Newberry NR, Nicoll RA. (1984). A bicuculline-resistant inhibitory post-synaptic potential in rat hippocampal pyramidal cells in vitro. *J Physiol* (Lond) 348: 239–254.

Newberry NR, Nicoll RA. (1985). Comparison of the action of baclofen with gamma-aminobutyric acid on rat hippocampal pyramidal cells in vitro. *J Physiol* (Lond) 360:161–185.

Nichols RA, Haycock JW, Wang JKT, Greengard P. (1987). Phorbol ester enhancement of neurotransmitter release from rat brain synaptosomes. *J Neurochem* 48:615–621.

Nichols RA, Sihra TS, Czernik JA, Nairn AC, Greengard P. (1990). Calcium/calmodulin-dependent protein kinase II increases glutamate and noradrenaline release from synaptosomes. *Nature* 343:647–651.

Nicoll A, Blakemore C. (1993). Single-fiber EPSPs in layer 5 of rat visual cortex. *Neurosci Rep* 4:167–170.

Nicoll A, Kim HG, Connors BW. (1993). Spatial organization of inhibitory synaptic responses onto pyramidal neurons of the rat neocortex. *Soc Neurosci Abstr* 19:1704.

Nicoll RA. (1988). The coupling of neurotransmitter receptors to ion channels in the brain. *Science* 241:545–551.

Nicoll RA, Kauer JA, Malenka RC. (1988). The current excitement in long-term potentiation. *Neuron* 1:97–103.

Nicoll RA, Malenka RC, Kauer JA. (1990). Functional comparison of neurotransmitter receptor subtypes in the mammalian central nervous system. *Physiol Rev* 70:513–565.

Nicoll RA, Eccles JC, Oshima T, Rubia FJ. (1975). Prolongation of hippocampal inhibitory postsynaptic potentials by barbituates. *Nature* 258:625–627.

Nikara T, Bishop PO, Pettigrew JD. (1968). Analysis of retinal correspondence by studying receptive fields of binocular single units in cat striate cortex. *Exp Brain Res* 6:353–372.

Nirenberg S, Cepko C. (1993). Targeted ablation of diverse cell classes in the nervous system in vivo. *J Neurosci* 13:3238–3251.

Noble D, Boyd CAR. (1993). The challenge of integrative physiology. In: Boyd CAR, Noble D, eds. *The Logic of Life.* Oxford: Oxford University Press: 1–13.

Noebels JL. (1979). Analysis of inherited epilepsy using single locus mutations in mice. *Fed. Proc.* 38:2405–2410.

Noebels JL. (1982). Spontaneous impulse generation in cortical axons. In: Culp WJ, Ochoa J, eds. *Abnormal Nerves and Muscles as Impulse Generators.* New York: Oxford University Press: 322–343.

Noebels JL. (1984a). Isolating single genes of the inherited epilepsies. *Ann Neurol* 163:S18–S21.

Noebels JL. (1984b). A single gene error of noradrenergic growth synchronizes central neurons. *Nature* 310:409–411.

Noebels JL. (1985). Tracing the cellular expression of neuromodulatory genes. *Trends Neurosci* 8:327–331.

Noebels JL, Prince DA. (1978a). Development of focal seizures in cerebral cortex: role of axon terminal bursting. *J Neurophysiol* 41:1267–1281.

Noebels JL, Prince DA. (1978b). Excitability changes in thalamocortical relay neurons during synchronous discharges in cat neocortex. *J Neurophysiol* 41: 1282–1296.

Noebels JL, Sidman RL. (1979). Inherited epilepsy: spike-wave and focal motor seizures in the mutant mouse tottering. *Science* 204:1334–1336.

Noebels JL, Sidman RL. (1989). Persistent hypersynchronization of neocortical neurons in the Mocha mutant mouse. *J Neurogen* 6:53–56.

Noebels JL, Marcom PK, Jalilian Tehrani MH. (1991). Sodium channel density in hypomyelinated brain increased by myelin basic protein gene deletion. *Nature* 352:431–434.

Nowak L, Bregestovski P, Ascher P, Herbet A, Prochiantz A. (1984). Magnesium gates glutamate-activated channels in mouse central neurones. *Nature* 307: 462–465.

Nussmeier NA, McDermott JP. (1988). Macroembolization: prevention and outcome modification. In: Hilberman M, ed. Boston: Martinus Nijhoff Publishing: 85–107.

Obenaus A, Mody I, Baimbridge KG. (1989). Dantrolene-Na (Dantrium) blocks induction of long-term potentiation in hippocampal slices. *Neurosci Lett* 98:172–176.

Obenaus A, Esclapez M, Houser CR. (1993). Loss of glutamate decarboxylase mRNA-containing neurons in the rat dentate gyrus following pilocarpine-induced seizures. *J Neurosci* 13:4470–4485.

Oberdick J, Schilling K, Smeyne R, Corbin JG, Bocchiaro C, Morgan JI. (1993). Control of segment-like patterns of gene expression in the mouse cerebellum. *Neuron* 10:1007–1018.

Ocorr KA, Schulman H. (1991). Activation of multifunctional Ca^{2+}/calmodulin-dependent kinase in intact hippocampal slices. *Neuron* 6:907–914.

Ohara PT, Havton LA. (1994). Dendritic architecture of rat somatosensory thalamocortical projection neurons. *J Comp Neurol* 341:159–171.

Ohishi H, Shigemoto R, Nakanishi S, Mizuno N. (1993a). Distribution of the mRNA for a metabotropic glutamate receptor (mGluR3) in the rat brain: an in situ hybridization study. *J Comp Neurol* 335:252–266.

Ohishi H, Shigemoto R, Nakanishi S, Mizuno N. (1993b). Distribution of the messenger RNA for a metabotropic glutamate receptor, mGluR2, in the central nervous system of the rat. *Neuroscience* 53:1009–1018.

Ojima H, Honda CN, Jones EG. (1991). Patterns of axon collateralization of identified supragranular pyramidal neurons in the cat auditory cortex. *Cereb Cortex* 1:80–94.

Ojima H, Honda CN, Jones EG. (1992). Characteristics of intracellularly injected infragranular pyramidal neurons in cat primary auditory cortex. *Cereb Cortex* 2:197–216.

Oka JI, Kobayashi T, Nagao T, Hicks TP, Fukuda H. (1993). $GABA_A$ receptor-induced inhibition of neuronal burst firing is weak in rat somatosensory cortex. *Neurosci Rep* 4:731–734.

Okazaki MM, McNamara JO, Nadler JV. (1989). N-methyl-D-aspartate receptor autoradiography in rat brain after angular bundle kindling. *Brain Res* 482: 359–364.

Okazaki MM, McNamara JO, Nadler JV. (1990). Kainate and quisqualate receptor autoradiography in rat brain after angular bundle kindling. *Neuroscience* 37:135–142.

Olbrich H, Braak H. (1985). Ratio of pyramidal cells versus non-pyramidal cells in sector CA1 of the human Ammon's horn. *Anat Embryol* 173:105–110.

O'Leary DM. (1992). Development of connectional diversity and specificity in the mammalian brain by the pruning of collateral projections. *Curr Opin Neurobiol* 2:70–77.

O'Leary DM, Koester SE. (1993). Development of projection neuron types, axon pathways, and patterned connections of the mammalian cortex. *Neuron* 10:991–1006.

Olney JW. (1986). Inciting excitototoxic cytocide among central neurons. *Adv Exp Med Biol* 203:631–645.

Olney JW, Labruyere J, Price MT. (1989). Pathological changes induced in cerebrocortical neurons by phencyclidine and related drugs. *Science* 244:1360–1362.

Olpe H, Ferrat T, Worner W, Andre P, Steinmann MW. (1992). Evidence for a "$GABA_B$ tone" in vivo: the effect of CGP 35348 on long-term potentiation, paired pulse inhibition and cortical excitability. *Pharmacol Commun* 2:21.

O'Rourke NA, Dailey ME, Smith SJ, McConnel SK. (1991). Diverse migratory pathways in the developing cerebral cortex. *Science* 258:299–302.

Osen-Sand A, Catsicas M, Staple JK, Jones KA, Ayala G, Knowles J, Grenningloh G, Catsicas S. (1993). Inhibition of axonal growth by SNAP-25 antisense oligonucleotides in vitro and in vivo. *Nature* 364:445–448.

Otis TS, Mody I. (1992a). Modulation of decay kinetics and frequency of $GABA_A$ receptor-mediated spontaneous inhibitory postsynaptic currents in hippocampal neurons. *Neuroscience* 49:13–32.

Otis TS, Mody I. (1992b). Differential activation of $GABA_A$ and $GABA_B$ receptors by spontaneously released transmitter. *J Neurophysiol* 67:227–235.

Otis TS, De Koninck Y, Mody I. (1993). Characterization of synaptically elicited $GABA_B$ responses using patch-clamp recordings in rat hippocampal slices. *J Physiol* (Lond) 463:391–407.

Otis TS, De Koninck Y, Mody I. (1994). Lasting potentiation of inhibition is associated with an increased number of γ-aminobutyric acid type A receptors activated during miniature inhibitory postsynaptic currents. *Proc Natl Acad Sci USA* 91:7698–7702.

Otis TS, Staley KJ, Mody I. (1991). Perpetual inhibitory activity in mammalian brain slices generated by spontaneous GABA release. *Brain Res* 545:142–150.

Ottersen OP, Storm-Mathisen J. (1984a). Immunocytochemical visualization of GABA fixed by glutaraldehyde in brain tissue. *Neuropharmacology* 23: 855–857.

Ottersen OP, Storm-Mathisen J. (1984b). Glutamate and GABA-containing neurons in the mouse and rat brain, as demonstrated with a new immunocytochemical technique. *J Comp Neurol* 229:374–392.

Palmini A, Andermann F, Tampieri D, Andermann E, Robitaille Y, Olivier A. (1992). Epilepsy and cortical cytoarchitectonic abnormalities: an attempt at correlating basic mechanisms with anatomo-clinical syndromes. *Epilepsy Res* 9(suppl):19–30.

Papadopoulos GC, Parnavelas JG, Cavanagh ME. (1987). Extensive co-existence of neuropeptides in the rat visual cortex. *Brain Res* 420:95–99.

Pape H. (1992). Adenosine promotes burst activity in guinea pig geniculocortical neurones through two different ionic mechanisms. *J Physiol* (Lond) 447: 729–753.

Pape H, McCormick DA. (1989). Noradrenaline and serotonin selectively modulate thalamic burst firing by enhancing a hyperpolarization-activated cation current. *Nature* 340:715–718.

Parfitt KD, Madison DV. (1993). Phorbol esters enhance synaptic transmission by a presynaptic, calcium-dependent mechanism in rat hippocampus. *J Physiol* 471:245–268.

Parnavelas JG, Uylings HBM. (1980). The growth of non-pyramidal neurons in the visual cortex of the rat: a morphometric study. *Brain Res* 193:373–382.

Parnavelas JG, Papadopoulos GC, Cavanagh ME. (1988). Changes in neurotransmitters during development. In: Peters A, Jones EG, eds. *Cerebral Cortex.* New York: Plenum: 177–209.

Partin KM, Patneau DK, Winters CA, Mayer ML, Buonanno A. (1993). Selective modulation of desensitization of AMPA versus kainate receptors by cyclothiazide and concanavalin A. *Neuron* 11:1069–1082.

Patlak J. (1991). Molecular kinetics of voltage-dependent Na^+ channels. *Physiol Rev* 71:1047–1080.

Patneau DK, Mayer ML. (1990). Structure-activity relationships for amino acid transmitter candidates acting at N-methyl-D-aspartate and quisqualate receptors. *J Neurosci* 10:2385–2399.

Pedersen SE, Sharp SD, Liu W, Cohen JB. (1992). Structure of the noncompetitive antagonist-binding site of the Torpedo nicotinic acetylcholine receptor. [^{3}H]Meproadifen mustard reacts selectively with the alpha-subunit Glu-262. *J Biol Chem* 267:10489–10499.

Pellegrini A, Gloor P, Sherwin AL. (1978). Effect of valproate sodium on generalized penicillin epilepsy in the cat. *Epilepsia* 19:351–360.

Pellock JM, Coulter DA. (1994). Tridione. In: Levy RH, Dreifuss FE, Mattson RH, Meldrum BS, Penry JK, eds. *Antiepileptic Drugs*, 4th ed. New York: Raven Press.

Penfield W, Erickson T. (1941). *Epilepsy and Cerebral Localization.* Springfield, IL: Charles C Thomas.

Penfield W, Jasper H. (1954). *Epilepsy and the Functional Anatomy of the Human Brain.* Boston: Little, Brown and Company.

Pennartz CMA, Boeijinga PH, Kitai ST, Lopes da Silva FH. (1991). Contribution of NMDA receptors to postsynaptic potentials and paired-pulse facilitation in identified neurones of the rat nucleus accumbens. *Exp Brain Res* 86: 190–198.

Pennefather P, Lancaster B, Adams PR, Nicoll RA. (1985). Two distinct Ca-dependent K currents in bullfrog sympathetic ganglion cells. *PNAS* 82:3040–3044.

Penny GR, Arsharpour S, Kitai ST. (1986). Substance P-immunoreactive neurons in the neocortex of the rat: a subset of the glutamic acid decarboxylase-immunoreactive neurons. *Neurosci Lett* 65:53–59.

People with Epilepsy. (1989). London: Central Health Service Council's Advisory Committee on the Health and Welfare of Handicapped Persons.

Perreault P, Avoli M. (1991). Physiology and pharmacology of epileptiform activity induced by 4-aminopyridine in rat hippocampal slices. *J Neurophysiol* 65:771–785.

Persohn E, Malherbe P, Richards JG. (1992). Comparative molecular neuroanatomy of cloned $GABA_A$ receptor subunits in the rat CNS. *J Comp Neurol* 326:193–216.

Peters A. (1987). Number of neurons and synapses in primary visual cortex. In: Jones EG, Peters A, eds. *Cerebral Cortex, Vol 6. Further Aspects of Cortical Function, Including Hippocampus.* New York: Plenum: 267–294.

Peters A, Fairen A. (1978). Smooth and sparsely-spined stellate cells in the visual cortex of the rat: a study using a combined Golgi-electron microscope technique. *J Comp Neurol* 181:129–172.

Peters A, Feldman M. (1973). The cortical plate and molecular layer of the late rat fetus. *Z Anat Entwickl-Gesch* 141:3–37.

Peters A, Kara DA. (1987). The neuronal composition of area 17 of rat visual cortex. IV. The organization of pyramidal cells. *J Comp Neurol* 260:573–590.

Peters A, Proskauer CC. (1980). Synaptic relations between a multipolar stellate cell and a pyramidal neuron in the rat visual cortex. A combined Golgi-electron microscope study. *J Neurocytol* 9:163–183.

Peters A, Sethares C. (1991). Organization of pyramidal neurons in area 17 of monkey visual cortex. *J Comp Neurol* 306:1–23.

Peters A, Miller M, Kimerer LM. (1983). Cholecystokinin-like immunoreactive neurons in rat cerebral cortex. *Neuroscience* 8:431–448.

Peters A, Proskauer CC, Ribak CE. (1982). Chandelier cells in rat visual cortex. *J Comp Neurol* 206:397–416.

Phillips CG. (1959). Actions of antidromic pyramidal volleys on single Betz cells in the cat. *Q J Exp Physiol* 44:1–25.

Phillis JW, Edstrom JP, Kostopoulos GK, Kirkpatrick JR. (1979). Effects of adenosine and adenine nucleotides on synaptic transmission in the cerebral cortex. *Can J Physiol* 57:1289–1312.

Pinault D, Pumain R. (1985). Ectopic action potential generation: its occurrence in a chronic epileptogenic focus. *Exp Brain Res* 60:599–602.

Pinault D, Pumain R. (1989). Antidromic firing occurs spontaneously on thalamic relay neurons: triggering of somatic intrinsic burst discharges by ectopic action potentials. *Neuroscience* 31:625–637.

Pinel JPJ, Rovner LI. (1978). Experimental epileptogenesis: kindling-induced epilepsy in rats. *Exp Neurol* 58:190–202.

Pini A. (1993). Chemorepulsion of axons in the developing mammalian central nervous stem. *Science* 261:95–98.

Piredda S, Gale K. (1985). A crucial epileptogenic site in the deep prepiriform cortex. *Nature* 317:623–625.

Pitler TA, Alger BE. (1992). Cholinergic excitation of GABAergic interneurons in the rat hippocampal slice. *J Physiol* (Lond) 450:127–142.

Placzek M, Tessier-Lavigne M, Yamada T, Dodd J, Jessell TM. (1990). Guidance of developing axons by diffusible chemoattractants. *Cold Spring Harbor Symp Quant Biol* 55:279–289.

Pollen DA, Trachtenberg MC. (1970). Neuroglia: gliosis and focal epilepsy. *Science* 167:1252–1253.

Pons TP, Garraghty PE, Mishkin M. (1988). Lesion-induced plasticity in the second somatosensory cortex of adult macaques. *Proc Natl Acad Sci USA* 85: 5279–5281.

Pons TP, Garraghty PE, Ommaya AK, Kaas JH, Taub E, Mishkin M. (1991). Massive cortical reorganization after sensory deafferentation in adult macaques. *Science* 252:1857–1860.

Price J, Thurlow L. (1988). Cell lineage in the rat cerebral cortex: a study using retroviral-mediated gene transfer. *Development* 104:473–482.

Prince DA. (1968a). The depolarization shift in "epileptic" neurons. *Exp Neurol* 21:467–485.

Prince DA. (1968b). Inhibition in "epileptic" neurons. *Exp Neurol* 21:307–321.

Prince DA. (1969a). Microelectrode studies of penicillin foci. In: Jasper WA, ed. *Basic Mechanisms of the Epilepsies*. Boston: Little, Brown & Co, Inc.

Prince DA. (1969b). Electrophysiology of "epileptic" neurons: spike generation. *Electroencephalogr Clin Neurophysiol* 26:476–487.

Prince DA. (1976). Cellular activities in focal epilepsy. In: Brazier MAB, Coceani F, eds. *Brain Dysfunction in Infantile Febrile Convulsions*. New York: Raven Press: 187–212.

Prince DA. (1978). Neurophysiology of epilepsy. *Annu Rev Neurosci* 1:395–415.

Prince DA, Connors BW. (1986). Mechanisms of interictal epileptogenesis. *Adv Neurol* 44:275–299.

Prince DA, Futamachi KJ. (1970). Intracellular recordings from chronic epileptogenic foci in the monkey. *Electroencephalogr Clin Neurophysiol* 29:496–510.

Prince DA, Huguenard JR. (1988). Functional properties of neocortical neurons. In: Rakic P, Singer W, eds. Neurobiology of Neocortex. New York: John Wiley & Sons: 153–176.

Prince DA, Tseng G. (1993). Epileptogenesis in chronically injured cortex: in vitro studies. *J Neurophysiol* 69:1276–1291.

Prince DA, Wilder BJ. (1967). Control mechanisms in cortical epileptogenic foci. "Surround" inhibition. *Arch Neurol* 16:194–202.

Prince DA, Wong RKS. (1981). Human epileptic neurons studied *in vitro*. *Brain Res* 210:323–333.

Prince DA, Lux HD, Neher E. (1973). Measurement of extracellular potassium activity in cat cortex. *Brain Res* 50:489–495.

Prince DA, Pedley TA, Ransom BR. (1978). Fluctuations in ion concentrations during excitation and seizures. In: Schoffeniels E, Franck JE, eds. *Dynamic Properties of Glia Cells: An Interdisciplinary Approach to Their Study in the Central and Peripheral Nervous System.* Oxford: Pergamon Press.

Puia G, Vicini S, Seeburg PH, Costa E. (1991). Influence of recombinant gamma-aminobutyric acid-A receptor subunit composition on the action of allosteric modulators of gamma-aminobutyric acid-gated Cl^- currents. *Mol Pharmacol* 39:691–696.

Pulsinelli WA. (1993). Strokes involving gray matter: studies on in situ models of cerebral ischemia. In: Waxman SG, ed. *Molecular and Cellular Approaches to the Treatment of Neurological Disease.* New York: Raven Press.

Pumain R. (1981). Electrophysiological abnormalities in chronic epileptogenic foci: an intracellular study. *Brain Res* 219:445–450.

Pumain R, Louvel J, Kurcewicz I. (1986). Long-term alterations in amino acid-induced ionic conductances in chronic epilepsy. *Adv Exp Med Biol* 203: 439–447.

Pumain R, Louvel J, Gastard M, Kurcewicz I, Vergnes M. (1992). Responses to N-methyl-D-aspartate are enhanced in rats with petit mal-like seizures. *J Neural Transm Suppl* 35:97–108.

Purpura DP, Housepian EM. (1961). Morphological and physiological properties of chronically isolated immature neocortex. *Exp Neurol* 4:377–401.

Purpura DP, Penry JK, Tower D, Woodbury DM, Walter R. (1972a). *Experimental Models of Epilepsy.* New York: Raven Press.

Purpura DP, Penry JK, Woodbury DM, Tower DB, Walter RD. (1972b). *Experimental Models of Epilepsy—A Manual for the Laboratory Worker.* New York: Raven Press.

Purves D, Riddle DR, LaMantia AS. (1992). Iterated patterns of brain circuitry. *Trends Neurosci* 15:362–367.

Qian Z, Gilbert ME, Colicos MA, Kandel ER, Kuhl D. (1993). Tissue plasminogen activator is induced as an immediate early gene during seizure, kindling and long term potentiation. *Nature* 361:453–457.

Qiao X, Noebels JL. (1991). Genetic heterogeneity of inherited spike-wave epilepsy: two mutant gene loci with independent cerebral excitability defects. *Brain Res* 555:43–50.

Qiao X, Noebels JL. (1992). $GABA_B$ receptor independent spike-wave epilepsy in the mutant mouse stargazer. *Pharmacol Abstr* 18:553.

Qiao X, Noebles JL. (1993). Developmental analysis of hippocampal mossy fiber outgrowth in a mutant mouse with inherited spike-wave seizures. *J Neurosci* 13:4622–4635.

Raastad M, Storm JF, Andersen P. (1992). Putative single quantum and single fibre excitatory postsynaptic currents show similar amplitude range and variability in rat hippocampal slices. *Eur J Neurosci* 4:113–117.

Racine RJ. (1972). Modulation of seizure activity by electrical stimulation. II. Motor seizure. *Electroencephalogr Clin Neurophysiol* 32:281–294.

Rakic P. (1972a). Extrinsic cytological determinants of basket and stellate cell dendritic pattern in the cerebellar molecular layer. *J Comp Neurol* 146:335–354.

Rakic P. (1972b). Mode of cell migration to the superficial layers of fetal monkey neocortex. *J Comp Neurol* 145:61–83.

Rall W. (1967). Distinguishing theoretical synaptic potentials computed for different soma-dendritic distributions of synaptic input. *J Neurophysiol* 30: 1138–1168.

Rapp M, Yarom Y, Segev I. (1992). The impact of parallel fiber background activity on the cable properties of cerebellar Purkinje cells. *Neural Comp* 4: 518–533.

Rauschecker JP, Hahn S. (1987). Ketamine-xylazine anesthesia blocks consolidation of ocular dominance changes in kitten visual cortex. *Nature* 326:183–185.

Raymond LA, Blackstone CD, Huganir RL. (1993). Phosphorylation and modulation of recombinant GluR6 glutamate receptors by cAMP-dependent protein kinase. *Nature* 361:637–641.

Raymond LR, Blackstone CD, Huganir RL. (1993). Phosphorylation of amino acid neurotransmitter receptors in synaptic plasticity. *Trends Neurosci* 16: 147–153.

Raynor K, Reisine T. (1992). Somatostatin receptors. *Crit Rev Neurobiol* 6:273–289.

Rayport M, Waller HJ. (1967). Technique and results of micro-electrode recording in human epileptogenic foci. *Electroencephalogr Clin Neurophysiol* 25: 143–151.

Recanzone GH, Merzenich MM, Dinse HR. (1992). Expansion of the cortical representation of a specific skin field in primary somatosensory cortex by intracortical microstimulation. *Cereb Cortex* 2:181–196.

Redburn DA, Schousboe A. (1975). *Neurotrophic Activity of GABA During Development.* New York: Alan R Liss.

Redman S. (1990). Quantal analysis of synaptic potentials in neurons of the central nervous system. *Physiol Rev* 70:165–198.

Regehr WG, Tank DW. (1990). Postsynaptic NMDA receptor-mediated calcium accumulation in hippocampal CA1 pyramidal cell dendrites. *Nature* 345: 807–810.

Regehr W, Kehoe JS, Ascher P, Armstrong C. (1993). Synaptically triggered action potentials in dendrites. *Neuron* 11:145–151.

Rehder V, Kater SB. (1992). Regulation of neuronal growth cone filopodia by intracellular calcium. *J Neurosci* 12:3175–3186.

Reid SA, Palovcik RA. (1989). Spontaneous epileptiform discharges in isolated human cortical slices from epileptic patients. *Neurosci Lett* 98:200–204.

Reiner O, Carrozzo R, Shen Y, Wehnert M, Faustinella F, Dobyns WC, Caskey CT, Ledbetter DH. (1993). Isolation of a Miller-Dieker lissencephaly gene containing G protein β-subunit-like repeats. *Nature* 364:717–721.

Reith AD, Bernstein A. (1991). Molecular basis of mouse developmental mutants. *Genes Dev* 5:1115–1123.

Ren J, Hamada J, Takeichi N, Fujikawa S, Kobayashi H. (1990). Ultrastructural differences in junctional intercellular communication between highly and weakly metastatic clones derived from rat mammary carcinoma. *Cancer Res* 50:358–362.

Represa A, LaSalle GLG, Ben-Ari Y. (1989a). Hippocampal plasticity in childhood epilepsy. *Neurosci Lett* 99:351–355.

Represa A, Tremblay E, Ben-Ari Y. (1989b). Transient increase of NMDA-binding sites in human hippocampus during development. *Neurosci Lett* 99:61–66.

Reuveni I, Friedman A, Amitai Y, Gutnick MJ. (1993). Stepwise repolarization from Ca^{2+} plateaus in neocortical pyramidal cells: evidence for nonhomogeneous distribution of HVA Ca^{2+} channels in dendrites. *J Neurosci* 13: 4609–4621.

Revah F, Galzi J, Giraudat J, Haumont P, Lederer F, Changeux J. (1990). The noncompetitive blocker [3H]chlorpromazine labels three amino acids of the acetylcholine receptor y-subunit: implications for the alpha-helical organization of regions MII and for the structure of the ion channel. *Proc Natl Acad Sci USA* 87:4675–4679.

Ribak CE. (1978). Aspinous and sparsely-spinous stellate neurons in the visual cortex of rats contain glutamic acid decarboxylase. *J Neurocytol* 7:461–478.

Ribak CE. (1985). Axon terminals of GABAergic chandelier cells are lost at epileptic foci. *Brain Res* 326:251–260.

Ribak CE. (1991). Epilepsy and the cortex. Anatomy. In: Peters A, Jones EG, eds. *Cerebral Cortex*. New York: Plenum: 427–483.

Ribak CE, Reiffenstein RJ. (1982). Selective inhibitory synapse loss in chronic cortical slabs: a morphological basis for epileptic susceptibility. *Can J Physiol Pharmacol* 60:864–870.

Ribak CE, Bradburne RM, Harris AB. (1982). A preferential loss of GABAergic, inhibitory synapses in epileptic foci: A quantitative ultrastructural analysis of monkey neocortex. *J Neurosci* 2:1725–1735.

Ribak CE, Harris AB, Vaughn JE, Roberts E. (1979). Inhibitory GABAergic nerve terminals decrease at sites of focal epilepsy. *Science* 205:211–214.

Ribak CE, Joubran C, Kesslak JP, Bakay RA. (1989). A selective decrease in the number of GABAergic somata occurs in pre-seizing monkeys with alumina gel granuloma. *Epilepsy Res* 4:126–138.

Rinzel J, Rall W. (1974). Transient response in a dendritic neuron model for current injected at one branch. *Biophys J* 14:759–789.

Robbins RJ, Brines MJ, Kim JH, Adrian T, De Lanerolle N, Welsh S, Spencer DD. (1991). A selective loss of somatostatin in the hippocampus of patients with temporal lobe epilepsy. *Ann Neurol* 29:325–332.

Roberts E. (1986). GABA: the road to neurotransmitter status. In: Olsen RW, Venter JC, eds. *Benzodiazepine/GABA Receptors and Chloride Channels: Structural and Functional Properties*. New York: Alan R Liss: 1–39.

Roberts E, Sherman MA. (1993). GABA—the quintessential neurotransmitter: electroneutrality, fidelity, specificity, and a model for the ligand binding site of $GABA_A$ receptors. *Neurochem Res* 4:365–376.

Robitaille R, Charlton MP. (1992). Presynaptic calcium signals and transmitter release are modulated by calcium-activated potassium channels. *J Neurosci* 12(1):297–305.

Robitaille R, Tremblay JP. (1991). Non-uniform responses to Ca^{2+} along the frog neuromuscular junction: effects on the probability of spontaneous and evoked transmitter release. *Neuroscience* 40:571–585.

Roche KW, Raymond LA, Blackstone C, Huganir RL. (1994). Transmembrane topology of the glutamate receptor subunit GluR6. *J Biol Chem* 269: 11679–11682.

Rogers CJ, Twyman RE, MacDonald RL. (1994). Benzodiazepine and b-carboline regulation of single $GABA_A$ receptor channels of mouse spinal neurones in culture. *J Physiol* (Lond) 475:69–82.

Ropert N, Miles R, Korn H. (1990). Characteristics of miniature inhibitory post-synaptic currents in CA1 pyramidal neurones of rat hippocampus. *J Physiol* (Lond) 428:707–722.

Rose D, Blakemore C. (1974). An analysis of orientation selectivity in the cat's visual cortex. *Exp Brain Res* 20:1–17.

Rosen AD, Vastola EF. (1971). Corticofugal antidromic activity in epileptogenic foci. *Trans Am Neurol Assoc* 96:297.

Rosen AD, Vastola EF, Hildebrand ZJM. (1973). Visual radiation activity during a cortical penicillin discharge. *Exp Neurol* 40:1–11.

Rosen AS, Morris ME. (1993). Anoxic depression of excitatory and inhibitory postsynaptic potentials in rat neocortical slices. *J Neurophysiol* 69:109–117.

Rosen JB, Cain CJ, Weiss SR, Post RM. (1992). Alterations in mRNA of enkephalin, dynorphin and thyrotropin releasing hormone during amygdala kindling: an in situ hybridization study. *Brain Res Mol Brain Res* 15:247–255.

Rosenmund C, Clements JD, Westbrook GL. (1993). Nonuniform probability of glutamate release at a hippocampal synapse. *Science* 262:754–757.

Ross CA, McInnis MG, Margolis RL, Li S. (1993). Genes with triplet repeats: candidate mediators of neuropsychiatric disorders. *Trends Neurosci* 16: 254–260.

Rossant J. (1990). Manipulating the mouse genome: implications for neurobiology. *Neuron* 2:323–334.

Rothman S. (1984). Synaptic release of excitatory amino acid neurotransmitter mediates anoxic neuronal death. *J Neurosci* 4:1884–1891.

Rothman SM, Olney JW. (1986). Glutamate and the pathophysiology of hypoxic-ischemic brain damage. *Ann Neurol* 19:105–111.

Rudy B. (1988). Diversity and ubiquity of K channels. *Neuroscience* 25:729–749.

Rutecki PA, Grossman RG, Armstrong D, Irish-Loewen S. (1989). Electrophysiological connections between the hippocampus and entorhinal cortex in patients with complex partial seizures. *J Neurosurg* 70:667–675.

Rutledge LT. (1969). Effect of stimulation on isolated cortex. In: Jasper HH, Ward AA Jr, Pope A, eds. *Basic Mechanisms of the Epilepsies*. Boston: Little, Brown: 349–355.

Rutledge LT, Ranck JB Jr, Duncan JA. (1967). Prevention of supersensitivity in partially isolated cerebral cortex. *Electroencephalogr Clin Neurophysiol* 23: 256–262.

Ruzdijic S, Pekovic S, Kanazir S, Ivkovic S, Stojiljkovic M, Rakic L. (1993). Temporal and spatial preferences of c-fos mRNA expression in the rat brain following cortical lesion. *Brain Res* 601:230–240.

Ryan SG, Buckwalter MS, Lynch JW, Handford CA, Segura L, Shiang R, Wasmuth JJ, Camper SA, Schofield P, O'Connell. (1994). A missense mutation in the gene encoding the α1 subunit of the inhibitory glycine receptor in the spasmodic mouse. *Nature Gen* 7:131–135.

Ryo Y, Miyawaki A, Furuichi T, Mikoshiba K. (1993). Expression of the metabotropic glutamate receptor mGluR1α and the ionotropic glutamate receptor GluR1 in the brain during the postnatal development of normal mouse and in the cerebellum from mutant mice. *J Neurosci Res* 36:19–32.

Sah P, Hestrin S, Nicoll RA. (1989). Tonic activation of NMDA receptors by ambient glutamate enhances excitability of neurons. *Science* 246:815–818.

Saint DA, Ju Y-K, Gage PW. (1992). A persistent sodium current in rat ventricular myocytes. *J Physiol* (Lond) 453:219–231.

Sakmann B, Brenner HR. (1978). Change in synaptic channel gating during neuromuscular development. *Nature* 276:401–402.

Sakurada K, Masu M, Nakanishi S. (1993). Alteration of Ca^{2+} permeability and sensitivity to Mg^{2+} and channel blockers by a single amino acid substitution in the N-methyl-D-aspartate receptor. *J Biol Chem* 268:410–415.

Salazar AM, Jabbari B, Vance SC, Grafman J, Amin D, Dillon JD. (1985). Epilepsy after penetrating head injury. I. Clinical correlates: a report of the Vietnam Head Injury Study. *Neurology* 35:1406–1414.

Salin PA, Parada I, Hoffman SN, Tseng GF, Prince DA. (1993). Axonal sprouting of adult rat neocortical pyramidal cells in chronic epileptogenic lesions. *Soc Neurosci Abstr* 19:2031.

Samulack DD, Lacaille JC. (1993). Hyperpolarizing synaptic potentials evoked in CA1 pyramidal cells by glutamate stimulation of interneurons from the oriens/alveus border of rat hippocampal slices. II. Sensitivity to GABA antagonists. *Hippocampus* 3:345–358.

Sanes JR. (1993). Topographic maps and molecular gradients. *Curr Opin Neurobiol* 3:67–74.

Sano K, Malamud N. (1953). Clinical significance of sclerosis of the cornu ammonis. *Arch Neurol Psychiatr* 70:40–53.

Sastry BR, Goh JW, Auyeung A. (1986). Associative induction of posttetanic and long-term potentiation in CA1 neurons of rat hippocampus. *Science* 232: 988–990.

Savage DD, McNamara JO. (1982). Kindled seizures selectively reduce a subpopulation of [^{3}H]quinuclidinyl benzilate binding sites in rat dentate gyrus. *J Pharmacol Exp Ther* 222:670–673.

Savage DD, Dashieff RM, McNamara JO. (1983). Kindled seizure-induced reduction of muscarinic cholinergic receptors in rat hippocampal formation: evidence for localization to dentate granule cells. *J Comp Neurol* 221:106–112.

Savage DD, Nadler JV, McNamara JO. (1984). Reduced kainic acid binding in rat hippocampal formation after limbic kindling. *Brain Res* 323:128–131.

Savage DD, Rigsbee LC, McNamara JO. (1985). Knife cuts of the entorhinal cortex: effects on development of amygdaloid kindling and seizure-induced decrease of muscarinic cholinergic receptors. *J Neurosci* 5:408–413.

Sayer RJ, Schwindt PC, Crill WE. (1990). High- and low-threshold calcium currents in neurons acutely isolated from rat sensorimotor cortex. *Neurosci Lett* 120:175–178.
Scharfman HE, Sarvey JM. (1985). Responses to t-aminobutyric acid applied to cell bodies and dendrites of rat visual cortical neurons. *Brain Res* 358: 385–389.
Scharfman HE, Sarvey JM. (1987). Responses to GABA recorded from identified rat visual cortical neurons. *Neuroscience* 23:407–422.
Scharfman HE, Sarvey JM. (1988). Physiological correlates of responses to gamma-aminobutyric acid (GABA) recorded from rat visual cortical neurons in vitro. *Synapse* 2:619–626.
Scharfman HE, Schwartzkroin PA. (1989). Protection of dentate hilar cells from prolonged stimulation by intracellular calcium chelation. *Science* 246:257–260.
Scharfman HE, Kunkel DD, Schwartzkroin PA. (1990). Synaptic connections of dentate granule cell and hilar neurons: results of paired intracellular recordings and intracellular horseradish peroxidase injections. *Neuroscience* 237:693–707.
Scheibel ME, Crandall PH, Scheibel AB. (1974). The hippocampal-dentate complex in temporal lobe epilepsy. *Epilepsia* 15:55–80.
Schlaggar BL, Fox K, O'Leary DD. (1993). Postsynaptic control of plasticity in developing somatosensory cortex. *Nature* 364:623–626.
Schlessinger J, Ullrich A. (1992). Growth factor signaling by receptor tyrosine kinases. *Neuron* 9:383–391.
Schmechel DE, Vickrey BG, Fitzpatrick D, Elde RP. (1984). GABAergic neurons of mammalian cerebral cortex: widespread subclass defined by somatostatin content. *Neurosci Lett* 47:227–232.
Schofield PR, Darlison MG, Fujita N, Burt DR, Stephenson FA, Rodriguez H, Rhee LM, Ramachandran J, Reale V, Glencorse TA. (1987). Sequence and functional expression of the GABA A receptor shows a ligand-gated receptor super-family. *Nature* 328:221–227.
Schröder H, Becker A, Lossner B. (1993). Glutamate binding to brain membranes is increased in pentylenetetrazole-kindled rats. *J Neurochem* 60:1007–1011.
Schulman H. (1993). The multifunctional Ca^{2+}/calmodulin-dependent protein kinases. *Curr Opin Cell Biol* 5:247–253.
Schulman H, Hanson PI. (1993). Multifunctional Ca^{2+}/calmodulin-dependent protein kinase. *Neurochem Res* 18:65–77.
Schulman H, Hanson PI, Meyer T. (1992). Decoding calcium signals by multifunctional CaM kinase. *Cell Calcium* 13:401–411.
Schwark HD, Jones EG. (1989). The distribution of intrinsic cortical axons in area 3b of cat primary somatosensory cortex. *Exp Brain Res* 78:501–513.
Schwartz JH, Greenberg SM. (1987). Molecular mechanisms for memory: second-messenger induced modifications of protein kinases in nerve cells. *Annu Rev Neurosci* 459–476.
Schwartzkroin PA. (1993). *Epilepsy: Models, Mechanisms and Concepts.* New York: Cambridge Press.
Schwartzkroin PA, Haglund MM. (1986). Spontaneous rhythmic synchronous activity in epileptic human and normal monkey temporal lobe. *Epilepsia* 27: 523–533.

Schwartzkroin PA, Knowles WD. (1984). Intracellular study of human epileptic cortex: in vitro maintenance of epileptiform activity? *Science* 223:709–712.
Schwartzkroin PA, Kunkel DD. (1985). Morphology of identified interneurons in the CA1 region of guinea pig hippocampus. *J Comp Neurol* 232:205–218.
Schwartzkroin PA, Mathers LH. (1978). Physiological and morphological identification of a nonpyramidal hippocampal cell type. *Brain Res* 157:1–10.
Schwartzkroin PA, Prince DA. (1976). Microphysiology of human cerebral cortex studied *in vitro*. *Brain Res* 115:497–500.
Schwartzkroin PA, Prince DA. (1978). Cellular and field potential properties of epileptogenic hippocampal slices. *Brain Res* 147:117–130.
Schwartzkroin PA, Prince DA. (1980a). Changes in excitatory and inhibitory synaptic potentials leading to epileptogenic activity. *Brain Res* 183:61–76.
Schwartzkroin PA, Prince DA. (1980b). Effects of TEA on hippocampal neurons. *Brain Res* 185:169–181.
Schwartzkroin PA, Stafsstrom CE. (1980). Effects of EGTA on the calcium-activated afterhyperpolarization in hippocampal CA3 pyramidal cells. *Science* 210:1125–1126.
Schwartzkroin PA, Wyler AR. (1980). Mechanisms underlying epileptiform burst discharge. *Ann Neurol* 7:95–107.
Schwartzkroin PA, Duijn H, Prince DA. (1974). Effects of projected cortical epileptiform discharges on unit activity in the cat cuneate nucleus. *Exp Neurol* 43:106–123.
Schwartzkroin PA, Scharfman HE, Sloviter RS. (1990). Similarities in circuitry between Ammon's horn and dentate gyrus: local interactions and parallel processing. In: Storm-Mathisen J, Zimmer J, eds. *The Hippocampal Region as a Model for Studying Brain Structure and Function.* Amsterdam: Elsevier: 269–286.
Schwartzkroin PA, Futamachi KJ, Noebles JL, Prince DA. (1975a). Transcallosal effects of a cortical epileptiform focus on ventrolateral nucleus of the cat. *Brain Res* 99:59–68.
Schwartzkroin PA, Mutani R, Prince DA. (1975b). Orthodromic and antidromic effects of a cortical epileptiform focus on ventrolateral nucleus of the cat. *J Neurophysiol* 38:795–811.
Schwartzkroin PA, Turner DA, Knowles WD, Wyler AR. (1983). Studies of human and monkey "epileptic" neocortex in the in vitro slice preparation. *Ann Neurol* 13:249–257.
Schwindt PC. (1992). Ionic currents governing input-output relations of Betz cells. In: McKenna T, Davis D, Zornetzer SF, eds. *Single Neuron Computation.* San Diego: Academic Press: 235–258.
Schwindt PC, Spain W, Crill WE. (1989). Long-lasting reduction of excitability by a sodium-dependent potassium current in cat neocortical neurons. *J Neurophysiol* 61:233–244.
Schwindt PC, Spain W, Crill WE. (1992a). Calcium-dependent potassium currents in neurons from cat sensorimotor cortex. *J Neurophysiol* 67:216–226.
Schwindt PC, Spain W, Crill WE. (1992b). Effects of intracellular calcium chelation on voltage-dependent and calcium-dependent currents in cat neocortical neurons. *Neuroscience* 47:571–578.
Schwindt PC, Spain W, Foehring RC, Stafstrom CE, Chubb MC, Crill WE. (1988). Multiple potassium conductances and their functions in neurons from cat sensorimotor cortex in vitro. *J Neurophysiol* 59:424–449.

Segal M. (1982). Multiple actions of acetylcholine at a muscarinic receptor studied in the rat hippocampal slice. *Brain Res* 246:77–87.
Segal M. (1987). Repetitive inhibitory postsynaptic potentials evoked by 4-aminopyridine in hippocampal neurons in vitro. *Brain Res* 414:285–293.
Segal M. (1990). A subset of local interneurons generate slow inhibitory postsynaptic potentials in hippocampal neurons. *Brain Res* 511:163–164.
Sejnowski TJ. (1977). Strong covariance with nonlinearly interacting neurons. *J Math Biol* 4:303–321.
Seress L, Ribak CE. (1983). GABAergic cells in the dentate gyrus appear to be local circuit and projection neurons. *Exp Brain Res* 50:173–182.
Serratosa JM, Weissbecker K, Delgado-Escueta AV. (1990). Childhood absence epilepsy: An autosomal recessive disorder? *Epilepsia* 31:651.
Servit Z, Strejckova A. (1970a). Epileptic focus in the frog forebrain. Triggering of the focal discharge with sensory stimuli. *Exp Neurol* 28:371–383.
Servit Z, Strejckova A. (1970b). An electrographic epileptic focus in the fish forebrain. Conditions and pathways of progagation of focal and paroxysmal activity. *Brain Res* 17:103–113.
Seubert P, Larson J, Oliver M, Jung MW, Baudry M, Lynch G. (1988). Stimulation of NMDA receptors induces proteolysis of spectrin in hippocampus. *Brain Res* 460:189–194.
Sharpless SK, Halpern LM. (1962). The electrical excitability of chronically isolated cortex studied by means of permanently implanted electrodes. *Electroencephalogr Clin Neurophysiol* 14:244–255.
Shatz CJ. (1990). Impulse activity and the patterning of connections during CNS development. *Neuron* 5:745–756.
Shaw C, Cameron L, March D, Cynader M, Zielinski B, Hendrickson A. (1991). Pre- and postnatal development of GABA receptors in Macaca monkey visual cortex. *J Neurosci* 11:3943–3959.
Sheardown MJ, Nieben EO, Hansen AJ, Jacobsen P, Honore T. (1990). 2,3-Dihydroxy-6-nitro-7-sulfamoyl-benzo (F) quinoxaline: a neuroprotectant for cerebral ischemia. *Science* 247:571–574.
Shelton DP. (1985). Membrane resistivity estimated for the Purkinje neuron by means of a passive computer model. *Neuroscience* 14:111–131.
Shen JM, Kriegstein AR. (1989). The development of bicuculline-induced epileptiform discharges in embryonic turtle cortex. *Neurosci Lett* 98:184–188.
Sheng M, Greenberg M. (1990). The regulation and function of c-fos and other immediate early genes of the nervous system. *Neuron* 4:477–485.
Sheng M, Thompson MA, Greenberg ME. (1991). A Ca^{2+}-regulated transcription factor phosphorylated by calmodulin-dependent kinases. *Science* 252: 1427–1430.
Sheng M, Liao YJ, Jan YN, Jan LY. (1993). Presynaptic A-current based on heteromeric K^+ channels detected *in vivo*. *Nature* 365:72–74.
Sheng M, Cummings J, Roldan LA, Jan YN, Jan LY. (1994). Changing subunit composition of heteromeric NMDA receptors during development of rat cortex. *Nature* 368:144–147.
Shiang R, Ryan RS, Zhu YZ, Hahn AF, O'Connell P, Wasmuth JJ. (1993). Mutations in the $\alpha 1$ subunit of the inhibitory glycine receptor cause the dominant neurologic disorder, hyperexplexia. *Nature Genet* 5:351–358.
Shigemoto R, Nakanishi S, Mizuno N. (1992). Distribution of the mRNA for a metabotropic glutamate receptor (mGluR1) in the central nervous system:

an in situ hybridization study in adult and developing rat. *J Comp Neurol* 322:121–135.

Shin C, McNamara JO, Morgan JI, Curran T, Cohen DR. (1990). Induction of c-fos mRNA expression by afterdischarge in the hippocampus of naive and kindled rats. *J Neurochem* 55:1050–1055.

Shin C, Pedersen HB, McNamara JO. (1985). gamma-Aminobutyric acid and benzodiazepine receptors in the kindling model of epilepsy: a quantitative radiohistochemical study. *J Neurosci* 5:2696–2701.

Shirasaki T, Munakata M, Akaike N. (1993). Heterogeneous distribution and developmental change of metabotropic glutamate receptors in the rat hippocampus CA1 pyramidal neurons. *Neurosci Lett* 157:191–194.

Shirokawa T, Nishigori A, Kimura F, Tsumoto T. (1989). Actions of excitatory amino acid antagonists on synaptic potential of layer 2/3 neurons of the cat's visual cortex. *Exp Brain Res* 78:489–500.

Shlaer R. (1971). Shift in binocular disparity causes compensatory change in the cortical structure of kittens. *Science* 173:638–641.

Sholl DA. (1956). *The Organization of the Cerebral Cortex.* New York: Hafner Publishing.

Siesjo BK, Bengtsson F. (1989). Calcium fluxes, calcium antagonists, and calcium-related pathology in brain ischemia, hypoglycemia, and spreading depression: a unifying hypothesis. *J Cereb Blood Flow Metab* 9:127–140.

Sigel E, Baur R, Trube G, Möhler H, Malherbe P. (1990). The effect of subunit composition of rat brain $GABA_A$ receptors on channel function. *Neuron* 5: 703–711.

Sigworth FJ. (1980). The variance of sodium current fluctuations at the node of Ranvier. *J Physiol* (Lond) 307:97–129.

Sillito AM. (1975). The contribution of inhibitory mechanisms to the receptive field properties of neurones in the striate cortex of the cat. *J Physiol* (Lond) 250:305–329.

Sillito AM. (1977). Inhibitory processes underlying the directional specificity of simple, complex and hypercomplex cells in the cat's visual cortex. *J Physiol* (Lond) 271:699–720.

Sillito AM, Versiani V. (1977). The contribution of excitatory and inhibitory inputs to the length preference of hypercomplex cells in layers II and III of the cat's striate cortex. *J Physiol* (Lond) 273:775–790.

Silva AJ, Paylor R, Wehner JM, Tonegawa S. (1992a). Impaired spatial learning in a calcium-calmodulin kinase II mutant mice. *Science* 257:206–211.

Silva AJ, Stevens CF, Tonegawa S, Wang Y. (1992b). Deficient hippocampal long-term potentiation in calcium-calmodulin kinase II mutant mice. *Science* 257:201–206.

Silva LR, Connors BW. (1992). Synchronized oscillations intrinsic to the neocortex. In: Speckmann EJ, Gutnick MJ, eds. *Epilepsy and Inhibition.* Munich: Urban & Schwarzenberg: 215–227.

Silva LR, Amitai Y, Connors BW. (1991a). Intrinsic oscillations of neocortex generated by layer 5 pyramidal neurons. *Science* 251:432–435.

Silva LR, Gutnick MJ, Connors BW. (1991b). Laminar distribution of neuronal membrane properties in neocortex of normal and reeler mouse. *J Neurophysiol* 66:2034–2040.

Silva LR, Chagnac-Amitai Y, Connors BW. (1994). Inhibitory synaptic input correlates with intrinsic membrane properties in neocortical pyramidal cells. *Soc Neurosci Abstr* 15:660.

Silver RA, Traynelis SF, Cull-Candy SG. (1992). Rapid-time-course miniature and evoked excitatory currents at cerebellar synapses *in situ*. *Nature* 355:163–166.

Silvia RC, Piercey MF, Hoffmann WE. (1987). U-74006F, an inhibitor of lipid peroxidations protects against lesion development following experimental stroke in the cat: histological and metabolic analysis. *Soc Neurosci Abstr* 13:1499.

Simmons PA, Pearlman AL. (1983). Receptive-field properties of transcallosal visual cortical neurons in the normal and reeler mouse. *J Neurophysiol* 50: 838–848.

Simon DK, Prusky GT, O'Leary DD, Constantine-Paton M. (1992). N-Methyl-D-aspartate receptor antagonists disrupt the formation of a mammalian neural map. *Proc Natl Acad Sci USA* 89:10593–10597.

Simon RP, Swan JH, Griffiths T, Meldrum BS. (1984). Blockade of N-methyl-D-aspartate receptors may protect against ischemic damage in the brain. *Science* 226:850–852.

Sivilotti L, Nistri A. (1991). GABA receptor mechanisms in the central nervous system. *Progr Neurobiol* 36:35–92.

Skene JHP. (1990). GAP-43 as a "calmodulin sponge" and some implications for calcium signaling in axon terminals. *Neurosci Res* 13:S112–S125.

Sladeczek F, Momiyama A, Takahashi T. (1993). Presynaptic inhibitory action of a metabotropic glutamate receptor agonist on excitatory transmission in visual cortical neurones. *J Neurochem* 61:S264.

Sloviter RS. (1983). "Epileptic" brain damage in rats induced by sustained electrical stimulation of the perforant path. I. Acute electrophysiological and light microscopic studies. *Brain Res Bull* 10:675–697.

Sloviter RS. (1987). Decreased hippocampal inhibition and a selective loss of interneurons in experimental epilepsy. *Science* 235:73–76.

Sloviter RS. (1991a). Permanently altered hippocampal structure, excitability, and inhibition after experimental status epilepticus in the rat: the "dormant basket cell" hypothesis and its possible relevance to temporal lobe epilepsy. *Hippocampus* 1:41–66.

Sloviter RS. (1991b). Feedforward and feedback inhibition of hippocampal principal cell activity evoked by perforant path stimulation: GABA-mediated mechanisms that regulate excitability *in vivo*. *Hippocampus* 1:31–40.

Sloviter RS. (1992). Possible functional consequences of synaptic reorganization in the dentate gyrus of kainate-treated rats. *Neurosci Lett* 137:91–96.

Smart TG. (1989). Excitatory amino acids: the involvement of second messengers in the signal transduction process. *Cell Mol Neurobiol* 9:193–206.

Smeyne RJ, Vendrell M, Hayward M, Baker SJ, Miao GG, Schilling K, Robertson LM, Curran T, Morgan JI. (1993). Continuous c-fos expression precedes programmed cell death in vivo. *Nature* 363:166–169.

Smith DO. (1980a). Morphological aspects of the safety factor for action potential propagation at axon branch points in the crayfish. *J Physiol* (Lond) 301: 261–269.

Smith DO. (1980b). Mechanisms of action potential propagation failure at sites of axon branching in the crayfish. *J Physiol* (Lond) 301:243–259.

Smith JMB, Kellaway P. (1964a). The natural history and clinical correlates of occipital foci in children. In: Kellaway P, Petersen I, eds. *Neurological and*

Electroencephalographic Correlative Studies in Infancy. New York: Grune & Stratton: 230–249.

Smith JMB, Kellaway P. (1964b). Central (rolandic) foci in children: an analysis of 200 cases. *Electroencephalogr Clin Neurophysiol* 17:451–472.

Smith MK, Colbran RJ, Soderling TR. (1990). Specificities of autoinhibitory domain peptides for four protein kinases. Implications for intact cell studies of protein kinase function. *J Biol Chem* 265:1837–1840.

Snyder EY, Deitcher DL, Walsh C, Arnold-Ridea S, Hartweig EA, Cepko CL. (1992). Multipotent neuronal cell lines can engraft and participate in development of mouse cerebellum. *Cell* 68:33–51.

Solbach S, Celio MR. (1991). Ontogeny of the calcium binding protein parvalbumin in the rat nervous system. *Anat Embryol* (Berl) 184:103–124.

Solis JM, Nicoll RA. (1992). Pharmacological characterization of $GABA_B$-mediated responses in the CA1 region of the rat hippocampal slice. *J Neurosci* 12:3466–3472.

Soltesz I, Mody I. (1994). Patch-clamp recordings reveal powerful GABAergic inhibition in dentate hilar neurons. *J Neurosci* 14:2365–2376.

Sommer B, Keinanen K, Verdoorn TA, Wisden W, Burnashev N, Herb A, Kohler M, Takagi T, Sakmann B, Seeburg PH. (1990). Flip and flop: a cell-specific functional switch in glutamate-operated channels of the CNS. *Science* 249: 1580–1585.

Somogyi P. (1977). A specific "axo-axonal" interneuron in the visual cortex of the rat. *Brain Res* 136:345–350.

Somogyi P. (1990). Synaptic organization of GABAergic neurons and $GABA_A$ receptors in the lateral geniculate nucleus and the visual cortex. In: Lam DM, Gilbert CD, eds. *Neural Mechanisms of Visual Perception.* Gulf Publishing: 35–62.

Somogyi P, Cowey A. (1981). Combined Golgi and electron microscopic study on the synapses formed by double bouquet cells in the visual cortex of the cat and monkey. *J Comp Neurol* 195:547–566.

Somogyi P, Cowey A. (1984). Double bouquet cells. In: Peters A, Jones EG, eds. *The Cerebral Cortex. Vol. I, Cellular Components of the Cerebral Cortex.* New York: Plenum: 337–360.

Somogyi P, Hodgson AJ, Smith AD. (1979). An approach to tracing neuron networks in the cerebral cortex and basal ganglia. Combination of Golgi staining, retrograde transport of horseradish peroxidase and anterograde degeneration of synaptic boutons in the same material. *Neuroscience* 4:1805–1852.

Somogyi P, Freund TF, Cowey A. (1982). The axo-axonic interneuron in the cerebral cortex of the rat, cat and monkey. *Neuroscience* 7:2577–2609.

Somogyi P, Cowey A, Halasz N, Freund TF. (1981). Vertical organization of neurons accumulating ^{3}H-GABA in the visual cortex of the rhesus monkey. *Nature* 294:761–763.

Somogyi P, Nunzi MG, Gorio A, Smith AD. (1983). A new type of specific interneuron in the monkey hippocampus forming synapses exclusively with the axon initial segments of pyramidal cells. *Brain Res* 259:137–142.

Somogyi P, Freund TF, Kisvarday ZF. (1984a). Different types of 3H-GABA accumulating neurons in the visual cortex of the rat. Characterization by combined autoradiography and Golgi impregnation. *Exp Brain Res* 54:45–56.

Somogyi P, Hodgson AJ, Smith AD, Nunzi MG, Gorio A, Wu J. (1984b). Different populations of GABAergic neurons in the visual cortex and hippocampus of cat contain somatostatin- or cholecystokinin-immunoreactive material. *J Neurosci* 4:2590–2602.

Somogyi P, Freund TF, Hodgson AJ, Somogyi J, Beuroukas D, Chubb IW. (1985). Identified axo-axonic cells are immunoreactive for GABA in the hippocampus and visual cortex of the cat. *Brain Res* 332:143–149.

Sonnenberg JL, Frantz GD, Lee S, Heick A, Chu C, Tobin AJ, Christakos S. (1991). Calcium binding protein (calbindin-D28k) and glutamate decarboxylase gene expression after kindling induced seizures. *Brain Res Mol Brain Res* 9:179–190.

Soong TW, Stea A, Hodson CD, Dubel SJ, Vincent SR, Snutch TP. (1993). Structure and functional expression of a member of the low voltage-activated calcium channel family. *Science* 260:1133–1136.

Spain W, Schwindt PC, Crill WE. (1987). Anomalous rectification in neurons from cat sensorimotor cortex *in vitro*. *J Neurophysiol* 57:1555–1576.

Spain W, Schwindt PC, Crill WE. (1991a). Post-inhibitory excitation and inhibition in layer V pyramidal neurones from cat sensorimotor cortex. *J Physiol* (Lond) 434:609–626.

Spain W, Schwindt PC, Crill WE. (1991b). Two transient potassium currents in layer V pyramidal neurones from cat sensorimotor cortex. *J Physiol* (Lond) 434:591–607.

Spencer SS, Kim J, Spencer DD. (1992). Ictal spikes: a marker of specific hippocampal cell loss. *Electroenceph Clin Neurophysiol* 83:104–111.

Spencer WA, Kandel ER. (1961). Electrophysiology of hippocampal neurons. Fast prepotentials. *J Neurophysiol* 24:272–285.

Sprengel R, Werner P, Seeburg PH, Mukhin AG, Santi MR, Grayson DR, Guidotti A, Krueger KE. (1989). Molecular cloning and expression of cDNA encoding a peripheral-type benzodiazepine receptor. *J Biol Chem* 264:20415–20421.

Stafstrom CE, Schwindt PC, Crill WE. (1982). Negative slope conductance due to a persistent subthreshold sodium current in cat neocortical neurons *in vitro*. *Brain Res* 236:221–226.

Stafstrom CE, Schwindt PC, Crill WE. (1984a). Repetitive firing in layer V neurons from cat neocortex in vitro. *J Neurophysiol* 52:264–277.

Stafstrom CE, Schwindt PC, Flatman JA, Crill WE. (1984b). Properties of subthreshold response and action potential recorded in layer V neurons from cat sensorimotor cortex in vitro. *J Neurophysiol* 52:244–263.

Stafstrom CE, Schwindt PC, Chubb MC, Crill WE. (1985). Properties of persistent sodium conductance and calcium conductance of layer V neurons from cat sensorimotor cortex in vitro. *J Neurophysiol* 53:153–170.

Staley KJ, Mody I. (1992). Shunting of excitatory input to dentate gyrus granule cells by a depolarizing $GABA_A$ receptor-mediated postsynaptic conductance. *J Neurophysiol* 68:197–212.

Stanley EF. (1993). Presynaptic calcium channels and the transmitter release mechanism. *Ann NY Acad Sci* 681:368–372.

Stasheff SF, Hines M, Wilson WA. (1993a). Axon terminal hyperexcitability associated with epileptogenesis in vitro. I. Origin of ectopic spikes. *J Neurophysiol* 70:961–975.

Stasheff SF, Mott DD, Wilson WA. (1993b). Axon terminal hyperexcitability associated with epileptogenesis in vitro. II. Pharmacological regulation by NMDA and $GABA_A$ receptors. *J Neurophysiol* 70:976–984.
Stasheff SF, Anderson WW, Clark S, Wilson WA. (1989). NMDA antagonists differentiate epileptogenesis from seizure expression in an in vitro model. *Science* 245:648–651.
Steinberg GK, Saleh J, Kunis D, DeLaPaz R. (1991). Protection after transient focal cerebral ischemia by the N-methyl-D-aspartate antagonist dextrorphan is dependent upon plasma and brain levels. *J Cereb Blood Flow Metab* 1015:1024.
Steinmeyer K, Kocke R, Ortland C, Gronemeier M, Jockusch H, Grunder S, Jentsch TJ. (1991). Inactivation of muscle chloride channel by transposon insertion in myotonic mice. *Nature* 354:304–306.
Stelzer A, Kay AR, Wong RKS. (1988). $GABA_A$ receptor function in hippocampal cells is maintained by phosphorylation factors. *Science* 241:339–341.
Stelzer A, Slater NT, ten Bruggencate G. (1987). Activation of NMDA receptors blocks GABAergic inhibition in an in vitro model of epilepsy. *Nature* 326:698–701.
Stent GS. (1973). A physiological mechanism for Hebb's postulate learning. *Proc Natl Acad Sci USA* 70:997–1001.
Steriade M, Deschenes M. (1984). The thalamus as a neuronal oscillator. *Brain Res Rev* 8:1–63.
Steriade M, Deschenes M. (1988). Intrathalamic and brainstem-thalamic networks involved in resting and alert states. In: Bentivoglio MA, ed. *Cellular Thalamic Mechanisms.* Amsterdam: Elsevier: 51–76.
Steriade M, Llinás RR. (1988). The functional states of the thalamus and the associated neuronal interplay. *Physiol Rev* 68:649–742.
Steriade M, McCarley RW. (1990). *Brainstem Control of Wakefullness and Sleep.* New York: Plenum Press.
Steriade M, Curro Dossi R, Nunez A. (1991). Network modulation of a slow intrinsic oscillation of cat thalamocortical neurons implicated in sleep delta waves: cortical potentiation and brainstem suppression. *J Neurosci* 11: 3200–3217.
Steriade M, McCormick DA, Sejnowski TJ. (1993). Thalamocortical oscillations in the sleeping and aroused brain. *Science* 262:679–685.
Steriade M, Parent A, Hada J. (1984). Thalamic projections of nucleus reticularis thalami of cat: a study using retrograde transport of horseradish peroxidase and double fluorescent tracers. *J Comp Neurol* 229:531–547.
Steriade M, Deschenes M, Domich L, Mulle C. (1985). Abolition of spindle oscillations in thalamic neurons disconnected from the nucleus reticularis thalami. *J Neurophysiol* 54:1473–1497.
Steriade M, Domich L, Oakson G. (1986). Reticularis thalami neurons revisited: activity changes during shifts in states of vigilance. *J Neurosci* 6:68–81.
Steriade M, Domich L, Oakson G. (1987a). The deafferented reticular thalamic nucleus generates spindle rhythmicity. *J Neurophysiol* 57:260–273.
Steriade M, Parent A, Pare D, Smith Y. (1987b). Cholinergic and non-cholinergic neurons of cat basal forebrain project to reticular and mediodorsal thalamic nuclei. *Brain Res* 408:373–376.
Steriade M, Datta S, Pare D, Oakson G, Curro Dossi R. (1990a). Neuronal activities in brain-stem cholinergic nuclei related to tonic activation processes in thalamocortical systems. *J Neurosci* 10:2541–2559.

Steriade M, Jones EG, Llinás RR. (1990b). *Thalamic Oscillations and Signaling.* New York: John Wiley and Sons.
Steriade M, Curro Dossi R, Pare D, Oakson G. (1991). Fast oscillations (20–40 Hz) in thalamocortical systems and their potentiation by mesopontine cholinergic nuclei in the cat. *Proc Natl Acad Sci USA* 88:4396–4400.
Stern P, Edwards FA, Sakmann B. (1992). Fast and slow components of unitary EPSCs on stellate cells elicited by focal stimulation in slices of rat visual cortex. *J Physiol* (Lond) 449:247–278.
Stillerman ML, Gibbs EL, Perlstein MA. (1952). Electroencephalographic changes in strabismus. *Am J Ophthalmol* 35:54.
Stratford K, Mason A, Larkman A, Major G, Jack J. (1989). The modelling of pyramidal neurons in the cat visual cortex. In: Durbin R, Mitchison G, eds. *The Computing Neuron.* Wokingham, England: Addison-Wesley.
Streit P. (1985). Glutamate and aspartate as transmitter candidates for systems of the cerebral cortex. In: Jones EG, Peters A, eds. *Cerebral Cortex, Vol 2. Functional Properties of Cortical Cells.* New York: Plenum: 119–143.
Stringer JL, Lothman EW. (1989). Maximal dentate gyrus activation: characteristics and alterations after repeated seizures. *J Neurophysiol* 62:136–143.
Strowbridge BW, Bean AJ, Spencer DD, Roth HR, Sheperd GM, Robbins RJ. (1992a). Low levels of somatostatin-like immunoreactivity in neocortex resected from presumed seizure foci in epileptic patients. *Brain Res* 587: 164–168.
Strowbridge BW, Masukawa LM, Spencer DD, Shepherd GM. (1992b). Hyperexcitability associated with localizable lesions in epileptic patients. *Brain Res* 587:158–163.
Stuart GJ, Sakmann B. (1994). Active propagation of somatic action potentials into neocortical pyramidal cell dendrites. *Nature* 367:69–72.
Stuart GJ, Dodt H-U, Sakmann B. (1993). Patch-clamp recordings from the soma and dendrites of neurons in brain slices using infrared video microscopy. *Pflügers Arch* 423:511–518.
Sucher NJ, Brose N, Deitcher DL, Awobuluyi M, Gasic GP, Bading H, Cepko CL, Greenberg ME, Jahn R, Heinemann SF, Lipton SA. (1993). Expression of endogenous NMDAR1 transcripts without receptor protein suggests post-transcriptional control in PC12 cells. *J Biol Chem* 268:22299–22304.
Sudhof TC, Camilli P, Niemann H, Jahn R. (1993). Membrane fusion machinery: insights from synaptic proteins. *Cell* 75:1–4.
Sudol M, Grant SGN, Maisonpierre PC. (1993). Protooncogenes and signaling processes in neural tissues. *Neurochem Int* 22:369–384.
Sugita S, Johnson SW, North RA. (1992). Synaptic inputs to $GABA_A$ and $GABA_B$ receptors originate from discrete afferent neurons. *Neurosci Lett* 134:207–211.
Sullivan HC, Osorio I. (1991). Aggravation of penicillin-induced epilepsy in rats with locus ceruleus lesions. *Epilepsia* 32:591–596.
Sutor B, Hablitz JJ. (1989a). EPSPs in rat neocortical neurons *in vitro.* I. Electrophysiological evidence for two distinct EPSPs. *J Neurophysiol* 61:607–620.
Sutor B, Hablitz JJ. (1989b). EPSPs in rat neocortical neurons *in vitro.* II. Involvement of N-methyl-D-aspartate receptors in the generation of EPSPs. *J Neurophysiol* 61:621–634.
Sutor B, Zielglgänsberger W. (1987). A low-voltage activated, transient calcium current is responsible for the time dependent depolarizing inward rectification of rat neocortical neurons *in vitro. Pflügers Arch* 410:102–111.

Sutula T, Steward O. (1987). Facilitation of kindling by prior induction of long-term potentiation in the perforant path. *Brain Res* 420:109–117.

Sutula T, Harrison C, Steward O. (1986). Chronic epileptogenesis induced by kindling of the entorhinal cortex: the role of the dentate gyrus. *Brain Res* 385: 291–299.

Sutula T, He XX, Cavazos J, Scott G. (1988). Synaptic reorganization in the hippocampus induced by abnormal functional activity. *Science* 239:1147–1150.

Sutula T, Cascino G, Cavazos J, Parada I, Ramirez L. (1989). Mossy fiber synaptic reorganization in the epileptic human temporal lobe. *Ann Neurol* 26:321–330.

Suzuki S, Rogawski MA. (1989). T-type calcium channels mediate the transition between tonic and phasic firing in thalamic neurons. *Proc Natl Acad Sci USA* 86:7228–7232.

Swann JW, Brady RJ, Smith KL, Pierson MG. (1988). Synaptic mechanisms of focal epileptogenesis in the immature nervous system. In: *Disorders of the Developing Nervous System: Changing Views on Their Origins, Diagnoses, and Treatments.* New York: Alan R Liss: 19–49.

Swann JW, Brady RJ, Martin DL. (1989). Postnatal development of GABA-mediated synaptic inhibition in rat hippocampus. *Neuroscience* 28:551–561.

Swann JW, Smith KL, Brady RL. (1991). Age-dependent alterations in the operations of hippocampal neural networks. *Ann NY Acad Sci* 627:264–276.

Swann JW, Smith KL, Brady RJ, Pierson MG. (1993). Neurophysiological studies of alterations of seizure susceptibility during brain development. In: Schwartzkroin PA, ed. *Epilepsy: Models, Mechanisms, and Concepts.* Cambridge: Cambridge University Press.

Szentágothai J. (1975). The "module-concept" in cerebral cortex architecture. *Brain Res* 95:475–496.

Szentágothai J, Arbib MD. (1974). Conceptual models of neural organization. *Neurosci Res Progr Bull* 12:306–510.

Takahashi K. (1965). Slow and fast groups of pyramidal tract cells and their respective membrane properties. *J Neurophysiol* 28:908–924.

Takahashi T, Nowakowski RN, Caviness VS Jr. (1993). Mode of cell proliferation in the developing mouse neocortex. *Proc Natl Acad Sci USA* 91:375–379.

Takahashi K, Tateishi N, Kaneda M, Akaike N. (1989). Comparison of low-threshold Ca^{2+} currents in the hippocampal CA1 neurons among the newborn, adult and aged rats. *Neurosci Lett* 103:29–33.

Takahashi T, Momiyama A, Hirai K, Hishinuma H, Akagi H. (1992). Functional correlation of fetal and adult forms of glycine receptors with developmental changes in inhibitory synaptic receptor channels. *Neuron* 9:1155–1161.

Tamaru M, Yoneda Y, Ogita K, Shimizu J, Nagata Y. (1991). Age-related decreases of the N-methyl-D-aspartate receptor complex in the rat cerebral cortex and hippocampus. *Brain Res* 542:83–90.

Tan S, Wenthold RJ, Soderling TR. (1994). Phosphorylation of AMPA-type glutamate receptors by calcium/calmodulin-dependent kinase II and protein kinase C in cultured hippocampal neurons. *J Neurosci* 14:1123–1129.

Tanabe Y, Masu M, Ishii T, Shigemoto R, Nakanishi S. (1992). A family of metabotropic glutamate receptors. *Neuron* 8:169–179.

Tancredi V, Hwa GGC, Zona C, Brancati A, Avoli M. (1990). Low magnesium epileptogenesis in the rat hippocampal slice: electrophysiological and pharmacological features. *Brain Res* 511:280–290.

Tanelian DL, Kosek P, Mody I, MacIver MB. (1993). The role of the $GABA_A$ receptor/chloride channel complex in anesthesia. *Anesthesiology* 78:757–776.

Tang C, Shi Q, Katchman A, Lynch G. (1991). Modulation of the time course of fast EPSCs and glutamate channel kinetics by aniracetam. *Science* 254:288–290.

Tang WJ, Gilman AG. (1991). Specific regulation of adenylyl cyclase by G protein β/γ subunits. *Science* 254:1500–1503.

Tasker JG, Peacock WJ, Dudek FE. (1992). Local synaptic circuits and epileptiform activity in slices of neocortex from children with intractable epilepsy. *J Neurophysiol* 67:496–507.

Tauck DL, Nadler JV. (1985). Evidence of functional mossy fiber sprouting in hippocampal formation of kainic acid treated rats. *J Neurosci* 5:1016–1022.

Taverna FA, Wang L, MacDonald JF, Hampson DR. (1994). A transmembrane model for an ionotropic glutamate receptor predicted on the basis of the location of asparagine-linked oligosaccharides. *J Biol Chem* 269:14159–14164.

Taylor CP, Dudek FE. (1984). Excitation of hippocampal pyramidal cells by an electrical field effect. *J Neurophysiol* 52:126–142.

Teillet M, Naquet R, Le Gal La Salle G, Merat P, Schuler B, Le Douari NM. (1991). Transfer of genetic epilepsy by embryonic brain grafts in the chicken. *Proc Natl Acad Sci USA* 88:6966–6970.

Telfeian AE. (1993). Pathways and mechanisms for the propagation of synchronized neural activity in neocortex. PhD dissertation, Brown University.

Telfeian AE, Wehr MS, Connors BW. (1990). Layer 5 is the preferred pathway for horizontal propagation of epileptiform discharges in neocortex. *Soc Neurosci Abstr* 16:21.

Teyler T, Aroniadou V, Berry RL, Borroni A, DiScenna P, Grover L, Lambert N. (1990). LTP in neocortex. *Sem Neurosci* 2:365–380.

Thalmann RH. (1984). Reversal properties of an EGTA-resistant late hyperpolarization that follows synaptic stimulation of hippocampal neurons. *Neurosci Lett* 46:103–108.

Thalmann RH. (1988). Evidence that guanosine triphosphate (GTP)-binding proteins control a synaptic response in brain: effect of pertussin toxin and GTPyS on the late inhibitory postsynaptic potential of hippocampal CA3 neurons. *J Neurosci* 8:4589–4602.

Thalmann RH, Peck EJ, Ayala GF. (1981). Biphasic response of hippocampal pyramidal neurons to GABA. *Neurosci Lett* 21:319–324.

Tharp BR. (1987). An overview of pediatric seizure disorders and epileptic syndromes. *Epilepsia* 28:536–545.

Thompson SM, Gähwiler BH. (1989a). Activity-dependent disinhibition. II. Effects of extracellular potassium, furosemide, and membrane potential on ECl- in hippocampal CA3 neurons. *J Neurophysiol* 61:512–523.

Thompson SM, Gähwiler BH. (1989b). Activity-dependent disinhibition. III. Desensitization and $GABA_B$ receptor-mediated presynaptic inhibition in the hippocampus in vitro. *J Neurophysiol* 61:524–533.

Thompson SM, Gähwiler BH. (1992). Effects of the GABA uptake inhibitor tiagabine on inhibitory synaptic potentials in rat hippocampal slice cultures. *J Neurophysiol* 67:1698–1701.

Thompson SM, Capogna M, Scanziani M. (1993a). Presynaptic inhibition in the hippocampus. *Trends Neurosci* 16:222–227.

Thompson SM, Capogna M, Scanziani M. (1993b). Presynaptic inhibition in the hippocampus. Reply. *Trends Neurosci* 16:396–397.

Thompson SM, Deisz RA, Prince DA. (1988). Relative contributions of passive equilibrium and active transport to the distribution of chloride in mammalian cortical neurons. *J Neurophysiol* 60:105–124.

Thomson AM. (1986). A magnesium-sensitive post-synaptic potential in rat cerebral cortex resembles neuronal responses to N-methylaspartate. *J Physiol* (Lond) 370:531–549.

Thomson AM. (1990). Augmentation by glycine and blockade by 6-cyano-7-nitroquinoxaline-2,3-dione (CNQX) of responses to excitatory amino acids in slices of rat neocortex. *Neuroscience* 39:69–79.

Thomson AM, West DC. (1993). Fluctuations in pyramid-pyramid excitatory postsynaptic potentials modified by presynaptic firing pattern and postsynaptic membrane potential using paired intracellular recordings in rat neocortex. *Neuroscience* 54:329–346.

Thomson AM, West DC, Lodge D. (1985). An N-methylaspartate receptor mediated synapse in rat cerebral cortex: a site of action of ketamine. *Nature* 313:479–481.

Thomson AM, Girdlestone D, West DC. (1988). Voltage-dependent currents prolong single axon postsynaptic potentials in layer III pyramidal neurons in rat neocortical slices. *J Neurophysiol* 60:1896–1907.

Thomson AM, Girdlestone D, West DC. (1989). A local circuit neocortical synapse that operates via both NMDA and non-NMDA receptors. *Br J Pharmacol* 96:406–408.

Thomson AM, West DC, Deuchars J. (1992). Local circuit, single axon excitatory postsynaptic potentials (EPSPs) in deep layer neocortical pyramidal cells. *Soc Neurosci Abstr* 18:1340.

Thomson AM, Deuchars J, West DC. (1993a). Large, deep layer pyramid-pyramid single axon EPSPs in slices of rat motor cortex display paired pulse and frequency-dependent depression, mediated presynaptically and self-facilitation, mediated postsynaptically. *J Neurophysiol* 70:2354–2369.

Thomson AM, Deuchars J, West DC. (1993b). Single axon excitatory postsynaptic potentials in neocortical interneurons exhibit pronounced paired pulse facilitation. *Neuroscience* 54:347–360.

Tingley WG, Roche KW, Thompson AK, Huganir RL. (1993). Regulation of NMDA receptor phosphorylation by alternative splicing of the C-terminal domain. *Nature* 364:70–73.

Titmus MJ, Faber DS. (1990). Axotomy-induced alterations in the electrophysiological characteristics of neurons. *Progr Neurobiol* 35:1–51.

Tombol T. (1978). Comparative data on the Golgi architecture of interneurons of different cortical areas in cat and rabbit. In: Brazier MA, ed. *Architectonics of the Cerebral Cortex*. New York: Raven Press: 59–76.

Tong G, Jahr CE. (1994). Multivesicular release from excitatory synapses of cultured hippocampal neurons. *Neuron* 12:51–59.

Traub RD, Jefferys JGR. (1994). Are there unifying principles underlying the generation of epileptic afterdischarges *in vitro*? In: Corner MA, ed. *Progress in Brain Research* 102:383–394.

Traub RD, Miles R. (1991). *Neuronal Networks of the Hippocampus.* Cambridge: Cambridge University Press.

Traub RD, Wong RKS. (1982). Cellular mechanism of neuronal synchronization in epilepsy. *Science* 216:745–747.

Traub RD, Wong RKS. (1983). Synaptic mechanisms underlying interictal spike initiation in a hippocampal network. *Neurology* 33:257–266.

Traub RD, Miles R, Wong RKS. (1989). Models of the origin of rhythmic oscillations in the hippocampal slice. *Science* 243:1319–1325.

Traub RD, Dudek FE, Snow RW, Knowles WD. (1985). Computer simulations indicate that electrical field effects contribute to the shape of the epileptiform field potential. *Neuroscience* 15:947–958.

Traub RD, Knowles WD, Miles R, Wong RKS. (1984). Synchronized afterdischarges in the hippocampus: simulation studies of the cellular mechanism. *Neuroscience* 12:1191–1200.

Traub RD, Knowles WD, Miles R, Wong RK. (1987a). Models of the cellular mechanism underlying propagation of epileptiform activity in the CA2-CA3 region of the hippocampal slice. *Neuroscience* 21:457–470.

Traub RD, Miles R, Wong RK. (1987b). Models of synchronized hippocampal bursts in the presence of inhibition. I. Single population events. *J Neurophysiol* 58:739–751.

Traub RD, Miles R, Wong RK, Schulman LS, Schneiderman JH. (1987c). Models of synchronized hippocampal bursts in the presence of inhibition. II. Ongoing spontaneous population events. *J Neurophysiol* 58:752–764.

Traynelis SF, Dingledine R. (1988). Potassium-induced spontaneous electrographic seizures in the rat hippocampal slice. *J Neurophysiol* 59:259–276.

Tremblay E, Represa A, Ben-Ari Y. (1985). Autoradiograph localization of kainic acid binding sites in the human hippocampus. *Brain Res* 343:378–382.

Tremblay E, Roisin MP, Represa A, Charriaut-Marlangue C, Ben-Ari Y. (1988). Transient increased density of NMDA binding sites in the developing rat hippocampus. *Brain Res* 461:393–396.

Trojaborg W. (1966). Focal spike discharges in children, a longitudinal study. Thesis, Copenhagen.

Trussel LO, Fischbach GD. (1989). Glutamate receptor desensitization and its role in synaptic transmission. *Neuron* 3:209–218.

Ts'o DY, Gilbert CD, Wiesel TN. (1986). Relationships between horizontal interactions and functional architecture in cat striate cortex as revealed by cross-correlation analysis. *J Neurosci* 6:1160–1170.

Tsaur M-L, Sheng M, Lowenstein D, Jan YN, Jan LY. (1992). Differential expression of K^+ channel mRNAs in the rat brain and down-regulation in the hippocampus following seizures. *Neuron* 8:1055–1067.

Tseng GF, Prince DA. (1990). Neuronal properties of identified corticospinal cells in vitro following spinal axotomy in vivo. *Soc Neurosci Abstr* 16:1163.

Tseng GF, Prince DA. (1991). Morphological and physiological characteristics of intracellularly filled axotomozed corticospinal neurons studied in vitro. *Soc Neurosci Abstr* 17:331.

Tseng GF, Prince DA. (1993). Heterogeneity of rat corticospinal neurons. *J Comp Neurol* 335:92–108.

Tsien RW, Ellinor PT, Horne HA. (1991). Molecular diversity of voltage dependent Ca^{2+} channels. *Trends Pharmacol* 12:349–354.

Tsumoto T. (1990). Excitatory amino acid transmitters and their receptors in neural circuits of the cerebral neocortex. *Neurosci Res* 9:79–102.

Tsumoto T. (1992). Long-term potentiation and long-term depression in the neocortex. *Progr Neurobiol* 39:209–328.

Tsumoto T, Eckart W, Creutzfeldt OD. (1979). Modification of orientation sensitivity of cat visual cortex neurons by removal of GABA-mediated inhibition. *Exp Brain Res* 34:351–363.

Tuff LP, Racine RJ, Adamec R. (1983). The effects of kindling on GABA-mediated inhibition in the dentate gyrus of the rat. *Brain Res* 277:79–90.

Turetsky DM, Goldberg MP, Choi DW. (1992). Kainate-activated cobalt uptake identifies a subpopulation of cultured cortical cells that are preferentially vulnerable to kainate-induced damage. *Soc Neurosci Abstr* 18:81.

Uchitel OD, Protti DA, Sanchez V, Cherksey BD, Sugimori M, Llinás R. (1992). P-type voltage-dependent calcium channel mediates presynaptic calcium influx and transmitter release in mammalian synapses. *Proc Natl Acad Sci USA* 89:3330–3333.

Ulas J, Monaghan DT, Cotman CW. (1990). Kainate receptors in the rat hippocampus: a distribution and time course of changes in response to unilateral lesions of the entorhinal cortex. *J Neurosci* 10:2352–2362.

Unwin N. (1993). Nicotinic acetylcholine receptor at 9Å resolution. *J Mol Biol* 229:1101–1124.

Urban L, Aitken PG, Friedman A, Somjen GG. (1990). An NMDA-mediate component of excitatory synaptic input to dentate granule cells in "epileptic" human hippocampus studied *in vitro*. *Brain Res* 515:319–322.

Urban L, Aitken PG, Crain BJ, Friedman AH, Somjen GG. (1993). Correlation between function and structure in "epileptic" human hippocampal tissue maintained *in vitro*. *Epilepsia* 34:54–60.

Uruno K, O'Connor MJ, Masukawa LM. (1994). Alterations of inhibitory synaptic responses in the dentate gyrus of temporal lobe epileptic patients. *Hippocampus* 4:583–593.

Valdes F, Dasheiff RM, Birmingham F, Crutcher KA, McNamara JO. (1982). Benzodiazepine receptor increases after repeated seizures: evidence for localization to dentate granule cells. *Proc Natl Acad Sci USA* 79:193–197.

Valtorta F, Benfenati F, Greengard P. (1992a). Structure and function of the synapsins. *J Biol Chem* 267:7195–7198.

Valtorta F, Greengard P, Fesce R, Chieregatti E, Benfenati F. (1992b). Effects of the neuronal phosphoprotein synapsin I on actin polymerization. I. Evidence for a phosphorylation-dependent nucleating effect. *J Biol Chem* 267: 11281–11288.

Valverde F. (1967). Apical dendritic spines of the visual cortex and light deprivation in the mouse. *Exp Brain Res* 3:337–352.

Valverde F. (1983). A comparative approach to neocortical organization based on the study of the brain of the hedgehog (Erinaceus europaeus). In: Gisiola S, Guerri, eds. *Ramon y Cajal's Contribution to the Neurosciences*. Amsterdam: Elsevier: 149–170.

Van Eden CG, Mrzljak L, Voorn P, Uylings HBM. (1989). Prenatal development of GABAergic neurons in the neocortex of the rat. *J Comp Neurol* 289: 213–227.

Vanderhaeghen JJ, Lotstra F, Demey J, Gilles C. (1980). Immunohistochemical localization of cholecystokinin-like and gastrin-like peptides in the brain and hypophysis of the rat. *Proc Natl Acad Sci USA* 77:1190–1194.

Vanni-Mercier G, Sakai K, Jouvet M. (1984). "Waking-state specific" neurons in the caudal hypothalamus of the cat. *Coll Roy Acad Sci Series* III 298 3: 195–200.

Vaughn JE. (1989). Fine structure of synaptogenesis in the vertebrate central nervous system. *Synapse* 3:255–285.

Veith G, Wicke R. (1968). Cerebrale Differenzierungsstörungen bei Epilepsie. *Westdtsch Ver* 515–534.

Verdoorn TA, Draguhn A, Ymer S, Seeburg PH, Sakmann B. (1990). Functional properties of recombinant rat $GABA_A$ receptors depend upon subunit composition. *Neuron* 4:919–928.

Verdoorn TA, Burnashev N, Monyer H, Seeburg PH, Sakmann B. (1991). Structural determinants of ion flow through recombinant glutamate receptor channels. *Science* 252:1715–1718.

Vergnes M, Marescaux C. (1992). Cortical and thalamic lesions in rats with genetic absence epilepsy. *J Neural Transm* 35 (suppl):71–83.

Vergnes M, Marescaux C, Depaulis A, Micheletti G, Warter J. (1988). Spontaneous spike-and-wave discharges in Wistar rats: A model of generalized nonconvulsive epilepsy. In: Avoli M, Gloor, P, eds. *Generalized Epilepsy.* Boston: Birkhauser Press.

Vergnes M, Marescaux C, Micheletti G, Depaulis A, Rumbach L, Warter J. (1984). Enhancement of spike and wave discharges by GABAmimetic drugs in rats with spontaneous petit-mal–like epilepsy. *Neurosci Lett* 44:91–94.

Vidal C, Changeux J. (1993). Nicotinic and muscarinic modulations of excitatory synaptic transmission in the rat prefrontal cortex in vitro. *Neuroscience* 56: 23–32.

Villarroel A, Sakmann B. (1992). Threonine in the selectivity filter of the acetylcholine receptor channel. *Biophys J* 62:196–205.

Vogt BA, Peters A. (1981). Form and distribution of neurons in rat cingulate cortex: areas 32, 24 and 29. *J Comp Neurol* 195:603–625.

von Bonin G. (1960). Introduction. In: Nowinski W, ed. *Some Papers on the Cerebral Cortex.* New York: Charles C Thomas: VII.

von der Marlsburg C, Schneider W. (1986). A neural cocktail party processor. *Biol Cyber* 54:29–40.

von Krosigk M, Bal T, McCormick DA. (1993). Cellular mechanisms of a synchronized oscillation in the thalamus. *Science* 261:361–364.

von Noorden GK. (1973). Experimental amblyopia in monkeys. Further behavioral observations and clinical correlations. *Invest Ophthalmol* 12:721–726.

von Noorden GK. (1985). Amblyopia: a multidisciplinary approach. *Invest Ophthalmol Vis Sci* 26:1704–1716.

von Noorden GK, Dowling JE. (1970). Experimental amblyopia in monkeys. II. Behavioral studies in strabismic amblyopia. *Arch Ophthalmol* 84:215–220.

Wafford KA, Bain CJ, Le Bourdelles B, Whiting PJ, Kemp JA. (1993). Preferential co-assembly of recombinant NMDA receptors composed of three different subunits. *Neuroreport* 4:1347–1349.

Wahl P, Schousboe A, Honore T, Drejer J. (1989). Glutamate-induced increase in intracellular Ca^{2+} in cerebral cortex neurons is transient in immature but permanent in mature cells. *J Neurochem* 53:1316–1319.

Wall JT, Kaas JH, Sur M, Nelson RJ, Felleman DJ, Merzenich MM. (1986). Functional reorganization in somatosensory cortical areas 3b and 1 of adult monkeys after median nerve repair: possible relationships to sensory recovery in humans. *J Neurosci* 6:218–233.

Wallace MN, Kitzes LM, Jones EG. (1991). Intrinsic inter- and intralaminar connections and their relationship to the tonotopic map in cat primary auditory cortex. *Exp Brain Res* 86:527–544.

Walsh C, Cepko CL. (1988). Clonally related cortical cells show several migration patterns. *Science* 241:1342–1344.

Walsh C, Cepko CL. (1992). Widespread dispersion of clonally related neurons across functional neocortical areas. *Science* 255:434–440.

Walsh C, Cepko CL. (1993). Clonal dispersion in proliferative layers of developing cerebral cortex. *Nature* 362:632–640.

Wang H, Kunkel D, Martin TM, Schwartzkroin PA, Tempel BL. (1993). Heteromeric K^+ channels in terminal and juxtaparnodal regions of neurons. *Nature* 365:75–79.

Wang L, Taverna FA, Huang X, MacDonald JF, Hampson DR. (1993). Phosphorylation and modulation of a kainate receptor (GluR6) by cAMP-dependent protein kinase. *Science* 259:1173–1175.

Wang Z, McCormick DA. (1993). Control of firing mode of corticotectal and corticopontine layer V burst-generating neurons by norepinephrine, acetylcholine, and 1S,3R-ACPD. *J Neurosci* 13:2199–2216.

Wang ZQ, Ovitt C, Grigoriades AV, Mohle-Steinien U, Ruther U, Wagner EF. (1992). Bone and hematopoietic defects in mice lacking c-fos. *Nature* 360: 741–745.

Ward AA Jr. (1972). Topical convulsant metals. In: Purpura DP, Penry JK, eds. *Experimental Models of Epilepsy.* New York: Raven Press: 13–35.

Ward AA Jr. (1978). Glia and epilepsy. In: Schoffeniels E, Franck JE, eds. *Dynamic Properties of Glia Cells.* Oxford: Pergamon Press.

Ward AA, Thomas LB. (1955). The electrical activity of single units in the cerebral cortex of man. *Electroencephalogr Clin Neurophysiol* 7:135–136.

Watanabe M, Inoue Y, Sakimura K, Mishina M. (1992). Developmental changes in distribution of NMDA receptor channel subunit mRNAs. *Neuroreport* 3:1138–1140.

Watkins J, Krogsgaard-Larsen P, Honore T. (1990). Structure-activity relationships in the development of excitatory amino acid receptor agonists and competitive antagonists. *Trends Pharmacol Sci* 11:25–33.

Wegner M, Cao Z, Rosenfeld MG. (1992). Calcium-regulated phosphorylation within the leucine zipper of C/EBPB. *Science* 256:370–373.

Wehmeier U, Dong D, Koch C, Van Essen D. (1989). Modeling the mammalian visual system. In: Koch C, Segev I, eds. *Methods in Neuronal Modeling. From Synapses to Networks.* Cambridge, MA: MIT Press: 335–360.

Weisskopf MG, Zalutsky RA, Nicoll RA. (1993). The opioid peptide dynorphin mediates heterosynaptic depression of hippocampal mossy fibre synapses and modulates long-term potentiation. *Nature* 362:423–427.

Westbrook GL. (1993). Glutamate receptors and excitotoxicity. In: Waxman SG, ed. *Molecular and Cellular Approaches to the Treatment of Neurological Disease.* New York: Raven Press.

Westenbroek R, Ahlijanian MK, Catterall WA. (1990). Clustering of L-type Ca^{2+} channels at the base of major dendrites in hippocampal pyramidal neurons. *Nature* 347:281–284.

Westenbroek R, Merrick DK, Catterall WA. (1989). Differential subcellular localization of the R_I and R_{II} Na^+ channel subtypes in central neurons. *Neuron* 3:695–704.

Westenbroek R, Hell JW, Warner C, Dubel SJ, Snutch TP, Catterall WA. (1992a). Biochemical properties and subcellular distribution of a N-type calcium channel alpha 1 subunit. *Neuron* 9:1099–1115.

Westenbroek R, Noebels JL, Catterall W. (1992b). Elevated expression of Type II Na^+ channels in hypomyelinated axons of shiverer mouse brain. *J Neurosci* 12:2259–2267.

Westerfield M, Joyner RW, Moore JW. (1978). Temperature-sensitive conduction failure at axon branch points. *J Neurophysiol* 41:1–8.

Westrum LE, White LE, Ward AA Jr. (1964). Morphology of the experimental epileptic focus. *J Neurosurg* 21:1033–1046.

White EL. (1989). Cell types. In: White EL, ed. *Cortical Circuits: Synaptic Organization of the Cerebral Cortex: Structure, Function and Theory.* Boston: Birkhauser: 19–43.

White EL, Hersch SM. (1981). Thalamocortical synapses of pyramidal cells which project from SmI to MsI cortex in the mouse. *J Comp Neurol* 198:167–181.

White EL, Hersch SM. (1982). A quantitative study of thalamocortical and other synapses involving the apical dendrites of corticothalamic projection cells in mouse SmI cortex. *J Neurocytol* 11:137–157.

White EL, Keller A. (1987). Intrinsic circuitry involving the local axonal collaterals of corticothalamic projection cells in mouse SmI cortex. *J Comp Neurol* 262:13–26.

White EL, Keller A. (1989). *Cortical Circuits: Synaptic Organization of the Cerebral Cortex Structure, Function and Theory.* Boston: Birkhauser.

White EL, Amitai Y, Gutnick MJ. (1994). A comparison of synapses onto somata of intrinsically bursting and regular spiking neurons in layer V of rat SmI cortex. *J Comp Neurol* 342:1–14.

Wickens JR, Abraham WC. (1991). The involvement of L-type calcium channels in heterosynaptic long-term depression in the hippocampus. *Neurosci Lett* 130:128–132.

Wiesel TN, Hubel DH. (1965). Comparison of the effects of unilateral and bilateral eye closure on cortical unit responses in kittens. *J Neurophysiol* 28: 1029–1040.

Wigström H, Gustafsson B, Huang Y-Y, Abraham WC. (1986). Hippocampal long-term potentiation is induced by pairing single afferent volleys with intracellularly injected depolarizing pulses. *Acta Physiol Scand* 126:317–319.

Williams BP, Read J, Price J. (1991). The generation of neurons and oligodendrocytes from a common precursor cell. *Neuron* 7:685–693.

Williams D. (1953). A study of thalamic and cortical rhythms in petit mal. *Brain* 76:50–69.

Williams K, Russell SL, Shen YM, Molinoff PB. (1993). Developmental switch in the expression of NMDA receptors occurs in vivo and in vitro. *Neuron* 10: 267–278.

Williams S, Lacaille JC. (1992). $GABA_B$ receptor-mediated inhibitory postsynaptic potentials evoked by electrical stimulation and by glutamate stimulation of interneurons in stratum lacunosum-moleculare in hippocampal CA1 pyramidal cells in vitro. *Synapse* 11:249–258.

Williamson A, Spencer DD, Shepherd GM. (1993). Comparison between the membrane and synaptic properties of human and rodent dentate granule cells. *Brain Res* 622:194–202.

Willshaw D, Dayan P. (1990). Optimal plasticity from matrix memories: what goes up must come down. *Neural Comp* 2:85–93.

Wilson CL, Isokawa M, Babb TL, Crandall PH. (1990). Functional connections in the human temporal lobe. *Exp Brain Res* 82:279–292.

Wilson SW, Placzek M, Furley AJ. (1993). Border disputes: do boundaries play a role in growth-cone guidance? *Trends Neurosci* 316:323.

Wilson WA, Wachtel H. (1974). Negative resistance characteristic essential for the maintenance of slow oscillations in bursting neurons. *Science* 196:932–934.

Winer JA. (1986). Neurons accumulating [^{3}H] gamma-aminobutyric acid (GABA) in supragranular layers of cat primary auditory cortex (AI). *Neuroscience* 19:771–793.

Winfield DA. (1981). The postnatal development of synapses in the visual cortex of the cat and the effects of eyelid closure. *Brain Res* 206:166–171.

Winguth SD, Winer JA. (1986). Corticocortical connections of cat primary auditory cortex (AI): laminar organization and identification of supragranular neurons projecting to area AII. *J Comp Neurol* 248:36–56.

Witcher DR, Kovacs RJ, Schulman H, Cefali DC, Jones LR. (1991). Unique phosphorylation site on the cardiac ryanodine receptor regulates calcium channel activity. *J Biol Chem* 266:11144–11152.

Witcher DR, Waard MD, Sakamoto J, Franzini-Armstrong C, Pragnell M, Kahl SD, Campbell KP. (1993). Subunit identification and reconstitution of the N-type Ca^{2+} channel complex purified from brain. *Science* 261:486–489.

Wolff JR. (1981). Evidence for a dual role of GABA as a synaptic transmitter and a promoter of synaptogenesis. In: DeFeudis FV, Mandel P, eds. *Amino Acid Neurotransmitters*. New York: Raven Press: 459–465.

Wolff JR, Balcar VJ, Zetzsche T, Bottcher H, Schmechel DE, Chronwall BM. (1984). Development of GABA-ergic system in rat visual cortex. *Adv Exp Med Biol* 181:215–239.

Wong ML, Smith MA, Licinio J, Doi SQ, Weiss SR, Post RM, Gold PW. (1993). Differential effects of kindled and electrically induced seizures on a glutamate receptor (GluR1) gene expression. *Epilepsy Res* 14:221–227.

Wong RKS, Miles R. (1993). Study of GABAergic inhibition and $GABA_A$ receptors in experimental epilepsy. In: Schwartzkroin PA, ed. *Epilepsy: Models, Mechanisms, and Concepts*. Cambridge: Cambridge University Press: 424–436.

Wong RKS, Prince DA. (1978). Participation of calcium spikes during intrinsic burst firing in hippocampal neurons. *Brain Res* 159:385–390.

Wong RKS, Prince DA. (1979). Dendritic mechanisms underlying penicillin-induced epileptiform activity. *Science* 204:1228–1231.

Wong RKS, Prince DA. (1981). Afterpotential generation in hippocampal pyramidal cells. *J Neurophysiol* 45:86–97.

Wong RKS, Prince DA, Basbaum AI. (1979). Intradendritic recordings from hippocampal neurons. *Proc Natl Acad Sci USA* 76:986–990.

Wong RKS, Schwartzkroin PA. (1982). Pacemaker neurons in the mammalian brain: mechanisms and function. In: Carpenter DO, ed. *Cellular Pacemakers, Vol 1. Mechanisms of Pacemaker Generation.* New York: John Wiley & Sons: 237–254.

Wong RKS, Stewart M. (1992). Different firing patterns generated in dendrites and somata of CA1 pyramidal neurones in guinea-pig hippocampus. *J Physiol* (Lond) 457:675–687.

Wong RKS, Traub RD. (1982). Synchronized burst discharge in disinhibited hippocampal slice. I. Initiation in CA2-CA3 region. *J Neurophysiol* 49:442–458.

Wong RKS, Watkins DJ. (1982). Cellular factors influencing GABA response in hippocampal pyramidal cells. *J Neurophysiol* 48(4):938–951.

Wong RKS, Prince DA, Basbaum AI. (1978). Intradendritic recordings from hippocampal neurons. *Proc Natl Acad Sci USA* 76:986–990.

Wong RKS, Traub RD, Miles R. (1986). Cellular basis of neuronal synchrony in epilepsy. *Adv Neurol* 44:583–592.

Worgotter F, Koch C. (1991). A detailed model of the primary visual pathway in the cat: comparison of afferent excitatory and intracortical inhibitory connection schemes for orientation selectivity. *J Neurosci* 11:1959–1979.

Wu CF, Ganetzky B, Haugland FN, Liu A. (1983). Potassium currents in Drosophila: different components affected by mutations of two genes. *Science* 220:1076–1078.

Wu K, Wasterlain C, Sachs L, Siekevitz P. (1990). Effect of septal kindling on glutamate binding and calcium/calmodulin-dependent phosphorylation in a postsynaptic density fraction isolated from rat cerebral cortex. *Proc Natl Acad Sci USA* 87:5298–5302.

Wuarin JP, Peacock WJ, Dudek FE. (1992). Single-electrode voltage-clamp analysis of the N-methyl-D-aspartate component of synaptic responses in neocortical slices from children with intractable epilepsy. *J Neurophysiol* 67:84–93.

Wuarin JP, Kim YI, Cepeda C, Tasker JG, Walsh JP, Peacock WJ, Buchwald NA, Dudek FE. (1990). Synaptic transmission in human neocortex removed for treatment of intractable epilepsy in children. *Ann Neurol* 28:503–511.

Wyler AR, Ward AA Jr. (1981). Neurons in human epileptic cortex—response to direct cortical stimulation. *J Neurosurg* 55:904–908.

Wyler AR, Fetz EE, Ward AA Jr. (1975). Firing patterns of epileptic and normal neurons in the chronic alumina focus in undrugged monkeys during different behavioral states. *Brain Res* 98:1–20.

Wyler AR, Burchiel KJ, Ward AA Jr. (1978). Chronic epileptic foci in monkeys: correlation between seizure frequency and proportion of pacemaker epileptic neurons. *Epilepsia* 19:475–483.

Wyler AR, Ojemann GA, Ward AA Jr. (1982). Neurons in human epileptic cortex: correlation between unit and EEG activity. *Ann Neurol* 11:301–308.

Xu Y, Baldassare M, Fisher P, Rathbun G, Oltz EM. (1993). LH-2, a LIM/homeodomain gene expressed in developing lymphocytes and neural cells. *Proc Natl Acad Sci USA* 90:227–231.

Yaari Y, Konnerth A, Heinemann U. (1986). Nonsynaptic epileptogenesis in the mammalian hippocampus *in vitro*. II. Role of extracellular potassium. *J Neurophysiol* 56:424–438.

Yamagata K, Andreason KI, Kaufmann WE, Barnes CA, Worley PF. (1993). Expression of a mitogen-inducible cyclooxygenase in brain neurons: regulation by synaptic activity and glucocorticoids. *Neuron* 11:371–386.

Yamamoto C, Higashima M, Sawada S. (1987). Quantal analysis of potentiating action of phorbol ester on synaptic transmission in the hippocampus. *Neurosci Res* 5:28–38.

Yarom Y, Spira ME. (1982). Extracellular potassium ions mediate specific neuronal interaction. *Science* 216:80–82.

Yeh G, Bonhaus DW, Nadler JV, McNamara JO. (1989). N-methyl-D-aspartate receptor plasticity in kindling: quantitative and qualitative alterations in the N-methyl-D-aspartate receptor-channel complex. *Proc Natl Acad Sci USA* 86:8157–8160.

Yovell Y, Abrams TW. (1992). Temporal asymmetry in activation of Aplysia adenylyl cyclase by calcium and transmitter may explain temporal requirements of conditioning. *Proc Natl Acad Sci USA* 89:6526–6530.

Yuste R, Katz LC. (1991). Control of postsynaptic Ca^{2+} influx in developing neocortex by excitatory and inhibitory neurotransmitters. *Neuron* 6:333–344.

Yuste R, Peinado A, Katz LC. (1992). Neuronal domains in developing neocortex. *Science* 257:665–668.

Yuste R, Gutnick MJ, Saar D, Delaney KR, Tank DW. (1994). Calcium accumulations in dendrites of neocortical pyramidal neurons: an apical band and evidence for two functional compartments. *Neuron* 13:23–43.

Zafra F, Hengerer B, Leibrock J, Thoenen H, Lindholm D. (1990). Activity dependent regulation of BDNF and NGF mRNAs in the rat hippocampus is mediated by non-NMDA glutamate receptors. *EMBO J* 9:3545–3550.

Zafra F, Castren E, Thoenen H, Lindholm D. (1991). Interplay between glutamate and gamma-aminobutyric acid transmitter systems in the physiological regulation of brain-derived neurotrophic factor and nerve growth factor synthesis in hippocampal neurons. *Proc Natl Acad Sci USA* 88: 10037–10041.

Zalutsky R, Nicoll RA. (1990). Comparison of two forms of long-term potentiation in single hippocampal neurons. *Science* 248:1619–1624.

Zhang L, Spigelman I, Carlen PL. (1991). Development of GABA-mediated, chloride-dependent inhibition in CA1 pyramidal neurones of immature rat hippocampal slices. *J Physiol* (Lond) 444:25–49.

Zhang SJ, Jackson MB. (1993). GABA-activated chloride channels in secretory nerve endings. *Science* 259:531–534.

Zheng JQ, Felder M, Connor JA, Poo MM. (1994). Turning of nerve growth cones induced by neurotransmitters. *Nature* 368:140–144.

Zheng X, Zhang L, Durand GM, Bennett MVL, Zukin RS. (1994). Mutagenesis rescues spermine and zinc potentiation of recombinant NMDA receptors. *Neuron* 12:811–818.

Zhou J, Potts JF, Trimmer JS, Agnew WS, Sigworth FJ. (1991). Multiple gating modes and the effect of modulating factors on the mI sodium channel. *Neuron* 7:775–785.

Zhu D, Caveney S, Kidder GM, Naus CCG. (1991). Transfection of C6 glioma cells with connexin 43 DNA: analysis of expression, intercellular coupling, and cell proliferation. *Proc Natl Acad Sci USA* 88:1883–1887.

Zorumski CF, Thio LL, Clark GD, Clifford DB. (1989). Calcium influx through N-methyl-D-aspartate receptors activates a potassium current in postnatal rat hippocampal neurons. *Neurosci Lett* 99:293–299.

Index